The maximum life span of multicellular organisms varies greatly: for a fruitfly it is about 30 days, for a dog about 20 years, and for a human about 100 years. Despite these differences, all animals show a similar pattern in their life spans – growth, adulthood, and aging, followed by death. The basic cause of aging in multicellular organisms (eukaryotes) lies at the level of the genes, although nutrition and various types of stresses do influence the rate and pattern of aging.

This book reviews the molecular biology of the gene in relation to aging. Until about a decade ago it was not possible to probe into the types of changes that occur in eukaryotic genes, due to their enormous complexity. The use of genetic engineering techniques, however, is beginning to unravel the changes that occur in the genes as an organism ages: the changing expression of specific genes under normal conditions and under various types of stress, the changes in the regulatory roles of the sequences in the promoter regions of genes, conformational changes that may occur in genes during aging, and the protein factors that are involved in the regulation of genes. The author presents basic information on eukaryotic genes and follows this with details of the changes that occur in their structure and function during aging. He reviews the latest studies being carried out in various laboratories, outlines the gaps and deficiencies in our present knowledge, and suggests the most profitable future areas of research.

Genes and aging is for all students and researchers interested in the molecular biology of aging.

Genes and aging

Genes and aging

M. S. KANUNGO

*Banaras Hindu University, Varanasi, and
Institute of Life Sciences, Bhubaneswar,
India*

CAMBRIDGE
UNIVERSITY PRESS

Published by the Press Syndicate of the University of Cambridge
The Pitt Building, Trumpington Street, Cambridge CB2 1RP
40 West 20th Street, New York, NY 10011-4211, USA
10 Stamford Road, Oakleigh, Melbourne 3166, Australia

First published 1994

Printed in the United States of America

Library of Congress Cataloging-in-Publication Data

Kanungo, M. S. (Madhu Sudan), 1927–
Genes and aging / M.S. Kanungo.

p.   cm.

Includes bibliographical references and index.

ISBN 0-521-38299-8

1. Aging – Genetic aspects.   2. Cells – Aging.   I. Title.
QP86.K335   1993
612.6'7 – dc20   93-2766
                                                             CIP

A catalog record for this book is available from the British Library.

ISBN 0-521-38299-8 hardback

Contents

v

Preface

The science of aging, though young in comparison to the science of development, has advanced considerably during the past three decades. Up until 1980, most of the research that had been done was on the physiological and biochemical aspects of aging. The information obtained from this research was covered in my book, *Biochemistry of Ageing,* published by Academic Press in 1980. Since the development in the early 1980s of genetic engineering technology, the research on aging has concentrated on the role of the gene and the changes that occur at the genetic level during the aging process. Much of this research is being focused on the genes that are expressed in either a greater or lesser degree in old age, the changes that occur in the promoters of genes, the trans-acting factors that alter with age, and the genes that vary in expression during aging under stress. However, with all the research that has been aimed at elucidating the changes in genes, we still do not know how many genes are involved in the aging process, nor do we know the specific genes responsible for aging. Also, whether the same set of genes undergoes alterations in expression in all organs during aging or whether there are tissue-specific alterations in the expression of genes remain to be answered.

In *Genes and Aging* I have attempted to assemble the information on age-related changes in genes that has appeared primarily in the last ten years, since it is during this period that biochemists and molecular biologists have concentrated on this problem. To make the book as comprehensive as possible I have also included information on the changes in gene products, the enzymes and other proteins, and chromatin as it relates to the structure of eukaryotic genes. This information is intended to provide new researchers in the field with the basic information necessary to study age-related changes in genes. At the same time I have tried also to point out the gaps and deficiencies in our knowledge of genes as they relate to aging.

ix

This basic information on age-related changes in genes is also intended to be useful to teachers, students, and researchers in this field. The book is not a compendium of data nor is it a catalog. But rather it is designed to be a reference book for researchers as well as a textbook for a course on genes and aging. I hope that this book will be an aid in the future research on aging. I will have fulfilled my objective if it stimulates young entrants to the field of aging research and generates an interest in some of the questions posed. A step toward our goal of understanding "why we age" will then have been taken.

When I wrote *Biochemistry of Ageing* in 1980, the farthest this science had advanced was the research on age-related changes in chromatin. Little did we realize then that genetic engineering technology would promote a major breakthrough in our studies on genes and aging – and this in only 10 years! It is an encouraging sign, and the next decade should see another breakthrough in our research on the biochemistry of aging.

Aging is a subject that is being taught in many universities all over the world, and an increasing number of researchers in the life sciences are entering this field. The interest in aging stems not only from the exponential increase in the number of elderly people and the recognition of the immensity of the problem, but also because aging as a discipline is a challenging biological problem for study. Fortunately, in this last decade of the twentieth century, molecular biologists have begun intensive studies on the gene, which may signal the final assault on understanding the changes that occur in genes. Hopefully, the next generation of scientists will be able to unravel the mysteries of the aging process and use their discoveries for the benefit of humanity.

I am conscious of the fact that despite my best efforts, I might have omitted certain data, and the work of some authors might not have been rightly interpreted. I am subject to correction on these scores. I am aware that in a work on genes, a lot of new information might have appeared between the date of submission of the final manuscript and the date of book publication. Indeed, as I write this preface, I am finding excellent new papers and reviews that I wish I could cite, but, in this rapidly moving field, such an attempt would become a never-ending task. To reiterate, my basic purpose in writing this book is to analyze the data at hand and to make projections for future work.

Readers may wonder why the aging of plants has been completely omitted in this book. I have neither carried out any work in this field nor have I kept up with the literature on plant aging. I thought it wise, therefore, not to enter an unfamiliar domain. Even so, there are several similarities between plant and animal aging, and the basic cause of aging, when known, may be found to be the same for both systems. Hence, students of plant aging should find the book informative and useful in that it deals with changes in genes, a subject that should be useful for their studies on plants.

The award of a Jawaharlal Nehru Fellowship during 1987–9 made it possible for me to write this book while at Banaras Hindu University. I am most grateful to the Jawaharlal Nehru Memorial Fund for this award. And I take this opportunity to express my continued esteem for the late Mr. Jawaharlal Nehru, who was a source of inspiration for me in my formative years, and for the late Mr. Rajiv Ghandhi, whom I had the honor of meeting soon after receiving the Nehru fellowship in 1986. I remember him with fondness and respect.

I would also like to thank the Council of Scientific and Industrial Research for the offer of an Emeritus Scientistship which I took up at Banaras Hindu University. A part of this book was written during my stay as Visiting Professor at the Tata Institute of Fundamental Research, Bombay.

I am grateful to Professor D. S. Kothari for his good will and encouragement during my academic career. The good wishes of Dr. K. L. Shrimali have been a constant source of encouragement. I also wish to thank Professor C. S. Jha, Vice-Chancellor of Banaras Hindu University, for his good will and for making my stay at the university a fruitful one.

I would like to express my gratitude to Mr. Biju Patnaik, Chief Minister of Orissa, for encouragement and support in setting up the Institute of Life Sciences at Bhubaneshwar. The library facilities of the Institute of Life Sciences, the university and medical libraries of Banaras Hindu University, and the Indian National Science Academy library have been of immense help in writing the book.

My Ph.D. students – several of them are now colleagues – have worked with me during the past 30 years to build in Banaras Hindu University a viable group devoted to research on aging. They have traveled and evolved with me, both in time and in depth, from enzymes to chromatin to genes, and made my research on aging an enjoyable and exciting experience. Especially, I wish to thank Dr. M. M. Chaturvedi for going through some sections of the manuscript and providing useful comments. The data reported from this laboratory are the work of Drs. S. N. Singh, S. P. Shukla, Behrose Gandhi, Sudha Rao, O. Koul, V. K. Moudgil, S. K. Patnaik, B. K. Ratha, T. C. James, G. B. Chainy, S. K. Srivastava, M. K. Thakur, Ratna Das, P. C. Supakar, M. M. Chaturvedi, B. R. Das, P. C. Rath, Y. K. Jaiswal, Anita Singh, Shweta Saran, and Sanjaya Singh. Sanjay Gupta, Rashmi Upadhyay, Monisha Mukherjee, R. N. Misra, and G. Mahendra have been of immense help in the preparation of this manuscript. My thanks are also due to A. R. Ganesh and R. Sharma for typing the manuscript, and to A. N. Singh for preparing the illustrations.

I am grateful to the Jawaharlal Nehru Memorial Fund, the Department of Science and Technology, the Government of India, the C.S.I.R., the University Grants Commission, and the PL-480 (U.S.A.) for the generous research grants that have supported my research on aging. My thanks are due to Cambridge University Press, especially to Dr. Robin Smith, and to Edith Feinstein and Kathryn Torgeson, for their help in publishing this book. I also wish to thank all the authors and publishers who have given me permission to reproduce the figures and tables from their publications.

I owe a special debt of gratitude to my wife, Sarat, for her understanding, encouragement, and patience that made the writing of this book possible. And the affection of my sons, Manas, Rajesh, and Tapas, has been a source of great satisfaction.

M. S. Kanungo

1
Introduction

Aging as a phenomenon in the life spans of organisms has intrigued mankind from time immemorial. Why and how is it that having attained a vigorous adulthood, all functions should undergo decay? The duration of this phase, which is referred to as aging or senescence, varies with species. It can be as short as a few days as in the female octopus which lays eggs only once, broods them, reduces its food intake, and dies soon after her young hatch. Among the so-called marsupial mice of Australia, the males live for only about a year. When they approach the end of their lives, they stop eating and engage in a competitive, brief but frantic mating. All males die shortly thereafter, perhaps due to hormonally induced stress. The females live long enough to suckle their young and wean them. Very few females live long enough to breed again. The female Pacific salmon also lays eggs once, and then dies soon after. These are sort of "sudden death" phenomena that occur soon after one-time reproduction, and the period of aging is too brief to be perceptible in these organisms.

In most species, however, such a phenomenon is not seen even when a large number of eggs or young are produced. Mice and rats give birth to a large number of young, take care of them during their weaning period, and are ready to breed again soon after. In higher mammals such as humans and elephants, only a few young are produced during the entire life span with long gaps, sufficient to take care of the young during the crucial early developmental period. These species have long life spans, and live long after reproduction has ceased. For example, female rats stop reproducing after about 1.5 years, but they may live thereafter for another 2 years. Human females usually do not reproduce after 45 years, but they may live up to 100 years. Thus, the duration for which an organism lives after attaining reproductive maturity varies widely among species.

Another important feature in the life span of animals is that in the

1

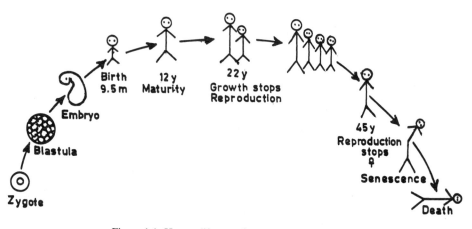

Figure 1.1. Human life span (m = months; y = years).

species which reproduce many times and for long periods, the initial reproductive rate is high, but it gradually decreases with age until it stops altogether. Concomitantly, the vigor and vitality of the animal also decrease. However, most of the functions go on late into life though decreasing gradually until the organism dies. This is a universal characteristic of all organisms.

Several questions arise. Why is an organism unable to maintain the vitality and vigor including the reproductive ability that it attained in the earlier age? Why does a mouse live for 3 years, a cat for 25 years, a horse for 45 years, an Indian elephant for 70 years, and a man for 100 years? Is it possible to maintain both vitality and vigor until the end of the life span? In other words, is it possible to add life into years and not merely years into life? This has been the wish of mankind. Scientists interested in this field of research wish to understand what changes occur after the attainment of reproductive maturity that cause deterioration of practically all functions including reproduction.

Life span is a continuum (Fig. 1.1). At one end is development and growth, and at the other end is deterioration of functions or senescence or aging. In between is the reproductive phase or adulthood. The time of onset, duration, and rate of senescence are dependent on the vigor and vitality of the reproductive or adult phase, and those of the reproductive phase are dependent on the vigor and vitality of the developmental phase. The three phases are thus interrelated. Hence, aging or senescence should not be considered as an isolated and independent phase of the life span. Information on developmental and reproductive phases may, therefore, help in our understanding of the mechanism of senescence.

Phases in life span

Development

The life span of a multicellular organism may be broadly divided into three phases: development (growth), reproductive period (adulthood), and senescence (aging). The developmental phase includes an increase in the number and size of cells, their differentiation to perform specialized functions, and formation of organs. At the molecular level, several genes that play specific roles at specific times have been identified for the developmental phase. These are, for example, *myo* D gene for differentiation of skeletal muscle cells, homeotic genes for segmentation of the body, and fibronectin gene for morphogenesis and organ formation. Several new proteins appear during this period, indicating the expression of hitherto inactive genes. The levels of proteins change as cells differentiate and organs form, indicating changes in the expression of corresponding genes. Concomitantly, the sizes of organs increase, which results in an increase in the size of the organism and its functional ability. These changes confer reproductive ability on the organism at a specific time when a certain stage of growth has been attained. Growth, however, generally continues even after reproductive ability is attained. For example, in humans, reproductive ability is attained at about age 12, but growth as measured by an increase in height continues up to about 20 years of age. In certain organisms like the poikilothermic vertebrates and invertebrates, growth continues long after reproductive maturity is attained. There is a good correlation between the time taken to reach reproductive maturity and the maximum life span, at least in mammals (Table 1.1). The exceptions seen in a few cases may be due to their adaptation to special habitats.

Reproductive phase

This phase is of special importance to the organism as it enables it to reproduce its own kind. The structures and functions that have evolved during natural selection in the species confer reproductive ability to the organism and aid in the perpetuation and evolution of the species. Certain genes that have been inactive up to this phase are now expressed,

Table 1.1. *Longevity and time to attain reproductive maturity for various mammals*

Scientific name	Common name	Maximum life span (months)	Length of gestation (months)	Age at puberty (months)
Homo sapiens	Man	1,380	9	144
Balaenoptera physalus	Finback whale	960	12	—
Elephus maximus	Indian elephant	840	21	156
Equus caballus	Horse	744	11	12
Pan troglodytes	Chimpanzee	534	8	120
Gorilla gorilla	Gorilla	472	9	—
Ursus arctos	Brown bear	442	7	72
Canis familiaris	Dog (domestic)	408	2	7
Bos taurus	Cattle (domestic)	360	9	6
Macaca mulatta	Rhesus monkey	348	5.5	36
Felis catus	Cat	336	2	15
Sus scrofa	Swine	324	4	4
Saimiri sciureus	Squirrel monkey	252	5	36
Ovis aries	Sheep	240	5	7
Capra hircus	Goat	216	5	7
Sciurus carolinensis	Gray squirrel	180	1.5	12
Oryctolagus cuniculus	European rabbit	156	1	12
Cavia porcellus	Guinea pig	90	2	2
Rattus rattus	House rat	56	0.7	2
Mesocricetus auratus	Golden hamster	48	0.5	2
Mus musculus	Mouse	42	0.7	1.5

Source: Rockstein, Chesky, and Sussman (1977).

as for example, the genes that code for the hormones FSH (follicle-stimulating hormone), LH (luteinizing hormone), in vertebrates, and for ovalbumin, vitellogenin, and lysozyme for egg formation in egg-laying vertebrates. Organisms that do not have a reproductive phase are of no importance for the perpetuation and evolution of the species.

Generally, the higher the number of offspring produced, or the faster the reproductive rate, or shorter the generation time of a species, the shorter is its maximum life span. For example, mice and rats reproduce faster and have shorter life spans than larger mammals like humans and elephants (Table 1.1). As mentioned above, there are some species that reproduce only once and die shortly thereafter. It appears as if reproduction depletes the organism of some essential substances, which are not replenished as fast as they are lost. Indeed, estradiol and testosterone, the two important hormones responsible for female and male fertility, respectively, decrease gradually after the initial reproductive phase. Also, the expression of genes responsible for the synthesis of egg proteins decreases, and egg laying also decreases.

As in the case of the developmental phase, the duration of the reproductive phase is also more or less defined, particularly in females. In humans, the reproductive phase is reached at about 12 years of age in both sexes. In females it ends at about 45 years when menopause occurs. Rats mature at 10 weeks, and the female rats stop reproducing at about 80 weeks.

Senescence

In senescence, which is a characteristic of all multicellular organisms, the functional abilities of all organs and the organism decrease. The decline becomes perceptible toward the later part of the reproductive phase. Thus, the reproductive phase smoothly merges into the senescence phase, unlike the transition from the developmental to the reproductive phase in which specific genes are expressed, and specific structures and functions appear that confer reproductive ability on the organism. One important feature of senescence is that reproduction does not occur in this phase. No special structures or functions appear; rather, those that are already present undergo change. So far no new genes have been shown to be expressed or repressed that may be responsible for aging. However, several genes that are already active have been shown to undergo changes in their expression.

The duration of senescence is not well defined because it is not known at what stage after the attainment of reproductive ability deterioration of functions begins. If cessation of reproductive ability is used as a criterion for senescence, then it begins around age 45 in human females. But it is common knowledge that in both males and females several functions, such as muscular activity and vital capacity of the lung, begin to decline from around age 30. The senescence phase is of little importance for the perpetuation and evolution of the species as the organism does not reproduce, except in humans and other higher vertebrates where parental care is important for the safety and growth of the offspring during its early developmental period. Among wild animals, only a few are able to live on to any appreciable period of the senescent phase as they are exterminated by predation, or die due to environmental changes or a lack of food.

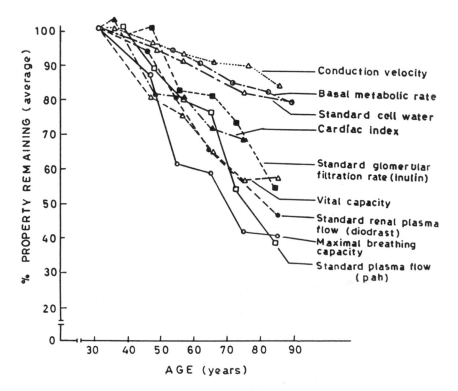

Figure 1.2. Decrease with age in some physiological activities in humans. Activities have been expressed as percent of mean value at 30 years of age. (Shock, 1959)

Functional changes during aging

Longitudinal studies carried out on various functions of human volunteers at increasing ages have shown that different functions decline at different rates (Fig. 1.2). These studies, which were conducted on 30-year-old volunteers, measured physiological functions. Therefore, they represent gross changes in the activities of cells and organs. It is likely that molecular changes leading to these physiological changes occur even earlier. Nevertheless, these studies show that (1) the rates of decline with age in the functions of different organs are different, and (2) different organs begin to decline in function at different times. There is great variability not only in the rates but also in the time at which various functions begin to decline in different individuals. There is no function or parameter that begins to decline at a fixed age and at a fixed rate in all individuals. This decline is also observed in experimental animals under given environmental conditions. So it has not been possible to define aging in terms of a specific parameter. The chronological age of an organism is the only parameter that is used by biologists, more for convenience, as an index of aging. However, it is common knowledge that this criterion can be very misleading since not infrequently one finds a man of 60 who may look like and be as active as a man of 40 or a man of 40 who may look like and be as inactive as a man of 60. In the absence of any definite parameter at any level – organ, cellular, and molecular – aging has been described as a process that causes a gradual decline in function and adaptability of an organism to environmental changes following the onset of reproductive maturity or a process that makes the organism more susceptible to diseases following the attainment of reproductive maturity.

Generally, the decline in overall activity of the organism accelerates with increasing age, which is evident in humans after 60 years of age. That is, the rate of decline in overall function between 70 and 80 is faster than that between 60 and 70, and that between 60 and 70 is faster than that between 50 and 60. This acceleration may be due to the cumulative effects of the decline in function of different organs or cells in the earlier phase. Hence, there is a continuous acceleration of deterioration of functions once it gets started. It is not known, however, when exactly the decline in function of each organ begins. The decline in function is obviously not uniform for any organ; it shows considerable variability among individuals of the same species, both in rate and magnitude, even in the same environmental conditions. Moreover, the rate and magnitude of decline differ in different environments. Those individuals that adapt better to the changes in the environment survive longer.

It is during this phase that adaptability to external and internal stresses decreases, and the homeostatic mechanisms deteriorate and susceptibility to diseases increases. Death occurs at some point during this phase not because all functions reach zero level, but because the functions of one or more organs so deteriorate due either to disease or other afflictions that life can no longer be sustained.

Life span

A survey of the life spans of various mammalian species shows that all individuals of a species have a more or less fixed maximum life

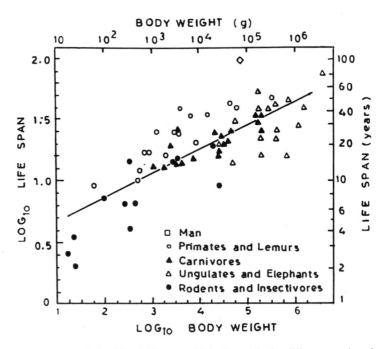

Figure 1.3. Relationship of life span with body weight for different species of mammals. (Redrawn from Sacher, 1959)

span (Table 1.1). This is also true of nonmammalian species. Not only that, various events in the life span of a species also occur at a more or less fixed time in the life span. For example, in humans, birth takes place after 9.5 months of gestation, reproductive maturity occurs at about 12 years, cessation of growth occurs usually at 20 years, and menopause usually at 45 years. Another interesting characteristic feature is that generally larger mammals have longer life spans, although there are exceptions, notably in humans. Also, there is an inverse relationship between metabolic rate and life span – the higher the metabolic rate, the shorter is the life span. This is to be expected because the surface areas of smaller mammals is larger than those of larger mammals on a per gram basis. So they lose heat faster than larger mammals. Hence, their metabolic rate as measured by oxygen consumption needs to be higher to maintain the body temperature and to counteract the loss of heat through the larger surface area. When life span and body weight are plotted on a log scale (Fig. 1.3; Sacher, 1959), a good correlation between the two parameters is found, with a few exceptions, especially

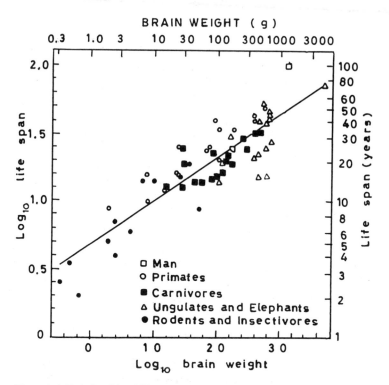

Figure 1.4. Relationship of life span with brain weight for different species of mammals. (Redrawn from Sacher, 1959)

for humans. According to this scale, human life span should be about 30, and less than that of horses. In general, primates have longer life spans than what is expected from this scale.

Most organs in a mammal have a close correlation with the total body weight, except the brain. Primates have relatively larger brains than those of other mammals. When weight of brain is plotted against life span in a log scale, a better correlation for the life span than with the body weight is found (Fig. 1.4; Sacher, 1959). A still better correlation with life span is found when both body weight and brain weight are taken into consideration using the following equation:

$$X = 0.636w + 0.198y = 0.471$$

where X = log life span; y = log body wt; w = log (brain weight/body weight$^{2/3}$).

It is conceivable that since a larger brain is expected to be more developed, the animal with a larger brain is able to sense changes in the environment and adapt to it better, and will have better homeostatic regulatory mechanisms than animals with small brains. Hence animals with larger brains will live longer. There are exceptions, however, even in this relationship. Even within a single species of mice, different strains with different body and brain size do not fit into this scheme (see Lamb, 1977). It is likely that natural selection favors larger brain size in long-lived species as these species have a greater learning capacity and have, therefore, a greater survival capacity. Therefore, selection pressures that favor increased longevity will also lead to an increase in brain size, since a larger and more developed brain enhances the chances of survival for longer periods.

An interesting relationship between gestation periods of mammals and their brain weight has been found by Sacher and Staffeldt (1974). Not only do mammals that have larger brains generally have longer life spans, they also produce smaller numbers of offspring. So it appears that selection for increased brain size results in a decrease in the reproductive rate. If this be so, a longer life span would result because for the perpetuation of such a species sufficient number of young have to be produced; this can occur only if the life span is long enough to compensate for the low reproductive rate. Hence, the long life span of mammals with a large brain is a secondary consequence of the selection for a larger brain. Such an argument, of course, does not hold good for lower vertebrates. The brain of a tortoise (*Testudo sumeiri*) is smaller than that of a rat, and yet it lives for over 150 years.

Longevity

That longevity within a species is determined by genetic factors has been well established. There is a positive correlation between longevity of parents and offspring. Abbott et al. (1974) found a direct relationship between the age at death of parents and the life span of their children (Table 1.2).

These data show that long-lived parents have long-lived children. Therefore, heredity (genes) of the parents contributes to longevity, though environmental factors including nutrition may also influence longevity of a species. That heredity is important for longevity is also evident from the finding that there is a greater similarity in the life spans of monozygotic twins than dizygotic twins (Kallmann & Jarvik, 1959). Since genomic content is identical in monozygotic twins, any differences ob-

Table 1.2. *Life span of progeny in families in which one parent lived for more than 90 years*

(1) Survivorship of offspring when father lived >90 years

Sex of offspring	Mother's age at death		
	<60	61–80	>81
Male	67.6	71.4	73.2
Female	73.8	74.1	77.2
Combined	70.9	72.8	75.1

(2) Survivorship of offspring when mother lived >90 years

Sex of offspring	Father's age at death		
	<60	61–80	>81
Male	67.0	69.3	70.9
Female	73.0	73.5	73.3
Combined	69.8	71.4	72.1

Source: Abbott et al., 1974.

served in the life spans of such twins are due to extrinsic factors. Since the genomic contents of dizygotic twins are different, any differences in their life spans are due to differences both in their genomes and extrinsic factors.

The findings on longevity from inbred strains of animals corroborate the importance of heredity in determining longevity. Matings of brother and sister and close relatives produce inbred strains of animals that are genetically more uniform than the outbred strains. Measurements of life spans of two short-lived inbred strains of *Drosophila subobscura* show no differences. When the two short-lived strains are mated, the F1 flies live for a far longer period (Fig. 1.5; Clarke & Maynard Smith, 1955). Similar results are obtained with inbred strains of mice. It is likely that the inbred strains have shorter life spans because they may be homozygous for certain recessive genes which may affect longevity adversely. The greater longevity and hybrid vigor seen in the F1 generation after mating of the two strains may be because the fly is heterozygous at most loci, especially at the loci that affect longevity. Similar results are obtained when mice strains with different longevities are inbred and

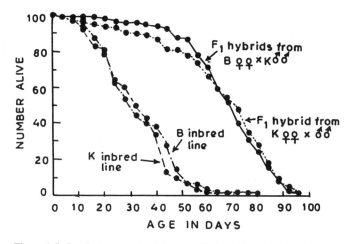

Figure 1.5. Survival curves for inbred and hybrid *Drosophila subobscura*. (Redrawn from Clarke & Maynard Smith, 1955)

outbred. At what stage of the life span the genes of a short-lived strain influence longevity is not known and requires further study.

Support for the role of genes in longevity determination has come from studies on fibroblast cells of patients suffering from premature aging syndromes. Two types of such diseases are known – Werner's syndrome and progeria. Werner's syndrome appears to be due to an autosomal recessive gene located on chromosome 8 (Goto et al., 1992). Patients with this disease have normal childhoods, healthwise. However, growth ceases during their teenage years; premature graying of hair and baldness occur; and the skin wrinkles and muscles atrophy. Other symptoms are hypogonadism, atherosclerosis, osteoporosis, soft-tissue calcification, juvenile cataracts, and a tendency toward diabetes and poor wound healing. Death, which occurs at a mean age of 47, appears to result from cardiovascular complications and/or malignancies. Fibroblast cells from these patients show a significantly lower number of population doublings in in vitro culture. The cells also have a significantly higher number of chromosomal abnormalities relative to the age-matched controls.

Progeria is probably caused by spontaneous and sporadic production of a dominant mutation in one of the parent's germ lines. The patients usually appear normal at birth, but begin to lose their hair and subcutaneous fat beginning from the first year of life. Growth slows and finally ceases. The skin becomes thinner, age spots appear on the body, bone resorption occurs, sexual development is limited, the nose is beaded, and the jaw is underdeveloped. No neurofibrillary tangles appear in the central nervous system. Surprisingly, the children have a normal to above-normal intelligence. The average age at death of these patients is 12 years, due generally to severe coronary artery disease. Their fibroblasts also undergo far fewer population doublings in in vitro culture than age-matched controls.

Although both diseases undoubtedly have a genetic basis, there are both similarities and differences in their respective symptoms. They appear to be more pathological than what is seen in normal aging. The symptoms are very varied, and therefore it is likely that the gene(s) has pleiotropic effects. Because the symptoms develop very fast, it is likely that a single gene initiates actions that bring about a cascade of effects involving several genes. Hence, these patients provide excellent experimental models for understanding gene action both in vivo and in vitro.

Several factors influence longevity. One of the best examples is the extension of the life span of rats by controlling nutrition (McCay, Crowell, & Mognard, 1935). Young rats fed a diet that was nutritionally adequate but low in calories had a slower growth rate, matured later than the rats fed ad libitum, and lived 30–40% longer than the controls. The diet restriction does not have any effect on the immune system, but retards the aging of collagen. It has been speculated that diet restriction decreases the metabolic rate, which may extend longevity (Arking, 1991).

Evolution of aging

The perpetuation and evolution of a species can occur through two strategies. In one, the organisms produce a large number of offspring, either once or a few times, and then die. Their young fend for themselves, and hence have a high mortality rate. A few live till reproductive maturity, reproduce, and die soon after. Such animals do not have long lives because if they do, the population would grow exponentially, and beyond a point would crash due to a shortage of food, competition among individuals, and a problem of space. In the second strategy, the reproductive rate is low and is linked with parental care. The mortality rate is low, and the individuals have long lives during which they reproduce several times, thereby ensuring perpetuation of the species.

Even though natural selection is believed to exert its influence by decreasing the frequency of genes that cause senescence and increasing the frequency of genes that prevent it, yet senescence persists. The most widely accepted evolutionary explanation for senescence is that it is beyond the influence of natural selection because predation, starvation, disease, and other environmental hazards cause the death of most wild animals before they reach the age at which bodily fitness starts to decline (Haldane, 1942; Medawar, 1957; Rose, 1985; Rose & Graves, 1990). That is, very few individuals live long enough to age. Since the number of individuals is very small, the force of natural selection is not sufficient to select the genes that cause senescence. It is believed that natural selection is weak at this level of organization. Therefore, it cannot sustain the traits that are detrimental to individuals, even if they otherwise benefit the individuals in some way (Williams, 1966). Even so, all animals undergo senescence and do not die immediately after reproducing once or twice, especially when environmental hazards are low. Therefore, increased longevity is likely to have evolved as a consequence of longevity assurance mechanisms (Sacher, 1980).

It is possible that senescence is a consequence of the effects of certain genes that are selected for their beneficial effects in adult life when the force of selection is the strongest. These genes are believed to be pleiotropic and have detrimental effects in later life (Williams, 1957). In other words, the senescence phase is a fall-out or late-acting effect of certain genes that were selected for some benefits to the adult. So there are no senescence-causing gene or genes. The pleiotropic effects that lead to senescence vary in duration. In species like salmon, which do not care for their young and in which the young have a high mortality rate during the interbreeding period, the deleterious effects of the pleiotropic genes are expressed suddenly because natural selection cannot oppose them, and death of females from senescence occurs within days (Williams, 1957). In species such as higher vertebrates that care for their young, the postreproductive period begins only after the young no longer benefit from parental care. Here, the deleterious pleiotropic effects of the genes that were selected for certain benefits to the adult, such as reproduction, generally begin to be expressed after the reproductive period is over.

In support of the above, Gustafsson and Pärt (1990) provide the example of the collared flycatcher (*Ficedula albicollis*) in which costly reproduction in early life accelerates senescence for fertility under natural conditions. The birds form an isolated population, and apparently are not exposed to environmental hazards in their natural habitat, unlike the birds that share the same ecological niche with other species. Yet their reproductive performance decreases with age, which may be due to some innate deterioration of the individual. These researchers compared birds breeding for the first time at the age of 1 year with those that started breeding at the age of 2 years. Then they compared individuals with experimentally enlarged brood size with those having reduced or unmanipulated brood size. The females that started breeding from the second year laid larger clutches throughout their subsequent lives than those that started breeding in the first year. So the former show late breeding and high fecundity, and the latter show early breeding and low fecundity. Furthermore, it was found that females with enlarged brood size laid smaller clutches later in life than those with reduced or unmanipulated broods. Thus, the cost of early reproduction is paid for in the form of reduced reproductive performance in later life. This is believed to be the result of selection for high early fertility during evolution (Williams, 1957). This is true for birds that breed in the first year and have a large clutch size.

Thus, it appears that reproduction itself induces senescence for fertility. It is likely that reproduction depletes certain essential factors that are necessary for maintaining reproduction itself, and the animal is unable to replenish these factors. It cannot be due to ecological risks because that should affect only survival (Partridge, 1989). Moreover, the collared flycatcher lives in an isolated population and, therefore, is subjected to minimal ecological hazards. Thus, the action of the pleiotropic gene, which is selected for conferring reproductive advantage by enabling early reproduction, has an effect on fertility itself later in life. So there is a link between early reproductive effect and late performance.

That such pleiotropic genes influencing senescence may exist has been inferred from breeding experiments in the laboratory. Selection for late reproduction in the female fruit fly increases longevity, decreases early fecundity, and increases late fecundity (Rose, 1984). Selection for early reproduction decreases longevity in *Tribolium* (Sokal, 1970). If pleiotropic genes have been selected by natural selection to confer benefit to the adult, but have deleterious effects that cause senescence later in life, then manipulation of such genes may extend the life span and prevent or postpone senescence. However, there is the danger that such manipulation may deprive the adult of the beneficial effects of the gene.

If one or more pleiotropic genes acting for the benefit of the adult have deleterious effects later, then why should the contributions of these genes vary so greatly in different species? Available data, though insufficient, suggest that pleiotropic effects are minimal in birds and are important in large mammals (Nesse, 1988). Despite the variations in the period of senescence, the declining force of natural selection with increasing age due to a sharply declining number of individuals may be the basis for the evolution of senescence.

Arking (1987) has analyzed the onset of senescence and longevity in *Drosophila* by generating long-lived strains through artificial selection. Both the mean and the maximum life spans are extended, the mean life span exceeding the maximum life span of the controls. This is due to a genetically based delay in the onset of senescence. The increase in the duration of the presenescent period is under both genetic and environmental control. Senescence itself is not under genetic control and appears to occur stochastically. Selection for decreased longevity is unsuccessful, which may be due to the requirement of a minimum species-specific life span.

Not only do genes that confer reproductive ability need to be selected through natural selection for the perpetuation of the species, but also it is likely that genes that control earlier events leading to reproductive maturity have also been selected; otherwise a mature organism will not be produced. However, any gene or genes that would perpetuate the juvenile stage and prevent maturation would be selected against. Thus, each stage of an organism leading to maturity is stabilized by the selection of specific genes, since the force of natural selection up until maturity is high because the number of individuals is high. Natural selection is unable to operate effectively after the attainment of maturity, since the number of individuals is considerably decreased. Hence, even if some of the traits of senescence, such as parental care, are advantageous, the possibility of selection of these traits is low.

Cell death

Two types of cell death occur during the life span: cell death in the embryo during the formation of organs, and cell death due to aging after adulthood. For the formation of an organ in the early embryo, many more cells are formed than are needed. The specific contour of the organ is then derived by the death of cells located at specific locations just as a sculptor chisels out a stone to produce a particular structure. The cells that must die at specific locations are earmarked, and the timing of their death is also fixed. The result is the production of an organ having a specific structure, which performs a specific function. Such cell death is seen during the formation of digits in the limb of a vertebrate embryo or the wing of a developing insect. Cell death that occurs during the embryonic period is not senescence because the death of these cells leads to an active and vigorous organ and an active organism.

Cells that die after the organs have fully formed and the organism has become reproductively mature are a sign of senescence or aging. Neurons, and cardiac and skeletal muscle cells are examples of such cells. These cells stop dividing soon after birth, become postmitotic, continue to perform specialized functions, and die at various stages after reproductive maturity is reached. Neither the timing of the death of these cells is fixed, nor are the cells that should die earmarked, unlike the death of cells during embryonic development. Since these cells are not replaced their death impairs the function of the organ. Therefore, such cell death is a sign of aging.

Bone marrow and epithelial cells are different from neurons and muscle cells since they continue to divide throughout the life span and hence are premitotic. Their cell-cycle time is short, but gradually becomes longer with age. The liver contains cells whose rate of division is very slow. At any given time, <0.01% of liver cells are in a mitotic state. However, if a part of the liver is excised, the remaining cells divide until the liver regains its normal size. The trigger for division of epithelial cells is apparently different from that for hepatic cells after hepatectomy. In the former, the signal is given at regular and short intervals. The liver cell, however, stays in the Go stage for a long period (days or weeks) before it divides again. What is that late-acting trigger? If hepatectomy is performed, all the cells immediately divide. This is apparently a stronger stimulus since it bypasses the stimulus that normally makes a hepatic cell divide.

Skeletal cells, cardiac muscle cells, and neurons lose the capacity to divide very early in the life span. They gradually die as a function of age. Muscle cells have the capacity to divide and regenerate only in the early part of the life span. When the muscle cells die, their place is taken by collagen and fibroblasts, and the function of the muscle gradually decreases. The neurons in mammals lose the capacity to divide soon after birth. Each neuron then establishes synaptic connections with other neurons through its axon and innumerable dendrites. These connections are required for the storage and retrieval of information and the control of various functions which last till their death. The division of neurons after these connections have been established would destroy these control mechanisms, because the two daughter cells that would result from the division of one cell have to re-establish all the connections. Besides, there will be the problem of space to accommodate the ever-increasing number of cells within the fixed skull. Therefore, the only alternative left is to keep the activities of the differentiated cells for as long as possible in order to postpone the onset of aging. This can be achieved by stabilization of the homeostatic and regulatory processes in the cell. That this has occurred is evident from the continuous increase in the life span and the reproductive period or adulthood of mammals during their evolution.

The science of aging

Aging is a universal phenomenon, just as development is. The two phenomena are at the two ends of the life span. Several important findings have been made in the elucidation of the problem of development, though much more remains to be explored for a proper understanding of the mechanism of development. Easier availability of experimental materials and well-timed events during development have greatly aided the progress in the field of development. These advantages are not available for studies on aging, although the techniques of genetic engineering and molecular biology that have been used for the past decade have increased our understanding of the types of changes that occur at the genetic level after adulthood. As mentioned above, not only the life span but also several events during the life span appear to be genetically controlled. However, the variations that are observed in the timing and duration of various phases such as development, adulthood, and aging are modulated by extrinsic factors such as nutrition, temperature, radiation, pollution, and psychological stress, as well as by intrinsic factors such as hormones and free radicals.

The elucidation of the basic cause of aging is an intellectually challenging and fundamental problem. The important questions that have confronted the biologists are: Why do all organisms undergo deterioration of function after attaining reproductive maturity? Why do all members of a species have a more or less fixed life span? At what age after attaining reproductive maturity does the process of aging begin? Is there a trigger or switch that sets in motion the process of deterioration in all organs? If so, how is it switched on? Is this process programmed, just as the process of development is believed to be? Is there a single universal cause of aging? Or does the cause differ with the organization of the system?

Answers to the above questions may help in devising methods for postponing the onset of the aging process and for the control of aging. These measures could conceivably prolong the active and youthful period from age 20 to 40 years to say 20 to 60 years or more. This will also greatly increase the period of work output in humans, besides giving them the psychological satisfaction of being youthful longer. The objective of all research on aging is to ensure better health for a longer period, and not merely an increase in the number of years lived. An extension of youth is likely to increase the average life span, and each individual may hope to live up to the maximum life span of about 100 years with a shorter period of senescence.

Even if all the diseases of old age are controlled, the organism will eventually die, because a living system, like any other dynamic system, is subject to the laws of nature. With the passage of time, the organism will progress toward a more probable and equilibrium state due to increase in entropy, and its homeostatic machinery will break down, resulting in the deterioration of vital functions and death. The prolongation of the adult period is expected to defer the onset of old-age diseases like Alzheimer's disease, cardiovascular and cerebrovascular diseases, cancer, and arthritis. The control of these diseases would be of great benefit to both the individual and society.

Several theories and models have been proposed to explain the cause of aging. They fall into two categories: (1) Theories that explain aging on the basis of changes occurring at the genetic level. These theories consider aging to be due to changes occurring at the primary site, that is, the gene. (2) Theories that consider aging to be due to changes in various gene products such as enzymes, collagen, and hormones; accumulation of free radicals; deterioration of structures such as cell membranes, lysosomes, and mitochondria; or changes in the immune system and homeostasis. It is obvious that these changes are secondary in nature because any change in the structural composition, regulation, accumulation, or depletion of factors would be initially due to changes in the functions of the specific genes responsible for the synthesis of various biochemical components.

Studies are being conducted on organisms of different complexities including humans, lower mammals, insects, nematodes, and protozoa. In vitro cell cultures are being used to answer specific questions at the genetic level. Although each model is useful in deriving some insight into the problem of aging, each has also certain drawbacks. Much progress has been made during the past three decades toward understanding the types of molecular and biochemical changes that occur as an animal ages. Alterations in the levels of enzymes and isoenzymes, and their induction by various hormones have given useful information about the expression of their genes. Since genes are complexed with chromosomal proteins to form chromatin, various techniques have been used to study the changes in the structure of chromatin during aging, since such changes influence the expression of genes. More recently, studies have been carried out on specific genes. Northern blot analyses have shown that the expressions of several genes decrease with age, whereas certain other genes are more expressed in old age. The expression of genes is altered by several factors such as hormones and temperature. These responses are generally lower in the older organism. Thus, the adaptability of the organism to changes in the internal and external environments appears to decrease with age. Since the expression of a gene is modulated by cis-acting regulatory sequences located at its 5′ flanking region, several workers have begun studying the trans-acting nuclear factors that bind to the cis-acting elements in the genes. This may show why the expression of a gene changes with age. It may be possible then to regulate the expression of a gene by altering the levels of trans-acting factors.

Data on the above aspects are discussed and analyzed; gaps in our knowledge that need to be filled are highlighted; and the possible approaches to the problem of aging that may answer key questions are examined in the following chapters. As will be evident, much more work remains to be done to derive answers to specific questions on the problem of aging. It is hoped that with the powerful tools of genetic engineering now available it will soon be possible to gain an insight into the basic mechanism of aging at the genetic level.

2
Phenotypic changes during aging

The genome in a eukaryote houses far more DNA and genes than it needs for its various functions. This is the inference one draws if one counts the various types of enzymes, structural proteins, mRNAs, tRNAs, and rRNAs that are coded by the genome. Only 3%–5% of the genome accounts for these molecules. Of the remaining, nearly 40% comprise repetitive DNA whose exact function is not known. What the rest of the DNA does is not known. Assuming that it is only the expressed fraction of the DNA or genes that plays a role in the aging process or holds the key to the aging process of an organism, one may look for changes that occur in this fraction during its life span, and hope to find a common pattern of changes in specific sets of genes. Such a finding may give an insight into the possible causes of aging since genes have been implicated in the following fixed-time life processes of mammals: gestation, attainment of maturity and growth completion, duration of reproduction and fertility, life span, and body size. These times may vary with the species of mammals. The variations seen in these characteristics among individuals within a species have been attributed to various extrinsic factors such as nutrition, temperature, radiation, and stress.

Biochemical research on aging commenced in the early 1950s. Until about 1970 the research was confined to mostly studies on changes in enzymes and structural proteins. It had two purposes: First, since enzymes catalyze all functions of the body, an understanding of their changes during the life span may throw light on the aging process. Second, since all proteins (including enzymes) are coded by genes, understanding them may give an insight into the types of changes that occur in genes as an organism ages. This approach was inevitable as methods had not developed to study eukaryotic genes. Beginning from the early 1970s, researchers looked into the transcription of various types of RNAs, but it was not until the 1980s that they started examining the way in which RNAs are transcribed by specific genes and the way the expression of genes is modulated. Nevertheless, studies on proteins that reveal how their levels may change and be modulated during the life span, as well as changes in isoenzyme patterns have given useful clues to the ways the expression of the corresponding genes change during the life span. Once the proteins are translated from their mRNAs, they undergo several types of posttranslational modifications. This type of change in proteins, however, does not reflect a role for genes unless one argues that the modifications are carried out by different enzymes which are coded by specific genes.

Since enzymes carry out all reactions in the body and are responsible for all structures, colors, forms, and sizes of the body, they are the main molecular phenotypes that represent specific genotypes of the organism. Hence, studies on the changes in enzymes and structural proteins during aging may give an insight into the types of changes that occur in genes as an organism ages. However, since only 3%–5% of the genotype of an organism is expressed as proteins, a large majority of the genotypic changes that may be occurring during aging remain unknown. Other approaches need to be taken to find out what happens to this vast, remaining amount of the genome as an organism ages. A large amount of data has accumulated on the various types of changes that occur in proteins during aging of animals. Certain representative examples of different aspects of proteins such as changes in their levels, induction, and isoenzyme patterns are described below. A more thorough treatment of these aspects may be found in *Biochemistry of Ageing* (Kanungo, 1980), which covers the literature up to 1978.

Changes in levels of enzymes

All biological reactions in an organism are catalyzed by enzymes. They are, therefore, essential for all functions and structures of the body. The initiation, duration, and termination of various phases in the life of an organism such as differentiation, development, and maturity may depend on various characteristics of enzymes such as their levels and isoenzyme patterns. Changes in their properties during aging may alter the activities of the organism. Much data have accumulated that show that the levels of certain enzymes decrease and others increase after reproductive maturity, while some enzyme levels remain the same. Each enzyme is coded by a gene, and for isoenzymes consisting of two types of subunits, two genes are involved. Hence, the study of various aspects of enzymes such as their levels, isoenzyme patterns, inducibility, and molecular properties may give insight into the mechanism of aging at the genetic level.

Several workers have compared the levels and activities of enzymes of adult animals with those in older animals. These data have several pitfalls. (1) Different authors have used different assay procedures which give different values for the same enzyme. (2) Assays of enzymes have generally been done at saturating substrate concentrations, a condition which does not occur in tissues under normal physiological conditions. (3) The levels of several enzymes have circadian rhythms. Hence measurement of their activities at different times of the day would give different values. (4) Different workers have expressed enzyme units differently: grams wet weight, grams dry weight, milligrams protein, and milligrams DNA. Since the levels of most of these parameters, especially wet weight and dry weight change as a function of age, a meaningful comparison of the activity of an enzyme at two different ages cannot be made. Also, no meaningful conclusion can be drawn on the activity if one enzyme of a metabolic path is expressed as units per milligram of protein and that of another of the same path as units per gram dry weight. (5) Different workers have considered different ages as old. For example, some workers have taken 60-week (15–16 months)-old rats as old. Others have regarded 100-week-old rats as old. (6) Two populations of animals of the same chronological age but kept on different diets may not have the same physiological condition. (7) Data for one sex are not comparable with those of another. (8) In general, the liver of mammals has been used for aging studies. Since approximately 0.01% of liver cells are in the process of division at any given time, this organ contains a heterogeneous mixture of dividing and nondividing cells. Therefore, data on mammalian liver cannot be compared with those of the brain, and skeletal and heart muscles whose cells stop dividing soon after birth.

Taking into account the above deficiencies in the studies on age-related changes in enzymes it is seen that no comparable data on all enzymes of even one metabolic path are available for any organism as a function of its age. Especially, data on the levels of regulatory enzymes of a metabolic path at different ages may be of greater significance than those of constitutive enzymes. For example, the study of the levels of the glycolytic enzymes, glucokinase, phosphofructokinase (PFK), and pyruvate kinase and the gluconeogenic enzymes, pyruvate carboxylase, fructose 1,6-diphosphatase (F-1,6-diPase), and glucose-6-phosphatase

(G-6-Pase), which are the key regulatory enzymes for glucose metabolism, would be of much value for understanding how glucose metabolism changes during aging.

Several reviews on changes in enzyme levels have been published (Kanungo, 1970; Finch, 1972; Wilson, 1973; Florini, 1975; Kanungo, 1980). Kanungo (1980) took a different approach by comparing the data on changes in enzymes during the aging of animals. Using the Enzyme Commission's system, he compiled the data on the six classes of enzymes in order to find out if enzymes of a particular class that catalyze a specific type of reaction show a specific type of change. To make the comparison meaningful, only data on enzymes whose activities were expressed as units per milligram of protein or units per milligram of DNA were included. The enzymes were also grouped according to their location in the cell. Data on female rats that were older than 80 weeks were used to represent old age, since rats stop reproducing after this age.

It was found that within each class of enzymes, the activities of some enzymes of old rats and mice are lower than those of adult animals, and some others are higher. Some enzymes do not show any difference in activity. Thus all the enzymes within a class do not show a specific type of change. Furthermore, certain enzymes that decrease in one tissue in old age increase in another. The number of enzymes studied are not large enough to make a broad generalization. In general, it appears that the enzymes responsible for oxidation decrease in activity after adulthood in postmitotic tissues such as the heart, skeletal muscle and brain. The activities of enzymes located in different compartments of a cell do not show any specific pattern of change.

The activities of enzymes of invertebrates show a somewhat definite pattern of change. Certain oxidoreductases decrease whereas others increase, and certain oxidoreductases do not show any change. However, all transferases studied so far decrease in activity, and a majority of hydrolases increase in activity. The studies on invertebrates have a major defect in that whole animals were used instead of specific tissues. So no definite conclusions can be drawn from these studies.

The changes in the activities of certain representative enzymes are described below. These data, however, do not throw much light on the types of changes that may occur at the level of the genes coding for these enzymes.

A general observation made by Reznick et al. (1985) is that the number of faulty enzyme molecules increase with age. They purified superoxide dismutase (SOD), glyceraldehyde-3P-dehydrogenase (Gly-

3P-DH), and aldolases B and C. Antisera were raised against the purified "native" enzymes and heat-denatured proteins. Western blot analyses showed that the levels of "native" active enzymes in all three cases were lower in old animals. The active-enzyme fraction present in the old animals had the same specific activity as that of the young animals. It appears that the enzymes undergo modifications with age before they are degraded. Also, these proteins have a longer half-life due either to a decrease in proteolytic enzymes or due to their denaturation.

Darnold, Vorbeck, and Martin (1990) have studied the enzymes involved in oxidative phosphorylation of adult and old rats. No age-related difference in the cytochrome C oxidase is seen. However, the activity of translocase declines. The age-related change in respiration is attributed to differences at the level of electron transport system, including its associated reactions. The ability of mitochondria of the old animal to respond to increased demand of oxidation is lower than that of the adult.

SOD protects cells against oxygen toxicity by removing superoxide radicals ($\cdot O_2^-$). It catalyzes the reaction which inactivates the superoxide radical:

$$\cdot O_2^- + \cdot O_2^- + 2H^+ \longrightarrow O_2 + H_2O_2$$

SOD is a dimer having two identical subunits and in eukaryotes has one copper and one zinc noncovalently attached to each subunit containing 153 amino acid residues. It has been postulated that free radicals increase with age and inactivate DNA, RNA, and proteins (Harman, 1981). Several workers have studied the activity of SOD as a function of age, because a decrease in the activity of the enzyme in old age would indicate an increase in the level of superoxide radical. In addition, glutathione peroxidase (GP) and vitamin C and E levels have been found to be lower in old age (Lebovitz & Siegel, 1980). SOD, GP, and vitamins C and E are scavengers of free radicals. Hence, it is reasonable to assume that the free-radical level increases with increasing age. If synthetic antioxidants are incorporated into the diet of mice, their mean survival time is significantly increased. There is good evidence that not only the activities of SOD and GP in old animals are lower than those of young animals, but also the levels of vitamins C and E are lower in old men, guinea pigs, and mice. Also, it has been reported that longer-lived species have higher levels of SOD (Tolmasoff, Ono & Cutler, 1980).

The cytochrome P-450s are a family of hepatic enzymes that metabolize drugs and toxic substances; they are induced by phenobarbitone

(PB). Sun, Lau, and Strobel (1986) examined the induction of six forms of cytochrome P-450s by PB and found a differential induction of the various forms during aging. Different cytochrome P-450s have different substrate specificities. Hence, the synthesis and induction of various forms of the enzyme during aging may affect the detoxification of toxic substances. Certain enzymes involved in protein synthesis also decline in activity. Gabius, Graupner and Cramer (1983) found in the nematode *Caenorhabditis elegans* that the activities of all aminoacyl-tRNA synthetases, except that for tryptophan, are lower in old nematodes. Several bases, particularly guanine and inosine of tRNAs, undergo methylation posttranscriptionally by tRNA methylase. The activity of tRNA methylase decreases after adulthood, and qualitative alterations in substrate specificities are seen.

Takahashi and Goto (1990) studied the effect of active oxygen on the activity of aminoacyl-tRNA synthetase of the rat liver as it has been suggested that superoxide and free radicals (HO·, ROO·) increase in old age. Treatment of leucyl-tRNA synthetase with Fe-ascorbate increases its heat lability. This inactivation is inhibited by free-radical scavengers such as mannitol and benzoate, which indicates that hydroxyl radicals (HO·) are responsible for the increased heat lability of the enzyme. Also, a considerable part of tyrosyl tRNA synthetase was converted to the heat-labile form under aerobic but not under anaerobic conditions even though Fe-ascorbate was not added. This indicates that heat lability of the enzyme is due to active oxygen, possibly generated by the reaction of dioxygen and transition metal ions present in the enzyme preparation. When the enzymes of young and old rats are treated as above, a higher percentage of heat-labile enzyme is seen in the older animals.

DNA polymerase activity and its fidelity have been an active area of study by several researchers on aging. Hanaoka et al. (1983) and Arai, Mitsui, and Yamada (1983) found a considerable decrease in the activity of DNA polymerase α of mouse liver following the neonatal period. Then it is maintained at a low level. This enzyme is located in the nucleus and is responsible for DNA replication. DNA polymerase β, which is also located in the nucleus and is responsible for DNA repair, does not show much change with age. DNA polymerase γ, which is located in mitochondria and is responsible for replication of mitochondrial DNA, increases in activity with age of the animal.

Carbonic anhydrase is present exclusively in cerebral glial cells (Maren, 1967), whose functional potentials are not well understood.

This enzyme has been used as a marker for aging in the rat since it shows a linear decrease in the cerebral cortex and cerebellum after adulthood (Koul & Kanungo, 1975) when expressed either as unit/milligrams of protein or unit/milligrams of DNA. Carbonic anhydrase III is also present in rat liver in which it decreases linearly after maturity (Wohlrab et al., 1988).

Creatine phosphokinase (CPK) is important for energy metabolism, especially of muscle. Cardiac CPK is a dimer made up of M and B subunits. The activity of CPK of the ventricle of the heart of old rats is lower than that of adult rats. Adrenalectomy of the rat lowers its activity in both adult and old rats. Hydrocortisone and cortisone lower it further in the adult, but increase it in old animals (Srivastava & Kanungo, 1980). In a related study, Srivastava and Kanungo (1982) studied the activity of skeletal myosin ATPase of the rat and found that both Ca- and actin-activation of the enzyme of old rats are lower than those of the adult, which may be due to a decrease in the titratable −SH groups of the enzyme. It is suggested that the enzyme molecule undergoes conformational changes in old age.

Sharma, Prasanna and Rothstein (1980) measured various parameters of phosphoglycerate kinase (PGK) in different rat tissue and found that the enzyme from old rats undergoes posttranslational modifications and is more heat stable. It has a slightly altered UV spectrum, but its K_m (Michaelis constant) and specific activity are not altered. PGK of young and old rats does not show any difference in isoelectric focusing. So the differences in some of the parameters may be due to conformational changes brought about by postsynthetic modifications. This is supported by the finding on enolase of *Turbatrix aceti* (Sharma, Prasanna & Rothstein, 1979) in which the conformation of the enzyme is first disrupted by guanidinium hydrochloride and is then allowed to fold back. The original conformation is regained. Gafni (1983) studied several enzymes involved in carbohydrate metabolism and found that Gly-3P-DH undergoes posttranslational modifications in old animals, resulting in lower activity of the enzyme. It is likely that deceleration of enzyme turnover due to a decrease in proteases, ubiquitin, and other factors may increase the possibility of covalent modifications of enzymes, resulting in differences in certain parameters of enzymes.

Not only does the decrease in the rate of transcription and translation decrease the level of an enzyme, but also its posttranslational modifications may decrease the number of active enzyme molecules that control specific metabolic steps. Fucci et al. (1983) found that ten key enzymes involved in metabolism of the liver are inactivated by a mixed-function oxidase system comprising NADPH cytochrome P-450 reductase and cytochrome P-450 isoenzyme II. The oxidase system inactivates the enzymes by oxidizing specific amino acid side chains, especially of histidine. These inactivated enzymes accumulate and are susceptible to proteolytic degradation. These observations are supported by the finding that inactive SOD molecules accumulate with increasing age in the human lens (Scharf, Dovrat & Gershon, 1987). Immunotitration using monospecific antibodies against SOD shows that catalytically inactive but antigenically reactive molecules accumulate with increasing age. These molecules are partially denatured and hence catalytically inactive.

The specific activity of guanylate cyclase has been measured in various cellular fractions of the heart and cerebral cortex of the rat during aging. Its activity in the microsomal and soluble fractions of the old heart is lower than that of the adult. Its activity is lower in the microsomal fraction of the brain of old rats, but is higher in the soluble fraction.

Ornithine decarboxylase (ODC) is a rate-limiting enzyme in the polyamine biosynthetic pathway. The polyamines – spermine, spermidine, and putrescine – have been implicated in cellular growth and differentiation, as well as RNA and protein synthesis. Being cations, these compounds bind to the phosphate groups in the DNA of chromatin and may alter thereby the structure of chromatin and its template activity. Studies of Das and Kanungo (1982) have shown that the levels of spermine and spermidine are far lower in the brain of old rats than adults. Also ODC activity is significantly lower. The ODC level is far higher in the heart and lung than in other tissues, and in both the tissues its level is lower in the old animals.

As mentioned earlier, only a few enzymes of any metabolic path have been studied as a function of age. Hence a general picture of the changes in a metabolic path cannot be discerned. One of the enzymes of the Krebs cycle that has been studied is isocitrate dehydrogenase (ICDH). NAD-linked ICDH is present only in mitochondria, but NADP-linked ICDH is present both in cytoplasm and mitochondria in different proportions. Yadav and Singh (1980, 1981) found that the activities of all three forms of ICDH are lower in the brain and liver of old rats as compared to those of adult rats. This may account for the decrease in oxygen consumption of the brain and liver of old rats.

Interesting changes in the circadian rhythm of enzyme levels are seen during aging. Moudgil and Kanungo (1973a) found that the specific activity of acetylcholinesterase (AChE) of the brain of immature (7 weeks) male rats is the highest at 6 a.m. and lowest at midnight. In the adult (35 weeks), the activity is the highest at midnight and lowest at 6 a.m. This diurnal rhythm of AChE, which is necessary for brain function, may be related to the behavior of rats. The adult animals are active in the night, which may necessitate a higher activity of the enzyme. Servillo, Della Fazia and Viola-Magni (1991) have also found a circadian rhythm in the expression of tyrosine aminotransferase (TAT). The peak activity of this enzyme in the liver of 3- and 12-month-old rats is at midnight; in the 24-month-old rat it is at 3 a.m.

Studies on the enzymes of cultured human fibroblasts have given useful information. Carlin et al. (1983) measured the tyrosine kinase activity of the epidermal growth factor (EGF) receptor of fibroblasts. EGF is a mitogen, which is needed for division of cells in the early phase, but not in old cells that have ceased to divide. Its receptor is a tyrosine-specific protein kinase that autophosphorylates its own tyrosine. This activity is lost in old cells. Takahashi and Zeydel (1982) found that γ-glutamyltranspeptidase (γ-GT) and glutaminase activities are greatly increased in old [49 population doublings (PD)] IMR-90 fibroblasts as compared to those of young (22 PD) cells. γ-GT is a membrane-bound enzyme and transfers the glutamyl moiety of glutathione to a variety of acceptors. A similar increase occurs in W1-38 fibroblasts. Also, alkaline phosphodiesterase is lower in old fibroblasts, whereas 5'-nucleotidase is higher.

The decline in protein synthesis during aging has been studied in relation to the nutritional status of animals. Several workers have shown that in animals fed restricted diets, the decrease in protein synthesis is retarded (Richardson & Semsei, 1987). How restricted diets influence gene expression is, however, not known. This observation is significant since dietary restriction also prolongs the life span (McCay, Crowell & Mognard, 1935; McCay, Sperling & Barnes, 1943; Guigoz & Munro, 1985). So it appears that there is a direct relationship between protein synthesis and longevity.

Makrides (1983) has reviewed the literature on protein synthesis and degradation during aging in animals. The general observations that may be made on the available literature are that protein synthesis decreases progressively with age which may be due to derangements in protein synthesis, including transcription and translation. Not all enzymes are altered during aging, and those that do change show posttranslational modifications that inactivate the enzyme molecules.

There is limited evidence to show that protein degradation slows down with age. This increases the $T_{1/2}$ (half life) of proteins, and may enhance the possibility of their posttranslational modification. It would be informative if efforts were made to find out if selected proteins show alterations in their synthesis and degradation. The information available so far is derived from studies on total proteins.

The overall fidelity of protein synthesis has been studied by various workers (Richardson & Semsei, 1987). No difference in fidelity is seen in the liver and brain of 7- and 33-month-old rats. The only difference is that the degree of incorporation of labeled amino acids into proteins is lower. Since the translation step has several substeps, it is not known at which substep(s) derangement occurs.

Kinetics and other parameters of enzymes

Proteins undergo several types of posttranslational modifications such as phosphorylation, methylation, acetylation, adenylation, and ADP-ribosylation. Although it is not known whether any change occurs in the rates of these modifications, it is reasonable to assume that the longer an enzyme stays in the cell, the greater the likelihood of its undergoing one or the other type of modification that may partially or completely decrease its activity. Thus, the fraction of inactive enzyme molecules will increase with the passage of time, which would adversely affect the activity of the cell. Such modifications would alter the K_m, K_i, M.W., electrophoretic mobility, antigenicity, and heat lability of enzymes.

The kinetic parameters of a few semipurified enzymes have been measured as a function of age. Studies on AChE (Moudgil & Kanungo, 1973b, c) and pyruvate kinase (Chainy & Kanungo, 1978a) of the brain; myosin ATPase (Kaldor & Min, 1975; Srivastava & Kanungo, 1979) and aldolase (Gershon & Gershon, 1973) of skeletal muscle; and SOD (Reiss & Gershon, 1976), cytoplasmic alanine aminotransferase (cAAT) (Patnaik & Kanungo, 1976), and aldolase (Gershon & Gershon, 1973) of mammalian liver show that in general there is no significant difference between the K_m, K_i, M.W., and electrophoretic mobility of enzymes purified from young and old animals. However, the specific activity of SOD (Reiss & Gershon, 1976) of the old rat is only 30%–70% of that of adult rats. The enzyme of old rats is also more thermolabile. These differences have been attributed to posttranslational modifications (Gershon, 1979).

That the structure of a protein synthesized in the old is the same as that of the young was first reported by Kanungo and co-workers using immunological (Kanungo & Gandhi, 1972) and peptide-mapping (Srivastava & Kanungo, 1979) techniques. Antisera produced against malate dehydrogenase (MDH) purified from the liver of young and old rats are immunologically the same. Tryptic maps of myosin ATPase purified from the skeletal muscles of young and old rats are identical. Later it was shown that tryptic maps of alcohol dehydrogenase purified from young and old *Drosophila* are identical (Subrahmanyam, Kannan & Reddy, 1984). PFK from skeletal muscle of adult and old rats has been purified and compared with respect to various parameters. Chromatographic elution patterns, electrophoretic mobility, K_m, K_i, pH, and temperature optima are the same. However, their K_m values for ATP in the presence of citrate, thermal stability, and inhibition by phosphoenol pyruvate are different. These changes have been attributed to the postsynthetic modifications of the enzyme molecules (Trigun & Singh, 1988).

Despite the fact that the molecular properties of only a few enzymes of young and old animals have been studied, the data so far obtained indicate that the amino acid sequences of the proteins synthesized in the old are apparently the same as those of the adult. Hence no changes in nucleotide bases seem to occur in a gene that codes for a protein as an animal ages. The only possible changes that seem to occur are the following: (1) Changes may occur at any one or more steps during the synthesis of a protein, such as transcription, processing of pre-mRNA into mRNA, and translation of the mRNA, which would change the level of the enzyme. (2) In an oligomeric enzyme which has isoenzymes, the expression of one of the genes coding for a subunit may change, leading to changes in the levels of isoenzymes, as was first reported for LDH (Kanungo & Singh, 1965) and cAAT (Patnaik & Kanungo, 1976).

An alternative view is that errors may occur as an organism ages in any of the steps leading to the synthesis of enzymes, resulting in the production of enzymes that have inappropriate amino acids. Such enzymes will have lower specific activities, and their molecular properties such as thermolability, specific activity, and antigenicity may be different (Orgel, 1963; Holliday, 1969; Holliday & Tarrant, 1972; see also Chapter 6). However, not all enzymes show a decrease in specific activity as a function of age. Whereas the specific activity of SOD decreases, that of myosin ATPase increases. In the latter case, both K_m and V_{max} also increase. Reiss and Gershon (1976), and Gupta and Rothstein (1976) have argued that the differences in the kinetic parameters of enzymes seen in old animals may be due to posttranslational modifications. This

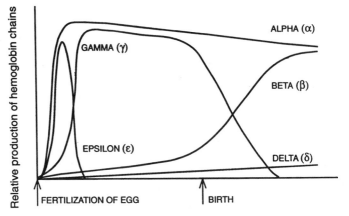

Figure 2.1. Synthesis of human globin chains during fetal development. (Zuck-erkandl, 1965; © 1965 Scientific American, Inc.)

could be a random event in which both new and old enzymes may be modified in such a way as to partially or fully inactivate a fraction of enzyme molecules.

Changes in multiple forms of proteins during aging

Another approach that has given useful information on the changes in gene expression during aging is the study of the changes in the isoenzyme patterns of enzymes. The lead came from the finding by Zuckerkandl (1965) that different types of hemoglobins appear during human development. These different molecular forms of hemoglobins have functional significance. Hemoglobin is a tetrameric protein having four globin chains. Hemoglobin of the 1- to 2-month-old fetus is of the $\alpha_2\epsilon_2$ type, which is replaced by $\alpha_2\gamma_2$ (fetal hemoglobin, HbF) for the rest of the gestation period. Hemoglobin of the newborn is $\alpha_2\beta_2$ (adult hemoglobin, HbA), which is present for rest of the life span (Fig. 2.1).

The α, ϵ, γ, and β chains are coded by separate genes. Hence the gradual shift in the hemoglobin pattern is due to the activation of certain genes and inactivation of other genes. For example, in the 1- to 2-month-old fetus genes for the α and ϵ chains ($\alpha_2\epsilon_2$) are active, but those for the γ and β chains are inactive. After the second month of fetal life, the γ gene gets activated, the α gene continues to be active, while the ϵ gene gets repressed. After birth, the β gene gets activated, the α gene continues to be active while the ϵ and γ genes remain repressed. This demonstrates that the duration for which a particular gene remains active is different. For example, the ε gene is active in the earliest period for only 2 months, whereas the α gene is active from the beginning till the end of the life span.

The oxygen-binding capacities of different hemoglobins are also different. HbF, which is present in the fetus, has a greater oxygen-binding capacity than HbA. This appears to be an adaptation of the fetus to the low oxygen content of the mother's blood, which is the sole vehicle for its oxygen supply. Whether the γ gene is expressed in the low oxygen environment of the fetus to produce HbF ($\alpha_2\gamma_2$) is not known. A related question would be: Does the higher oxygen content of the air that a newborn breathes repress the γ gene and cause the expression of the β gene? It would be of interest to study the expression of α, β, γ, and ε genes of people who live in high altitudes and are reported to have longer life spans.

Like the different molecular forms of hemoglobin, several enzymes have different molecular forms called isoenzymes, which have two or more subunits or chains. A well-studied enzyme that has isoenzymes is lactate dehydrogenase (LDH), which catalyzes the reversible conversion of pyruvate to lactate. Wieland and Pfleiderer (1957) found that LDH has five isoenzymes. They are made up of two types of subunits, H and M, which assemble in different proportions to give five types of active tetrameric isoenzymes – H4, H3M1, H2M2, H1M3, and M4. The H and M subunits differ in their amino acid sequences, and are coded by two separate genes (Shaw & Barto, 1963). Markert and Moller (1959) reported that the LDH isoenzymes are tissue-specific.

That oxygen level is important for the expression of genes for H and M subunits is evident from the LDH patterns in mammalian and avian embryos. The mammalian embryo that develops in an anaerobic environment has greater amounts of M4 and H1M3. As the embryo develops, a shift toward H4 and H3M1 occurs, indicating an increasing expression of the gene for H and a gradual repression of the M gene. The chick embryo, on the other hand, grows in an aerobic environment and has greater proportions of H4 and H3M1; it shows a shift toward M4 and H1M3 as it develops. It has been shown in tissue-culture experiments that the cells synthesize more M4-LDH under anaerobic conditions (Goodfriend, Sokal & Kaplan, 1966).

The above findings raise the question of whether such changes in isoenzyme patterns are confined to the developmental period or continue throughout the life span. If one finds changes occurring beyond the developmental stage, it would show that the expression of the genes coding for different subunits undergoes alterations that are responsible for the changes in the isoenzyme patterns. This was first examined by Kanungo and Singh (1965) and Singh and Kanungo (1968) using LDH from different tissues of rats of various ages. They showed that not only the total activity of LDH decreases, but also the proportion of M4-LDH in the heart, brain, and skeletal muscle of a 96-week-old rat is considerably lower than that of a 30-week-old rat. M4-LDH catalyzes the conversion of pyruvate to lactate faster than H4-LDH. Therefore, a larger proportion of M4-LDH would be of advantage in a tissue for deriving energy in the anaerobic condition when oxygen supply is cut off temporarily or is inadequate. The decrease in M4-LDH in old age may decrease the capacity of the tissue to withstand anaerobic conditions, which may result in its damage due to lack of sufficient energy. This correlates well with the higher frequency of cardiac failure and an inability to sustain longer periods of muscular exercise in old age.

It may be inferred from the above data that the lower proportion of M4-LDH is due to a decrease in the synthesis of the M subunit, which in turn is due to lower expression of the gene for the M subunit. It is also likely that the mRNA for the M subunit may be degraded faster. In either case, the synthesis of M subunit will decrease, resulting in a decrease in M4-LDH. This is likely to make the tissues more aerobic and increasingly dependent on the Krebs cycle. If the shift towards the decrease in the synthesis of M subunit in the old can be stopped, its ability to produce energy during anaerobic conditions can be maintained as in the adult. It has been shown that 17β-estradiol preferentially increases the synthesis of M subunits in the uterus of immature rats and rabbits, but the synthesis of H subunits is not altered (Goodfriend & Kaplan, 1964). Preferential synthesis of M subunits of LDH for increasing the level of M4-LDH and of other useful enzymes by administration of hormones and inducers may be a desirable method to prevent the decline in various activities of an organism after attainment of adulthood.

The findings on isoenzymes of pyruvate kinase (PK) have relevance in the light of the studies on LDH. PK has four isoenzymes, PK-1, PK-2, PK-3, and PK-4. In the skeletal muscle of the rat, PK-4 is the predominant form at birth, but by day 14, it is replaced by PK-3, the presence of which continues up to 52 weeks of age (Osterman, Fritz & Wuntch, 1973). Similar changes are seen in cardiac muscle. Moreover, the level of PK decreases in the brain, heart, and skeletal muscle of both male and female rats till old age (Chainy & Kanungo, 1976, 1978a,

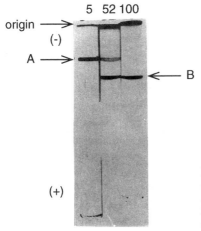

Figure 2.2. Polyacrylamide gel electrophoresis of cytoplasmic alanine aminotransferase of the liver of 5-, 52-, and 100-week-old rats. (Kanungo & Patnaik, 1975)

1978b). PK catalyzes the conversion of phosphoenol pyruvate (PEP) to pyruvate. The K_m of PK-3 for PEP is lower (0.75×10^{-4} M) than that for PK-4 (4.0×10^{-4} M) (Imamura, Taniuchi & Tanaka, 1972). The shift toward PK-3 would make the tissue more aerobic and more dependent on the Krebs cycle. This, together with the decrease in M4-LDH, would make the tissue more susceptible to damage when oxygen supply is cut off.

Studies on cytoplasmic alanine aminotransferase (cAAT) of rat liver show that the phenomenon of sequential changes in isoenzyme pattern extends to old age (Kanungo & Patnaik, 1975; Patnaik & Kanungo, 1976). cAAT is a dimer made up of two subunits, A and B, and has two active isoenzymes, cAAT-A and cAAT-B. Polyacrylamide gel electrophoresis of purified cAAT of the liver of 5-, 52-, and 100-week-old female rats shows that the liver of the immature rat has only cAAT-A, and the liver of old rats has only cAAT-B. The adult has both the isoenzymes, but the level of cAAT-A is lower (Fig. 2.2). The A and B subunits are under the control of two separate genes (Chen & Giblett, 1971). Hence, the sequential appearance and disappearance of the two isoenzymes during the life span of the rat is due to the sequential expression and repression of the two genes responsible for the synthesis of their subunits. What factors regulate the expression of the two genes are not known, but it is likely that, as in the case of hemoglobin, endogenous factors that activate or repress genes may be involved. The levels of these factors may change during aging and cause appearance or disappearance of corresponding protein subunits by influencing the expression of their genes. PFK has two isoenzymes, I and II. PFK II isoenzyme is lower in the liver and heart of old rats. Skeletal muscle PFK is absent in old rats (Trigun & Singh, 1987). These differences in PFK II and PFK levels may be due to lower expression of the gene.

Useful information has come from studies on *Drosophila*. Starch gel electrophoresis of hexose-P-isomerase of both male and female *Drosophila* shows that, of the five isoenzymes, the proportions of three slow-moving ones are considerably higher in the old insects (67 days). In the female, the two fast-moving isoenzymes that are present in young flies disappear in old flies (Hall, 1969). Three isoenzymes of esterase that are not present in young *Drosophila* appear in the old (49 days). Of the five isoenzymes of alcohol dehydrogenase (ADH), ADH3 and ADH5 are present throughout the adult life, but ADH2 is present only up to day 12 (Dunn, Wilson & Jacobson, 1969). It is likely that ADH2 is kinetically preferred at early age. Unfortunately, these experiments were done with the whole insect, and hence the changes seen are the sum total of the changes that may be occurring in all organs. Nevertheless, the appearance and disappearance of an isoenzyme at particular stages of the life span indicate changes in the expression and repression of specific genes.

Changes similar to those of isoenzymes during aging of animals have been reported for certain senescence marker proteins of rat liver (Chatterjee, Surendranath & Roy, 1981). In vitro translation of the mRNAs of the liver of 30-, 500-, and 850-day-old rats was studied by ^{35}S-methionine incorporation into proteins, followed by gel electrophoretic studies. Three unique proteins of M.W. 28.5, 26.3, and 19.5 K_d were observed, which showed age-related changes. The immature animals had only the 26.3 K_d protein, the young had 28.5 and 19.5 K_d proteins, but the 26.3 K_d protein was absent. The 26.3 K_d protein was found in old rats, but the 28.5 and 19.5 K_d proteins had disappeared.

Induction of enzymes

The above data show not only that the activities of several enzymes change with age, but also that the isoenzyme patterns of several enzymes change. The following questions arise from these data:

1. Are these changes irreversible?
2. Can we manipulate their levels?
3. What are the factors that are responsible for their changes?

Since proteins are coded by genes, such studies should provide an insight into the types of changes that occur at the level of the respective genes coding for these enzymes.

Attempts have been made by several workers to induce enzymes at various phases of the life span using different hormones. It is seen that under identical conditions, the induction in the old organism is lower than that of the adult, as for example, G-6-Pase, F-1,6-diPase, PFK, and hexose-P-isomerase (Singhal, 1967a, 1967b; Singhal, Valadares & Ling, 1969); malate dehydrogenase (Kanungo & Gandhi, 1972); AChE (Moudgil & Kanungo, 1973c); and choline acetyltransferase (CAT) (James & Kanungo, 1978). There are a few enzymes whose induction is the same both in the adult and the old animal, as for example, tryptophan pyrrolase (Gregerman, 1959), AAT (Patnaik & Kanungo, 1976), and TAT (Ratha & Kanungo, 1977). Induction of arginase in the liver (Rao & Kanungo, 1974), on the other hand, is greater in the old. Glucokinase and TAT of mouse liver are induced by fasting (Adelman, 1975). Hepatic TAT is induced by exposure of the rat to cold temperatures (Finch, Foster & Mirsky, 1969). In these studies it was found that induction of the enzyme is the same in both young and old animals, except that the time required for induction in the old is longer.

Class-wise the induction of oxidoreductases decreases, and that of transferases increases or is not altered in old age. The induction of lyases and hydrolases decreases in the liver of mammals in old age. In the brain, the induction of both transferases and hydrolases decreases in old age. However, since the number of enzymes studied is small, a definitive conclusion cannot be drawn concerning the pattern of induction of specific classes of enzymes.

Another observation on induction of enzymes is that it varies from organ to organ. For example, the induction of pyruvate kinase is higher in the heart of old rats, whereas in the brains of the same rats, it is lower than that of the adults (Chainy & Kanungo, 1976, 1978a, 1978b). In general, however, the induction of several enzymes is impaired in old age. Hence, the adaptability or response to a specific stimulus or stress is impaired in old age (Adelman, 1975). Such impairment may involve:

1. A longer lag period for induction in the old, as for example, glucokinase (Adelman, 1970a), TAT (Adelman, 1970b), and serine dehydratase (Rahman & Peraino, 1973).
2. A degree of induction by the same dose of inducer that is either lesser or higher in old age, as for example, cytoplasmic malate dehy-

drogenase (cMDH) (Kanungo & Gandhi, 1972), G-6-Pase (Rahman & Peraino, 1973) and ODC (Ferioli, Ceruti & Comolli, 1976).

3. Alterations that may occur both in the magnitude and the time required for induction, such as in thymidine kinase (Roth et al., 1974).
4. A response of the enzyme to the inducer that may be the same in both the adult and the old as in the case of TAT (Adelman, 1975), and mitochondrial glycerophosphate dehydrogenase (Bulos, Shukla & Sacktor, 1972).

In general, it is well established that the impairment of induction occurs either in magnitude or in the lag period or both. Most induction studies have been made using steroid hormones, the reason being that they act at the genetic level. Since these hormones are lipids, they pass through the target cell membrane and bind to specific cytosol protein receptors to form a H-R complex. The H-R complex then enters the nucleus and binds to specific cis-acting elements in the promoter regions of target genes to stimulate their transcription. The binding of the H-R complex to the DNA may be mediated by distinct tissue-specific trans-acting protein factors, as it is found that a hormone may stimulate the expression of one gene in one tissue but fails to induce the same gene in another tissue. For example, 17β-estradiol stimulates the expression of the ovalbumin gene in the oviduct of egg-laying vertebrates, but the same gene in the liver is not stimulated. 17β-Estradiol stimulates the expression of the vitellogenin gene in the liver, but not in the oviduct. This tissue-specific expression may involve specific trans-acting factors in addition to the estradiol receptor.

Induction of enzymes by steroid hormones gives useful information on the changes in the activities of corresponding genes because they act at the level of specific genes (Fig. 2.3). Kanungo, Patnaik and Koul (1975) showed that 17β-estradiol induced AChE in the cerebral hemisphere of immature and adult ovariectomized rats, but not in old rats. One possible reason for this lack of induction in old rats may be a decrease in the estradiol-binding receptor protein. To verify this, homogenates of the cerebral hemisphere of 7-, 44-, and 108-week-old ovariectomized rats were incubated with ^{3}H-estradiol, and the estradiol-binding protein was assayed (Kanungo, Patnaik & Koul, 1975). It was found that the affinity of this protein for 17β-estradiol is the highest in the cerebral hemisphere of immature rats in which the induction of AChE is also the highest. The affinity is lower in old rats. So, the molecular basis for the impairment of induction of AChE by estradiol

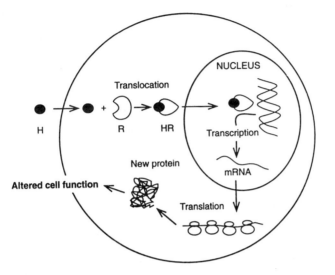

Figure 2.3. Mechanism of action of steroid hormone.

in the brain of old rats appears to be a decrease in the estradiol-binding protein. This was corroborated by the finding of Carmickle, Kalimi and Terry (1979), who measured the level of glucocorticoid receptor in different regions of the brain of 4-, 14-, and 29-month-old male rats after adrenalectomy. They observed a 50% decrease in ^{3}H-dexamethasone binding in the hypothalamus and hippocampus, but there was no change in the cerebral cortex. They also found a 65% decrease in the binding of ^{3}H-estradiol in the hypothalamus and cerebral cortex of ovariectomized rats.

Haji et al. (1984) and Haji and Roth (1984) measured the ability of nuclear and cytosolic 17β-estradiol receptors of the uterus of the rat to stimulate RNA polymerase II activity on incubation with estradiol. Nuclei from 6- to 8-month-old rats are three times more efficient than 20-month old rats in inducing the enzyme. It was further shown that the ability of estradiol-receptor complex to bind to acceptor sites in uterine nuclei is reduced by 50% in old rats (Chuknyiska & Roth, 1985). However, when 17β-estradiol is administered to young rats, RNA polymerase II activity is stimulated within 6 hours, peaking at 12 hours. In old rats, stimulation is observed after 12 hours, peaking at 18 hours. Furthermore, whereas 1 μg estradiol/100 g body weight could produce stimulation of the enzyme in young rats, more than 3 µg/100 g body weight was needed for old rats (Chuknyiska & Roth, 1985; Chuknyiska et al., 1985a, b).

Insulin receptor is a kinase which undergoes autophosphorylation. In the absence of insulin, autophosphorylation of the membrane-bound receptor is low. Insulin stimulates this activity, which is greatly reduced in old age (Carrascosa, Ruiz & Martinez, 1989). This may be the reason for the greater insulin resistance in old age.

Wohlrab et al. (1988) have reported that in the liver of adult male rats three specific proteins – D2, D3 and D4 – are present. Their levels decrease linearly with increasing age. They are absent in the newborn male and have the highest level in 60-day-old male rats. The female rats do not have these proteins at any age; because of this they may be used as biomarkers for male aging. These proteins have high degree of homology with carbonic anhydrase III. If senescent male rats are administered dihydroxytestosterone (DHT), D4 is restored, but D2 and D3 only partially. It appears that the serum testosterone level and testosterone receptors in hepatic cells are responsible for the decrease in the three biomarkers in old age.

Another protein that may be used as a senescence marker is the $\alpha_2\mu$ globulin of the liver of male rats. Richardson et al. (1987) found that its synthesis in hepatocytes of rats decreases by 90% as they age from 6 to 22 months. Its M_r is 18,700. It is synthesized in parenchymal cells, secreted into the blood, and finally excreted in urine. Estradiol inhibits its synthesis, but androgen, T3, glucocorticoids, and growth hormone stimulate its synthesis. Its synthesis increases up to 9–12 weeks and then gradually decreases. When rats are given a calorie-restricted diet (40% total calories), its synthesis increases by about twofold.

It is known that estradiol stimulates the synthesis of its own receptor. Is the decrease in the level of specific estradiol receptor in the brain of female rats due to a decrease in the estradiol level that is known to occur in old age? Androgen level also declines in male rats as they grow old. Does it have any relationship with the AChE level of the cerebral hemisphere? AChE is vital for brain function. Administration of estradiol after adulthood may prevent the decline in its level. A related study was done by Levin et al. (1981). When rats were kept under restricted feeding after weaning their dopaminergic receptor level at 24 months was the same as that of 3- to 6-month-old rats and 50% more than that of normal 24-month-old rats. Lee, Paz and Gallop (1982) found that high-affinity, low-density lipoprotein (LDL) surface receptors on fibroblast cells in culture need more LDL to induce the receptors in older cells.

Attempts to increase the levels of certain enzymes of the brain of old rats by administration of estradiol hormone have given encouraging results. Castration of old male and female rats lowers the levels of AChE and cholineacetyltransferase (CAT) in the cerebral hemisphere of rats. AChE hydrolyzes acetylcholine (ACh), and CAT is responsible for the synthesis of ACh. Administration of 10 μg testosterone/100 g body weight raises the levels of these enzymes in both the sexes. Testosterone is more effective than estradiol in inducing CAT (James & Kanungo, 1978). It is relevant here to mention that the synthesis of ACh is significantly lower in old rats as studied by the incorporation of labeled choline (Gibson, Peterson & Jenden, 1981).

Cells contain an enzyme that converts testosterone to estradiol. It is not known whether the induction of AChE and CAT is due to testosterone or estradiol. Nevertheless, this finding has given a lead for the manipulation of the levels of specific enzymes by administration of appropriate amounts of their inducers. Similar results were obtained for pyruvate kinase of the brain and heart (Chainy & Kanungo, 1978a, 1978b).

The induction of NAD- and NADP-linked ICDH by hydrocortisone and estradiol in the brain and liver of rats of various ages has been studied by Yadav and Singh (1980, 1981). Adrenalectomy lowers the level of NAD-linked ICDH considerably. Administration of hydrocortisone to the rat raises the activity significantly in the adult, but not old animals. These effects are not so pronounced for NADP-linked ICDH. Ovariectomy lowers the level of both forms of ICDH. Administration of estradiol raises their levels only in the adult, but not in the old.

Another approach to the restoration of induction of enzymes in an organ is by stimulating the organ to regenerate. Kanungo and Gandhi (1972) showed that the level of mitochondrial MDH decreases in the liver of young rats after adrenalectomy, but not in old adrenalectomized rats. Administration of cortisone to these rats causes induction of the enzyme in young rats but not in old rats. However, if the liver of old rats is partially hepatectomized and allowed to regenerate for 3 days, and then cortisone is administered, the enzyme is induced. So the impairment of induction in old age is repaired. This may be because after hepatectomy, the remaining quiescent liver cells begin to divide, their genes become activated, and the factors required for induction begin to appear. Since liver cells are premitotic by nature and retain the ability to divide throughout the life span, the liver is able to regenerate. Tissues in the brain, heart, and skeletal muscle cannot regenerate in higher vertebrates, as their cells lose the capacity to divide at an early stage of development and become postmitotic. It has not been possible to make these cells divide in vitro. Once a way is found to make them divide, it may be feasible to activate and regenerate these tissues.

These studies have shown that factors like hormones, their receptors, and the tissue-specific trans-acting factors needed for expression of specific genes are important for the maintenance of the levels and adaptive response of enzymes. Changes in the levels of such factors in old age impair metabolic functions and contribute to the aging process.

Protein synthesis can also be induced by heat, as has been shown in several types of organisms beginning from *Escherichia coli* to mammalian cells. When *Drosophila* are kept at higher temperatures for brief periods, four types of heat shock proteins (HSP) are induced. Though the roles of these proteins are not well understood, it has been shown that both transcription and translation are enhanced. Fargnoli et al. (1990) subjected cultures of fibroblast cells derived from the lung and skin of rats to a temperature of 42.5°C and found that HSP 70 mRNA and HSP 70 protein (a 70 K_d protein) are synthesized, but the level of their synthesis is lower in confluent old cells. When freshly excised lung tissues of young (5 month) and old (24 month) rats were subjected to heat shock, the synthesis of HSP 70 was lower in the old animals. How temperature induces the expression of the HSP 70 gene is an interesting problem for study.

Changes in collagen

Collagen is a structural protein that is fibrous in nature. It constitutes 25%–30% of the total protein of the animal body. It is synthesized in all cell types, but is deposited extracellularly in all tissues. Tendon and ligament are predominantly made up of collagen. Verzar and his co-workers first reported that the solubility of collagen decreases, but its tensile strength increases with age (Verzar, 1955, 1958; Verzar & Thoenen, 1960). This indicates that the polypeptide chains of collagen become increasingly cross-linked with age. Verzar proposed that, with the passage of time, increasing amounts of insoluble collagen accumulate in the extracellular space, preventing the flow of nutrients and oxygen to cells and causing them to starve and die. This may contribute to aging (see Kanungo, 1980 for details). It is significant that the amount of insoluble collagen in human tendon increases with age. The increase is so predictable that it is possible to determine human age by measuring the amount of collagen which is resistant to collagenase.

Extracellular collagen is present as fibrils, which are made up of several monomers. Each monomer is a triple helix of three polypeptide chains that are coiled around each other. Two types of polypeptide chains are found in collagen of different tissues, $\alpha 1$ and $\alpha 2$. $\alpha 1$ has four variants, $\alpha 1(I)$, $\alpha 1(II)$, $\alpha 1(III)$, and $\alpha 1(IV)$. The four variants of $\alpha 1$ and that of $\alpha 2$ differ in amino acid sequence and are coded by different genes. Each α chain has 1,050 amino acid residues. Approximately 15–25 residues at both N- and C-terminal regions are noncollagenous and nonhelical. The central region, consisting of ~1,000 residues, has a repeating Gly-X-Y sequence. Proline generally occurs at the X position and hydroxyproline at the Y position. The various α chains differ from each other in the amino acids at the X and Y positions. The glycine content and its occurrence at every third position is a constant feature in all chains, but proline and hydroxyproline contents vary among α chains.

Four types of collagens – I, II, III, and IV – having four different types of triple helical assembly are present in various tissues. Therefore, collagen exhibits molecular polymorphism, as do several enzymes that are present as isoenzymes. There are five types of α chains. Therefore, 35 different triple helices are possible. Why only four types of triple helices are formed is not known.

Rigid tissues like bone, tendon, and skin have type I collagen, which is made up of two $\alpha 1(I)$ and one $\alpha 2$ chains. It is the most abundant collagen in vertebrates. Type II is present in hyaline cartilage, eye lens, cornea and the nucleus pulposus of intervertebral disks. It is made up of three $\alpha 1(II)$ chains. Type III collagen is found in the skin, blood vessels, smooth muscle, and synovial membrane. It has three $\alpha 1(III)$ chains. Type IV is predominantly present in the basement membrane and has three $\alpha 1(IV)$ chains.

Heikkinen and Kulonen (1964) extracted collagen by neutral salt from the skin of rats ranging from 1.5 to 24 months, and found that the amount extracted is lower in old age (Table 2.1). Especially, the amount of extractable single α chains (monomer) decreases rapidly and that of α chain trimers (γ-collagen) increases. No change is seen in α chain dimers (β-collagen).

A significant increase occurs in the covalent cross-linking among the

Table 2.1. *Proportions of extractable collagen of rat skin as a function of age*

Type of collagen	Age (months)				
	1.5	3	6	12	24
Monomer (α-chain)	40	43	37	22	19
Dimer (β-collagen)	54	54	54	54	54
Trimer (γ-collagen)	2	3	9	12	27

Source: Heikkinen and Kulonen (1964).

three chains of the triple helix and between triple helices. This is the cause of the gradual increase in the tensile strength of collagen as a function of age. Such changes are secondary in nature and have been dealt with in detail by Kanungo (1980).

Not only do the types of collagen change in tissues during the early developmental period, but there is also evidence that switching may occur in the types of collagen being synthesized in certain tissues as the animal grows old. Chondrocytes of cultured cartilage cells synthesize type II collagen, but when they grow old type I is synthesized (Mayne, Vail & Miller, 1975). The turnover of collagen is high during the developmental period, which is due partly to a high level of collagenase that cleaves the Gly-Leu bond in collagen leading to its degradation. Decreases in the level of collagenase and several other enzymes necessary for maturation of collagen may lead to structural and functional changes in collagen during aging. It is necessary to determine what factors are responsible for switching of collagen types both in cell cultures and in vivo.

Turnover of proteins

In general, the rate of protein synthesis decreases by 40%–79% in mammals and by 60%–90% in insects (Makrides, 1983). Limited evidence suggests that protein degradation may slow down with age in mammals and nematodes. If the protein degradation system is inefficient in older animals, proteins will tend to remain longer and their $T_{1/2}$ will increase. In that case they will be subjected to more posttranslational modifications and undergo alterations. This may alter their conformation, kinetic parameters, electrophoretic mobility, and thermostability.

These phenotypic changes in proteins are, however, independent of the corresponding genes. It is desirable to study the rates of synthesis and degradation of individual proteins to get an insight into this aspect of their turnover.

The turnover of proteins has several important implications for the cell and the organism. If an enzyme is turned over fast it is degraded rapidly, and is synthesized equally fast so that the desired number of active molecules remain at any given time to carry out a specific function. If degradation is faster than synthesis, the number of molecules will decline, and the activity of the metabolic path of which it is a part will decline. This rapid degradation may be due to an increase in proteases and the enzymes needed for ubiquitin-mediated degradation (Hershko, 1988). Protein degradation is a random process. Apparently the degradation machinery in the cell does not distinguish between the newly synthesized protein and the old protein from the pool of proteins. The proteins that are covalently modified by the short 76-amino acid polypeptide, *ubiquitin,* are marked for degradation. Ubiquitination is an energy-dependent enzymic process (Hershko, 1988, 1991; Rechsteiner, 1991). So a decrease in the levels of ubiquitin, ATP, or the enzymes involved in the degradation pathway, and proteases may decrease the rate of degradation of proteins. Also, it is known that an enzyme bound to its cofactor degrades more slowly. Sharma et al. (1979) found that the rates of synthesis and degradation of enolase and its turnover slow down in *Turbatrix aceti* with age. The $T_{1/2}$ for its synthesis is 73 hours and degradation 58 hours in the young (5-day-old), and 163 hours and 161 hours in the old (22- to 36-day-old). The slowdown of protein turnover in old age may result in protein modification. Rothstein, Coppens and Sharma (1980) also studied the K_m, electrophoretic mobility, heat sensitivity, and specific activity of enolase of skeletal muscle, heart, and liver of the rat, and found no change in the enzyme of young and old animals.

It is reasonable to expect a relationship between the rates of protein synthesis and degradation, and the protein content of a tissue. When protein synthesis was measured by incorporation of ^{3}H-valine into postmitochondrial proteins of the kidney in 4.5-, 7.5-, 22-, and 31-month-old rats, a 73%–87% decrease was observed between 4.5-week- and 31-month-old rats. There was no difference in the fidelity of poly-U translation by ribosomes of the kidney in young and old rats. This shows that the fidelity of the translational machinery does not decline.

Microtubules (MT) are important for the maintenance of the structure of cells, in addition to their role in spindle formation and cell division. Two MT assembly proteins – MAPI and II – promote MT assembly. They are degraded by the protease, cathepsin D, which is found to increase with age (Matus & Green, 1987), a condition that may cause defective MT assembly and degeneration of cell structure.

Modifications of bases

Transfer RNAs have several modified bases, one of them being pseudouracil (ψ). Ribosomal RNA also has ψ, but its content is very low. Pseudouracil is formed enzymatically after the tRNA is transcribed. So the excretion of ψ may be used as a measure of the turnover of tRNA, since it is not reutilized. Luch et al. (1979) found that ψ excretion per day in the urine is higher in old men. Its excretion in a 25-year-old man was 28 mg/24 hours, but it was 83 mg/24 hours in a 90-year-old man. Hence tRNAs are degraded more rapidly in the older person.

Modification of guanine to queuine (G → Q) also occurs posttranscriptionally and is seen in tRNA (Watson et al., 1987). Singhal et al. (1981) found that the highest level of queuine is found in 9-month-old rats which then declines. Such a modification at the anticodon of a tRNA alters its capacity to recognize the appropriate codon on an mRNA (see Watson et al., 1987).

Methylation of cytosines at position 5 (5mC) in DNA is a well-established modification in all eukaryotes except insects. Approximately 2% of cytosines occur as 5mC, especially in CpG doublets and in –CCGG– sequences. Methylation of cytosine occurs after replication by a methylase present in the chromatin. Cytosine has been implicated in regulation of gene expression (for details see Chapter 5). Adenine is also methylated at position 7 (7mA), but it is not found in eukaryotes.

Another interesting modification is 7m guanine (7mG) found in both mitochondrial and nuclear DNA. Park and Ames (1988) analyzed mitochondrial and nuclear DNA of rat liver by HPLC. They found that in 6-month-old rats the 7mG content is 1/31,000 bases in the mitochondrial DNA, and 1/105,000 bases in nuclear DNA. It is nearly 2.5 times higher in 24-month-old rats. If DNA is exposed to methylating carcinogens, the guanines and adenines are methylated to produce 7mG, 6mG, and 3mA. 7mG lies exposed in the major groove of DNA. Since its enzymatic repair is very slow, it remains in DNA for a long time. Though it does not interfere with replication, its role in transcription is an interesting aspect that needs to be studied.

Changes in RNA

Turnover of RNA is a measure of its synthesis and degradation, both of which are controlled by several factors. Messenger RNAs are translated into proteins. Their turnover is highly variable and is dependent on several factors. Transfer RNAs and rRNAs have longer $T_{1/2}$ and are essential for translation of mRNAs.

Transfer RNA

Transfer RNAs (tRNAs), like several enzymes, have isoacceptors, which have different specificities for the corresponding codon and amino acid. The redundancy of the genetic code is variable. For example, arginine and serine have six codons each, whereas tryptophan has just one. Hence, there are multiple tRNAs having matching anticodons for the same amino acid. Therefore, studies on isoacceptors of tRNAs may give useful information on the changes in the expression of their genes. The tRNA content of the skeletal muscle of the old rat is lower than that of adult or young rats. The arginyl-, glutamyl-, and tyrosyl-tRNAs of the liver do not show either qualitative or quantitative changes with age (Manjula & Sundari, 1980). On the other hand, of the four isoacceptor tRNAs for arginine I, II, III, and IV, the proportions of III and IV are higher and that of II is lower in the old (Manjula & Sundari, 1981).

Glutamyl-tRNA has three isoacceptors, but there is no change in their proportions during aging. Prolyl-tRNA has three isoacceptors – I, II, and III. There is a distinct increase in I and II in the old animal. Such changes in the proportions of isoacceptor tRNAs may affect the rate of translation. The patterns of isoacceptor tRNAs have been reported to change in nematodes (Reitz & Sanadi, 1972) and *Drosophila* (Hosbach & Kubli, 1979; Owenby, Stulberg & Jacobson, 1979).

Transfer RNAs have several modified bases which are formed post-transcriptionally. It is likely that such bases have a role in the aminoacylation of tRNAs and translation. Studies on the changes in the levels of modified bases of tRNA with increasing age, as seen for the conversion of G→Q, may provide important clues to the changes seen in protein synthesis during aging.

Both the contents of tRNAs and their rates of synthesis decrease in the heart and kidney of mice (Neumeister & Webster, 1981). However, no such change is seen in the liver of the rat (Cook & Buetow, 1982), nor are any differences seen in the extent of aminoacylation of the cytoplasmic tRNAs, total tRNA synthetase activity, and the rate of aminoacylation of individual tRNAs of the liver. The tRNA system (tRNAs plus aminoacyl-tRNA synthetases) isolated from the liver of old rats is less efficient in in vitro cell-free protein synthesis than that of adult rats. Also, in heterologous assays, adult tRNA synthetases are more active in charging isoleucine, methionine, proline, glutamine, phenylalanine, and glutamic acid than old enzymes. Another possibility of changes in tRNA may be in their splicing. Changes in the levels of tRNA splicing enzymes may contribute to changes in the levels of functional tRNA and in turn influence translation.

Mays-Hoopes et al. (1983) purified tRNAs from the liver of young and old rats and studied their capacity to be acylated by specific aminoacyl-tRNA synthetases. Specifically, tRNA^{phe} was studied. Apparently, the tRNA of old rats is modification deficient, and their acylation is reduced, but it is unlikely that they produce errors in proteins. When mRNAs for hemoglobin and ovalbumin and tRNAs from young and old rats are used for translation in vitro, the amount of protein synthesized is lower with tRNA derived from the liver of old rats.

Ribosomal RNA

Even though only four types of ribosomal RNA (rRNA) (5.0S, 5.8S, 18S, and 28S) are required to form ribosomes in eukaryote cells, they form the highest proportion of cellular RNA. Especially in dividing cells that require synthesis of large amounts of proteins, the proportion of rRNA far exceeds those of tRNA and mRNA. In such cells the proportion of rRNA may be about 80% of the total cellular RNA. The cell is able to synthesize a large amount of rRNAs due to the presence of multiple copies of rRNA genes (rDNA). The number of copies of genes for rRNA in *Xenopus* and human cells is about 20 per haploid genome. Whether the rate of transcription of rRNA decreases as the cells stop dividing and differentiate, or whether only a few of the genes are transcribed is not known.

Ribosomal RNAs are transcribed from two separate sequences of rDNA (ribosomal DNAs that transcribe rRNA). One of the rDNA is transcribed by RNA polymerase I into a long 45S precursor rRNA in the nucleolus. This rRNA is processed and cleaved into one each of 18S, 5.8S, and 28S rRNA in the order mentioned. This gene is tandemly repeated. The other rDNA is located elsewhere in the genome, and is transcribed by RNA polymerase III (which also transcribes tRNA) into a 5.0S rRNA consisting of 120 nucleotides. The 28S, 5.8S, and 5S rRNA assemble together with about 60 ribosomal proteins to form a 60S ribosomal subunit. The 18S rRNA and about 30 ribosomal proteins form the 40S ribosomal subunit. The two subunits together form an 80S ribosome to which mRNA binds for translation into proteins. Accurate processing of 45S pre-rRNA, assembly of rRNAs, and ribosomal proteins are important for the accurate translation of mRNAs.

A decrease in transcription of rDNAs may decrease the level of rRNAs and hence of protein synthesis. Johnson and Strehler (1972) and Johnson, Chrisp and Strehler (1972) found a tissue-specific loss of ribosomal genes in aging beagles, the loss being more pronounced and rapid in highly specialized tissues like the brain, heart and skeletal muscle, and less so in the kidney, liver, and spleen. Later, Gaubatz and Cutler (1978) found a similar loss of rDNA in the brain, spleen, and kidney of mice as a function of age. Also, Strehler and Chang (1979) showed a loss of rDNA in human myocardium and cerebral cortex, and Shmookler-Reis and Goldstein (1980) found a loss of reiterated sequences during serial passage of human diploid fibroblast cells. However, Peterson, Cryar, and Gaubatz (1984) did not find any decrease in the copy number of rDNAs in in vivo aging of mouse myocytes or in in vitro aging of W1-38 human fibroblasts. These studies, therefore, need to be done using the recent technique of RNA–DNA hybridization.

The rate of transcription of rDNA appears to decrease during aging. Buys, Osinga and Anders (1979) reported a decrease in transcription of rRNA in human lymphocytes and fibroblasts. Also, an age-related decrease in rRNA synthesis occurs in the macronucleus of *Paramecium* (Heifetz & Smith-Sonneborn, 1981). Unfortunately, so far no one has studied the age-related changes in the rDNA genes, their transcription, processing of the pre-rRNAs, and assembly of rRNAs to form ribosomes. The occurrence of multiple copies of rDNA genes and the abundance of copies of only three species of rRNAs generated from a single pre-rRNA offer a unique advantage to study the changes that occur at the level of ribosomal genes and the processing of their transcripts.

Messenger RNA

There are only four genes for rRNA, and about 60 genes for tRNA. However, mRNAs are generally coded by single-copy genes. It is estimated that a mammalian cell may have 50,000–1,00,000 genes that code for proteins. Hence, there are as many types of mRNA. Except a few tRNAs like those for glycine, proline, arginine, and histidine which are present in large number for the synthesis of collagen in specialized cells, and histones in dividing cells, the levels of most of the other tRNA and their isotypes are more or less equal. Likewise, all rRNA occur in equimolar amounts for the formation of ribosomes. On the other hand, not only the types of mRNA are enormous, their levels also vary greatly. Depending on the types and conditions of cells, the number of copies of mRNAs may vary from zero to millions. Despite this high heterogeneity, only 5%–7% of the total RNA content of a liver cell consists of mRNAs, 10%–15% of tRNAs, and 70%–80% of rRNAs. Moreover, both the tRNA and rRNA have a far longer $T_{1/2}$ than mRNAs.

Most protein-coding genes are split by a varying number of introns, except histone and protamine genes which have many copies that are clustered in specific regions of chromosomes. They do not have poly-A^+ tails at their 3′ end. The genes that are split produce longer transcripts (hnRNAs), which have a 7mG cap at the 5′ end. The hnRNAs undergo processing in the nucleus during which poly-A^+ tails having nearly 200 adenylate residues are added at the 3′ end, and the regions of the hnRNAs corresponding to the introns are removed. This generates mRNAs with the 5′ cap and 3′ poly-A^+ tail. They are then translocated to the cytoplasm for translation. Transcripts of genes that have a large number of introns, such as those for collagen, vitellogenin, and fibronectin, undergo more complex processing as compared to those of genes for globin or *myc* which have only two introns to be removed. It is conceivable that the efficiency of the hnRNA processing machinery may decrease as a function of age, especially for those hnRNAs that have several introns. For instance, transfer of information from the transcript of a histone gene should be faster and more accurate than from a gene that has several introns. Again this is an area in which we have little knowledge and one that needs to be explored using modern techniques.

Another important structural aspect of mRNA is its poly-adenylated 3′ end. Poly-A^+ tails consisting of nearly 200 adenylates are added at the 3′ end of hnRNAs before their introns are excised. Poly-A^+ mRNAs are translocated across the nuclear membrane to the cytoplasm where they are translated. It is seen that the tail gradually shortens as the mRNA undergoes repeated cycles of translation. Muller, Zahn and Arendes (1979), Muller et al. (1979), Arendes, Zahn and Muller (1980), and Bernd et al. (1982) found that the poly-A^+ tails of the mRNAs of the oviduct of senescent Japanese quails are shorter than those of young birds. The rate of poly-A$^+$ synthesis decreases in hepatocytes by 68% between 6 and 30 months of age in rats. The rate of synthesis of total RNA also decreases, but the rate of decline in the synthesis of rRNA and tRNA is less rapid than that of poly-A$^+$ mRNA in hepatocytes as a function of age of the rat (Richardson et al., 1982). Semsei, Szeszak and Zs-Nagy (1982) also found a decrease in the total RNA and mRNA synthesis in the cerebral cortex of rat brain as a function of age.

Birchenall-Sparks et al. (1985) studied several properties of poly-A$^+$ mRNAs of hepatocytes of 4- and 30-month-old rats. Newly synthesized poly-A$^+$ mRNAs have poly-A$^+$ tails of about 150 nucleotides, and those isolated from the steady-state pool of the cytoplasm have about 70 nucleotides. There is no age-related difference; neither is there any difference in their translational activity as tested in vitro using wheat germ extract and rabbit reticulocyte lysate. Nor is there any difference in the 5' cap structure of these poly-A$^+$ mRNAs.

Turnover of RNA

Ribosomal RNAs have long half-life, in the range of 5 to 10 days. This is because they are bound to ribosomal proteins, and hence they may not be accessible to RNases. Their half-lives are tissue-specific, but do not show age-related changes. The half-lives of transfer RNAs are more variable and range between 1 and 11 days (Menzies, Mishra, & Gold, 1972). The existence of isotypes of tRNAs makes it an interesting problem for further study. Changes in isotypes, as in the case of isoenzymes, would affect the efficiency of incorporation of amino acids into mRNAs during translation, and thereby signify changes in the expression of the corresponding genes.

Very little information is available on the half-lives of mRNAs in relation to aging. The early studies using the inhibitor of mRNA synthesis, actinomycin D, showed that the $T_{1/2}$ of mRNAs of old *Drosophila* is shorter. It is necessary to determine the $T_{1/2}$ of specific mRNAs to get a general idea of the turnover of mRNAs. It is also necessary to find out if the levels of RNases change with age.

DNA polymerases

There are three steps through which information stored in the DNA is transferred: DNA → DNA, DNA → RNA, and RNA → protein. The crucial enzymes that catalyze the transfer of information are DNA polymerases, RNA polymerases, and aminoacyl tRNA synthetases. Accurate transfer of information requires that these enzymes have high fidelity. Also, it is necessary to have a mechanism to remove any error in the product whenever it occurs. Both these accuracy-requiring mechanisms operate in eukaryotes, but despite the high fidelity of the enzymes and error-removing mechanisms, mistakes occur and perpetuate from cell to cell, and from generation to generation. This has had the desirable effect that it has permitted evolution of organisms. Another desirable aspect of this system is that it is not too error prone, because if it were, possibly viable organisms would not have appeared, and the course of evolution would have been undefined.

The activity of DNA polymerase and its fidelity in relation to aging have been subjects of active studies by several workers. Hanaoka et al. (1983) found a considerable decrease in the activity of DNA polymerase α of the mouse liver after the neonatal stage till maturity, at which point it is maintained at a low level. This enzyme is located in the nucleus and is responsible for DNA replication. DNA polymerase β, which is also located in the nucleus, is responsible for DNA repair and does not show much change with age. DNA polymerase γ, which is a mitochondrial enzyme responsible for replication of mitochondrial DNA, increases in activity.

The first step in the information-transfer system is the replication of DNA by DNA polymerases. Several workers have studied the fidelity of these enzymes as a function of age. Hopfield (1974) reported that the replication machinery has high fidelity. Even if any errors occur in replication, they are removed by proofreading. Particularly, accuracy in the germ line is essential for gene survival, though it may not be so essential for somatic cells. However, Fry and Weisman-Shomer (1976), Linn, Kairis & Holliday (1976), Murray and Holliday (1981), and Krauss and Linn (1982) found that the fidelity of DNA polymerases α, β, and γ purified from fibroblast cells in culture is lower in late-passage cells. On the contrary, Agarwal, Tuffner and Loeb (1978) and Fry, Loeb & Martin (1981) reported that the fidelity of DNA polymerase α from human lymphocytes and that of β from mouse liver does not change with age. This was supported by the finding that when DNA polymerases, α and β, of regenerating liver of young and old mice are tested for their fidelity, they are found to carry out polymerization accurately. They also do not show any differences in their thermolability. Moreover, the fidelity of enzymes from normal liver of young and old mice is the same.

These studies question the validity of extrapolating the data from cells in in vitro culture to an in vivo system. It appears that the fidelity of DNA polymerase is not lower in old age. Even if errors occur during replication, the enzymes involved in proofreading remove such errors to ensure faithful copying of DNA from generation to generation. However, there is still another step at which errors can become incorporated, that is, when damaged or nicked DNA is repaired. Nicks occur randomly, and may involve variable lengths of the DNA. We do not have information on the accuracy of this repair step. Furthermore, DNA polymerase α is induced in regenerating liver. Studies on repair of DNA and induction of the gene for DNA polymerase α may give useful information on this interesting facet of aging.

In contrast to DNA polymerase α, DNA polymerase β of mice neurons is reported to be highly error-prone. The error frequency is about 1 in 7000, which is not significantly different in young and old animals (Rao, Martin & Loeb, 1985). This enzyme is involved in DNA repair and is the predominant enzyme in the neurons that stop dividing soon after birth in mammals.

Conclusions

It is clear from the above information that both the types of macromolecules, proteins, and RNA that are coded by genes undergo changes in their levels. The general trend is a decrease in their levels. Though these decreases may largely be due to a decrease in transcription, derangements at other steps that follow cannot be ruled out. For protein coding genes, processing of the hnRNA in the nucleus to produce an mRNA, translocation of the mRNA to the cytoplasm where it binds to ribosomes in the rough endoplasmic reticulum, and the translation process are steps where derangements may occur. For rRNAs and tRNAs, in addition to transcription, processing of transcripts and modification of bases are important steps for producing functional molecules. In recent years much attention has been focused on transcription, which is by itself a complex process, particularly because several regulatory steps influence transcription. From the available data on phenotypic changes, no clear picture has emerged as to the types of molecules that are crucial or important in aging.

Perhaps the only part of these studies that has given some lead as to the types of changes that may be occurring at the genetic level is the research on the changes in isoenzymes and the induction of enzymes. It is clear from these studies that switching of genes occurs during the life span, though it is not obvious what factors are responsible. Isotypes of tRNAs also show a similar pattern, but more work needs to be done on tRNAs to arrive at a definite conclusion.

3
Chromatin

In multicellular organisms, the problems of transmitting information stored in the DNA from cell to cell and from generation to generation are enormous. A diploid (2N) human cell has about 7 picograms (pg) of DNA which is distributed among 23 pairs of chromosomes. The total length of the DNA of 23 pairs of chromosomes is about 2 meters. Therefore, it has to be highly compacted in order to be accommodated within a nucleus whose diameter may range from 5 to 8 μ. In the highly condensed metaphase chromosomes of a liver cell, the DNA is compacted about 10,000-fold (Pienta, Getzenberg & Coffey, 1991). Approximately 3×10^9 base pairs (bp) of DNA are distributed among 23 chromosomes. Only 50,000 to 100,000 genes that are coded by nearly 5×10^7 bp of DNA or 3%–5% of the total DNA are estimated to be functional. So nearly 95%–97% of the DNA in a human cell has no known function.

The DNA in eukaryote chromosomes exists as a complex, unlike that in prokaryotes, mitochondria, and plastids, with two types of chromosomal proteins, histones and nonhistone chromosomal (NHC) proteins, forming a supramolecular complex called chromatin, the genetic apparatus of eukaryotes. The three components – DNA, histones, and NHC proteins – are present approximately in equal proportions. Various types of RNA are also found in the chromatin, but they occur as transcription products of DNA and are not structural components of the chromatin. In order to be compacted and accommodated within the nucleus the DNA, which is negatively charged at physiological pH due to its phosphate groups, binds to two molecules each of four types of positively charged core histone molecules (H2A, H2B, H3, and H4). This forms nucleosomes which are connected with each other by the DNA strand to form a 10 nm nucleosome fiber that has a bead-on-a-string structure. This is the first-order organization of the chromatin which compacts the DNA about sixfold. The fifth histone, H1, which

55

is bound to the DNA in the internucleosomal region, then coils the 10-nm fiber to achieve the second-order organization which is a 30-nm filament. This compacts the DNA by another six- to seven-fold. Further higher-order organizations of the DNA–histone complex are achieved by coiling and looping of the 30-nm filament which is aided by NHC proteins. This forms the interphase chromatin. Such organization of chromatin accomplishes two important purposes: It accommodates 2-m long DNA within a nucleus of ~ 8 μ diameter and it facilitates selective expression and repression of genes in a cell- and tissue-specific manner.

The problems that arise for the DNA to carry out its functions while in a very compact structure are varied and enormous. Some of these problems are: (1) How does the DNA extricate itself from this complex during cell division and get precisely replicated during the S phase, which lasts only for a short period of the cell cycle? (2) How does the DNA assume its original complex once again in the daughter cells? (3) How do certain cells in multicellular organisms differentiate into specialized cells that perform specific functions? Certain dividing cells gradually stop dividing, begin to express certain genes more and more, and other genes less and less, stop dividing altogether, become postmitotic, and perform special functions. Neurons and skeletal and cardiac muscle cells exhibit this type of behavior. Epithelial cells continue to divide, and liver cells retain the ability to divide throughout the life span. A considerably fewer number of genes are expressed in these cells. How does a postmitotic human cell select and express only a few genes that are distributed in the 23 pairs of chromosomes existing as complexes with histones and NHC proteins?

Some somatic cells do not synthesize certain proteins normally, but can be made to produce them when subjected to induction. Despite the complexity of the chromosome and its compaction with histones and NHC proteins, the inducing substance finds its target gene either directly or indirectly. The differentiation state of a cell illustrates that certain genes are silenced or repressed permanently, and certain genes remain active throughout an organism's life. Once differentiated, the cells do not revert to their premitotic state. How are these two opposite states of genes, active and inactive, brought about? How does the DNA after replication form protein complexes so that only those genes that were active in the parent cell remain active also in the daughter cells to maintain cell lineage? Knowledge of the basic structure of the chromatin is necessary to understand how the expression of genes and its modulation are carried out and how these functions change as an organism ages. Several reviews

Table 3.1. *Characteristics of calf thymus histones*

Type	M.W.	No. of residues	Lys: Arg	N–terminal residue	C–terminal residue	α-Helix content	β-Pleated sheet	Net charge
H1[a]	22,500	224	22.7	Ac–Ser[b]	Lys	55	5	+58
H2A	13,960	129	1.17	Ac–Ser	Lys	40	20	+15
H2B	13,774	125	2.5	Pro	Lys	35	20	+19
H3	15,273	135	0.72	Ala	Ala	39	15	+20
H4	11,236	102	0.78	Ac–Ser	Gly	28	31	+16

[a]For rabbit thymus H1 (van Holde, 1989).
[b]Ac = acylated.

have been published on the structure and function of chromatin (Elgin & Weintraub, 1975; Kornberg, 1977; Chambon, 1978; Kanungo, 1980; Felsenfeld & McGhee, 1986; Widom, 1989; Van Holde, 1989; Elgin, 1990; Svaren & Chalkley, 1990; Grunstein, 1990; Pienta, Getzenberg & Coffey, 1991; Konnberg & Lorch, 1992; Felsenfeld, 1992).

Histones

Histones, which were first discovered by Stedman and Stedman in 1943, are proteins of small molecular weight present in the chromatin of all eukaryotes. Devoid of tryptophan, they are rich either in lysine or arginine residues; hence they are basic in nature, and are positively charged at physiological pH. Because of this property they bind to DNA, which has negatively charged phosphate groups linearly arranged in its backbone. They are present in approximately 1:1 ratio with DNA. Generally, five major types of histones are present in all types of cells: H1, H2A, H2B, H3, and H4, which are classified on the basis of the ratio of lysine: arginine. Various histones are compared in Table 3.1.

Histones are easily dissociated from the chromatin in 1.5 M NaCl and can be resolved by polyacrylamide gel electrophoresis (Fig. 3.1). An important characteristic of all histones is that their positively charged lysine or arginine residues are clustered in specific regions of the polypeptide chain. H1 has lysine residues clustered in its C-terminal domain, and the other four histones have the basic amino acids clustered in their N-terminal domain. This accounts for the long stretches of β-pleated sheets in the secondary structure of the chain. It is obvious that these positively charged regions would bind to the negatively charged phos-

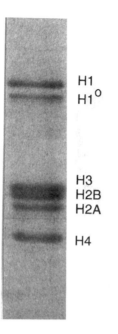

H1
H1^{o}

H3
H2B
H2A

H4

Figure 3.1. Rat liver histones resolved by 18%
SDS-polyacrylamide gel electrophoresis. His-
tones were stained by coomassie brilliant blue
R.

phate groups in the DNA more firmly than the other regions. Another
important feature is that two molecules each of H2A, H2B, H3, and
H4 form an octamer around which the DNA is coiled 1.75 times to form
a nucleosome. One molecule of H1 is bound to the DNA that links two
nucleosomes. Nucleosomal histones are more extended than H1 as they
have higher content of β-pleated sheet. A comparison between H1 and
nucleosomal histones is given in Table 3.2.

H1 histone

This is a very basic histone with approximately 25% of its residues being lysine. So it is more readily dissociated from DNA than
other histones. It has many isotypes, as several species of H1 with
variable amino acid residues are present in a single tissue. Rat thymus
and liver have five isohistones of H1. Also, the relative proportions of
the subfractions of H1 vary among tissues of the same organism (Kincade
& Cole, 1966a, 1966b) and during the cell cycle (Hohmann & Cole,
1971). Different subfractions of H1 are synthesized at various stages of

Table 3.2. *Comparison of H1 and nucleosomal histones*

Property	H1	H2A, H2B, H3, and H4
Location	Internucleosomal	Nucleosomal
Diversity	Several variations	Less variable
Stoichiometry	1 molecule 200 bp^{-1}	2 molecules 200 bp^{-1}
M.W.	22,500	11,000–15,000
Amino acids		
N-terminal	Cationic	Highly cationic
Central	Hydrophobic	Hydrophobic
C-terminal	Highly cationic	Low cationic
Amino acid sequence	Several variations	H2A, H2B – less variable
		H3, H4 – nearly invariant

Source: Kanungo (1980).

development of the sea urchin egg. Changes in the H1 isotypes occur during transition from blastula to gastrula, and at hatching of the sea urchin egg. Hence, it is likely that different H1 isotypes may have different functions.

H1 histone has a major basic C-terminal domain in contrast to the other four histones. The N-terminal region (1–40) has a basic domain (24–39) and a random coil. Most of the variations in amino acids in H1 isohistones are seen in the N-terminal domain. Functional differences of H1 isohistones such as interaction with different regions of DNA and with NHC proteins may depend on these variations. The central region, 39 ± 4 to 116 ± 4, has apolar amino acids in addition to most of the acidic amino acids and two aromatic amino acids. It is globular, largely invariant, and highly conserved. It is this domain that plays a role in binding to the nucleosomes in inactive chromatin. In active chromatin H1 no longer remains associated to nucleosomes through its central globular region but only through its C-terminal and N-terminal domains (Nacheva et al., 1989). The C-terminal is lysine rich and, therefore, highly basic. Since it is also highly conserved within a species, it may have a common role in all H1 histones. It has a random coil and is the main DNA-binding region.

Micrococcal nuclease (MNase), an endonuclease, specifically cleaves DNA in between nucleosomes. H1 is bound to 20–60 bp of this linker DNA that connects two adjacent nucleosomes. The length of the linker DNA varies among tissues, and also changes during development and differentiation. H1 does not participate in the formation of nucleosome structure, is more basic than other histones, is the first histone to be displaced from chromatin by acid or alkali and is more susceptible to degradation by protease when still bound to chromatin (Bartley & Chalkley, 1970). When H1 is added to H1-depleted chromatin, the latter contracts (Bradbury, 1975). When H1 is added to double-stranded DNA in solution, doughnut-shaped structures are formed (Hsiang & Cole, 1977), unlike other histones that form globular aggregates resembling nucleosomes. Thus, H1 is involved in the formation of higher order structure of chromatin, that is coiling of the string of nucleosomes to form a superhelical structure, as well as stabilizing the nucleosome. H5, a variant of H1 found in erythrocytes of birds, also forms similar structures with DNA.

Another important characteristic of H1 histone that is different from other histones is that it has a more rapid turnover in cell cultures (Appels & Ringertz, 1974). Whereas the synthesis of the other four histones is coupled to DNA synthesis during the S phase, that of H1 has been shown to occur at G1 in Friend and HeLa cells (Zlatanova & Swetly, 1978).

Core histones – H2A, H2B, H3, and H4

Digestion of chromatin by MNase yields globular structures called nucleosomes. Analysis of nucleosomes shows that they contain H2A, H2B, H3, and H4 histones. Their structures have been more conserved than H1. Furthermore, H3 and H4 are more conserved than H2A and H2B. H3 has a cysteine at 110 which has been conserved throughout evolution. H3 forms dimers by a disulfide bridge (Otero & Felsenfeld, 1977). It is phosphorylated at the G2/M boundary, and is rapidly dephosphorylated during G1. So phosphorylation precedes disulfide bridge formation.

Purified H3 and H4 form tetramers in solution, and it is the C-terminal domains that are involved in this interaction. Reconstitution experiments using partially cleaved H3 and H4 show that the first 41 and 31 residues from their N-terminals, respectively, are not essential for the tetramer formation. Removal of 45 and 18 residues from the C-terminals of H3 and H4, respectively, prevents the tetramer formation. The regions important for tetramer formation are 42–120 of H3, and 38–102 of H4 (Bohm et al., 1977).

H5 histone

Besides the five types of histones that are present in all types of nucleated cells, the nucleated erythrocytes of nonmammalian vertebrates contain another type of histone, H5, which is similar to H1 in many respects. It was first detected in chick erythrocytes by Neelin and Butler (1961). It has 197 amino acid residues, shows an electrophoretic band near H1, and has a M.W. of 23,000. H5 shows polymorphism and has an internucleosomal location. It binds to A-T-rich regions of DNA and has a stabilizing role on chromatin like H1. It is also lysine rich, lysine residues being clustered at the C-terminal region and constituting 23% of the amino acid residues. Nuclear magnetic resonance (NMR) studies show that its conformation is different from that of H1 and, therefore, it may have evolved independently from H1 (Chapman et al., 1978). The lysine residues of H5 are more acetylated than that of H1, but it is less phosphorylated than H1. It has a large number (21) of serine residues. Unlike H1, no clustering of basic amino acids is seen at its N-terminal and its N-terminal region is globular, also unlike that of H1.

The mRNA of H5 also does not have a poly-A$^+$ tail at its 3' end as in other histones. The H5 content in the early phase of maturation of avian erythroid cells is low, but as the maturation process advances, its content increases, and concomitantly the overall transcription activity of the chromatin decreases, even though the RNA polymerase level does not change. In the nondividing mature erythrocytes, synthesis of H5 histone continues even though the other histones are not synthesized (Sung et al., 1977). If H5 is removed from the chromatin, the repression of transcription is reversed. Its synthesis is neither coordinated with that of other histones, nor is it synchronized with DNA synthesis. Rather, H5 is synthesized after the synthesis of other histones. Since the gradual appearance of H5 in developing erythrocytes parallels repression of transcription, it is likely that it does so by condensing the chromatin leading to its inactivation. If H5 is introduced into other cells, their transcription is also repressed. It is significant that the newly synthesized H5 in developing erythroid cells is phosphorylated, and subsequently they get dephosphorylated with concomitant maturation of cells and decrease in transcription. Thus H5 appears to be important for maintaining the highly repressed state of nucleated chromatin. It is noteworthy that the gene for H5 is expressed only in erythroid cells at a specific stage, but how its expression is triggered and programmed is not known.

Protamines

These are basic proteins of small molecular weight that take the place of histones in the chromatin of sperm. They appear at the spermatid stage of spermatogenesis and replace histones in the chromatin. They show polymorphism. For example, trout sperm has three types of protamines, which have 31–33 amino acid residues. Mammalian protamines are longer (45 amino acids), are rich in arginine, but have no lysine or tryptophan. Two-thirds of the amino acids in protamines are arginine which is clustered together in long stretches and bound to the spermatid DNA; this binding completely represses transcription. When protamines are removed, the chromatin assumes a beaded appearance and becomes susceptible to MNase. Replacement with protamines causes loss of this structure and the chromatin becomes resistant to MNase. The serine residues undergo phosphorylation and dephosphorylation, and this covalent modification may also be involved in their association with DNA besides arginine (Dixon et al., 1977). The primary structure of a fish protamine is

$$_2HN-Pro-Arg4-Ser-Arg-Pro-Val-Arg5-Pro-Arg2-Pro-Arg2$$
(1)

$$HOOC-Arg4-Gly-Arg6-Ser-Val$$
(33)

Protamines, like histones, are synthesized in the cytoplasm and then translocated into the nucleus. Their mRNAs are short and are translated on diribosomes. The RNA has a poly-A$^+$ tail at its 3′ end unlike histone mRNA, and its mRNA is capped at its 5′ end by 7-methylguanine (m G). Iatrou, Spira and Dixon (1978) have shown that, although protamine is synthesized at the spermatid stage in trout testis, its mRNA is transcribed much earlier, that is, at the primary spermatocyte stage. An analogous situation is seen for histones of *Xenopus*. Its oocyte has maternal mRNA for histones in inactive form which are activated and translated as the egg divides. Protamines are present only in spermatocytes, but how their genes are expressed only in these cells, and what triggers their expression at a specific stage of development of these cells remain to be elucidated.

Histone genes

Histones are synthesized during the S phase of the cell cycle, which has helped the isolation of histone mRNA from rapidly dividing embryos for identifying and locating their genes by molecular hybridization and cloning techniques (Kedes, 1976). During the early cleavage stage of the sea urchin embryo, histones constitute 25%–30% of the newly synthesized proteins, and histone mRNA constitute nearly 70% of the total mRNA. Also, histone mRNA hybridize with DNA several hundred-fold faster than other types of mRNA. This indicates that there are multiple copies of histone genes. In the sea urchin, their genes are reiterated 300–1,000 times per haploid genome (Kedes, 1976). In *Drosophila, Xenopus,* chickens, and humans, the reiteration is 100, 10–20, 10, and 10–20 times, respectively. This large variation in reiteration may be related to the respective requirement of histones during early cleavage. The *Xenopus* egg contains a large amount of maternal histones, whereas the sea urchin egg has negligible amounts. Hence the former need not synthesize much histone in the early rapidly dividing stage, whereas the latter has to rapidly synthesize histones so that the cells may divide in the early stage of development. The large number of reiterated histone genes helps this process.

Histone genes are located in chromosome 2 in *Drosophila*. The five genes for the histones are G–C rich and are tandemly repeated. They are separated from each other by A–T rich spacer regions which are not transcribed. The entire coding region of histone genes is contained in 6–7 kbp of DNA. The arrangement of histone genes in trout, newt, sea urchin, and *Drosophila* along with their transcription are

Trout	H4	H2B	H1	H2A	H3
	→	→	→	→	→
Newt	H1	H3	H2B	H2A	H4
	→	→	←	→	→
Sea urchin	H1	H4	H2B	H3	H2A
	→	→	→	→	→
Drosophila	H3	H4	H2A	H2B	H1
	←	→	←	→	←

The structural genes for histones do not have introns unlike those for globin, ovalbumin, etc. (Schaffner et al., 1978). The spacer regions between the individual histone genes do not have repetitive sequences unlike those for rRNA and 5S RNA genes. Though the direction of transcription of histone genes is the same for all sea urchins and trout, those for *Drosophila* and newts are different (Lifton et al., 1978). Since transcription always takes place in the $5' \to 3'$ direction, the transcription of H2B and H4 genes occurs in opposite strands in *Drosophila*.

Histone variants

Histone genes are present in multiple copies. The genes are arranged tandemly, each repeat carrying all five genes, though there are differences in their arrangements and direction of transcription.

It appears that nonallelic variations of the genes have arisen during the evolution of the species. Kincade and Cole (1966a, 1966b) found that calf thymus H1 has several variants. Panyim and Chalkley (1969) found a variant of H1 which has a histidine unlike other histones; they named it H1°. Mouse somatic cells have six variants of H1 (H1°, H1a, H1b, H1c, H1d, and H1e). H3 and H4 are highly conserved. Only two variants of H4 have been reported; these are found in sea urchins. There are several variants of H2A and H2B. H5 is an extreme variant of H1, which is found in nucleated erythrocytes of birds, reptiles, amphibians, and fish. It is present in the internucleosomal region like H1. H5 is present in erythrocytes that are not only terminally differentiated but are also transcriptionally inactive. H1° is present in cells that are terminally differentiated.

It is interesting to speculate on how one type of H1 variant is replaced by another during mammalian spermatogenesis. During this process, spermatogonia become spermatocytes which in turn become spermatids. Spermatogonia contain H1a and H1c, unlike other cells which have H1b, H1d, and H1e. In spermatids histones are replaced by arginine-rich protamines. Seyedin and Kistler (1980) found that before protamine replacement occurs, H1 is replaced by a testis-specific H1t which is present in mammalian testis. It is present specifically in the late leptotene and pachytene periods of the meiotic prophase. It has been proposed that the presence of H1a and H1c may render the higher-order structure of the chromatin less stable, and thus permit genetic recombination that occurs at the pachytene period.

Histone variants found in somatic cells are different from those found in developing spermatocytes. The histones of spermatogonia are the same as those of somatic cells. The testis-specific variants replace the somatic variants during the meiotic prophase of spermatogenesis. During the transition from spermatocytes to spermatids, the testis-specific variants are replaced by protamines. Another interesting feature of histones is that there are tissue-specific differences in their variants, and their relative proportions are different (Zweidler, 1984).

Modifications of chromosomal proteins

Covalent modifications of chromosomal proteins have been shown to influence chromatin conformation and hence gene activity. Modifications of histones and NHC proteins may alter their binding to DNA, and thereby alter the accessibility of a gene to RNA polymerase II and transcription factors, that bind to transcription start site, and trans-acting factors that bind to cis-acting elements in its promoter region. Four types of covalent modifications are known to occur in chromosomal proteins – acetylation, phosphorylation, ADP-ribosylation, and methylation.

Acetylation

Phillips (1963) reported the occurrence of acetyl groups in histones. Allfrey, Faulkner and Mirsky (1964) then showed that histones in isolated nuclei can be acetylated. Two types of acetylated amino acid residues occur in histones: (1) Acetylated N-terminal serine is present in H1, H2A, and H4. It is an irreversible postsynthetic modification catalyzed by a cytosolic enzyme. (2) Acetylated lysyl residues are present in histones as a result of postsynthetic reaction that occurs in the cystosol (Ruiz-Carrillo, Wangh & Allfrey, 1975), and in the nucleus after the histones are translocated from cytosol into the nucleus and bind to DNA (Candido & Dixon, 1972a). Internal acetylation of H1 is negligible. Acetyltransferase catalyzes this energy-dependent reaction using ace-tyl~CoA as the donor of the acetyl group. The enzyme is present as a component of NHC proteins. The $-NH_2$ groups of internal lysyl residues located in the N-terminal half of the core histones are acetylated to form $-N-$acetyllysine (Gershey, Vidali & Allfrey, 1968; Allfrey, 1970). Up to four acetyl groups may be present in a histone molecule. Deacetylation is carried out by deacetylase, also present in the NHC protein fraction. The reaction is

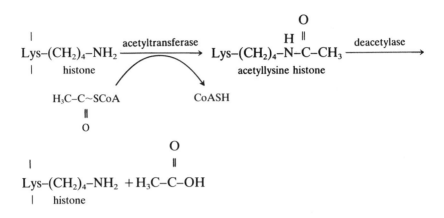

Acetylation occurs in internal lysyl residues 9, 14, 18, and 23 of H3, and 5, 8, 12, and 16 of H4 histones (DeLange et al., 1969; Candido & Dixon, 1972b). These sites have been conserved during evolution (Waterborg, Fried & Mathews 1983; Chicoine et al, 1986). Thorne et al. (1990) have shown that H3 has a strict order of acetylation sites: Lys-14, Lys-23, and Lys-18. In H4, Lys-16 is acetylated first, followed by acetylation of Lys 12, 8, and 5. H2B is acetylated at Lys-12 and Lys-15. Lys-5 and Lys-20 are also acetylated. H2A has only one site for acetylation.

Acetylation of internal lysyl residues is a reversible process and shows rapid turnover with a $T_{1/2}$ of only about 3 minutes (Nelson et al., 1978). Unlike phosphorylation, cyclic AMP has no effect on acetylation. Maximum acetylation occurs at interphase, and as cells enter mitosis, acetylation decreases. Minimum acetylation of histones is seen at prophase and metaphase, especially in H4 when chromosomes are condensed (D'Anna et al., 1977). RNA synthesis is minimal during these phases. As the cells enter telophase and begin to become extended, a concomitant increase in H4 acetylation occurs (Berkovic & Mauritzen, 1977). Acetylation of H3 and H4 decreases in avian erythroblasts as they undergo maturation to form erythrocytes in which the chromatin is highly condensed and transcriptionally inactive (Ruiz-Carrillo et al., 1976). Hence histone deacetylation may be correlated with the inhibition of transcription.

The above conclusions are experimentally corroborated by the findings that stimulation of RNA synthesis in lymphocytes by mitogens

(Pogo, Allfrey & Mirsky, 1966), in liver after partial hepatectomy (Pogo et al., 1968), and in target tissues by hormones (Libby, 1968) follows histone acetylation. Marushige (1976) showed that transcription of calf thymus chromatin increases if the nucleosomal histones are acetylated. Ruiz-Carrillo, Wangh and Allfrey (1975) found that acetylation of histones precedes the increase in RNA synthesis.

Further evidence for the role of acetylation of histones in transcription comes from the finding that in ciliates the heterochromatin is transcriptionally inactive and has a low level of acetylation. The euchromatin is transcriptionally active and highly acetylated (Lipps, 1975). The transcriptionally active macronucleus of *Tetrahymena pyriformis* has acetylated histones, but the repressed micronucleus does not (Gorovsky et al., 1973). In the sea urchin spermatid which does not synthesize RNA, H4 is completely deacetylated although it is acetylated in the embryo in which extensive gene activity is seen (Burdic & Taylor, 1976). Chromatin having highly acetylated histones is more easily digested by DNase I (Nelson et al., 1978).

The finding that butyrate stimulates acetylation of histones (Riggs et al., 1977; Candido, Reeves & Davie, 1978) has greatly facilitated studies on the role of acetylation in chromatin function. This is due to the inhibition of histone deacetylase by butyrate. Butyrate does not affect the rate of acetylation (Boffa et al; 1978; Reeves & Candido, 1978). Butyrate extensively acetylates histones, especially of H3 and H4 in HeLa cells (Simpson, 1978). Nucleosomal DNA is then digested five to ten times faster by DNase I. Nelson et al. (1978) and Sealy and Chalkley (1978) showed that DNase I preferentially digests chromatin regions that are highly acetylated. This indicates that nucleosomal regions containing acetylated histones undergo conformational changes and may loosen up to make the DNA more accessible to RNA polymerase and transcription factors. Therefore, acetylation of histones may be necessary for transcription, which requires movement by RNA polymerase along the template DNA (Davie & Candido, 1978).

The laboratory of Kanungo has carried out studies on the role of acetylation on transcription (Thakur, Das & Kanungo, 1978; Kanungo & Thakur, 1979b; Das & Kanungo, 1979). These researchers have shown a direct correlation between acetylation and transcription (Kanungo, 1980). Slices of cerebral cortex were incubated with ^{14}C-acetate for 60 minutes, nuclei were then purified and the incorporation of ^{3}H-UMP into RNA was studied for 30 minutes. Not only does acetylation significantly increase transcription, prior treatment of slices with 17 β-estradiol and butyric acid stimulates transcription further. Studies of Perry and Chalkley (1981) corroborate these findings. It was suggested that acetylation of histones decreases their net positive charge and dissociates them from the DNA, making it available for transcription (Kanungo & Thakur, 1979b; Kanungo, 1980). Particularly, regional modulation of histones is of importance in this context. Of the five histones, it is the core histones that are predominantly acetylated, particularly at their N-terminal domains containing lysine and arginine residues. It is probable that these regions are located at the transcription start site and cis-acting elements. Their dissociation may permit the binding of RNA polymerase, transcription factors, and trans-acting factors for transcription to be carried out.

Allan et al. (1982) have suggested that hyperacetylation modifies the higher-order structure of chromatin and loosens it, enabling transcription to occur. This is supported by the finding that histones of spermatids of rainbow trout (Christensen, Rattner & Dixon, 1984), rooster (Oliva & Mezquita, 1982), and rat (Grimes & Henderson, 1984) are maximally acetylated before they are replaced by protamines. When nucleosomes are reconstituted with acetylated histones derived from butyrate-treated HeLa cells, partial unwinding of DNA is seen (Norton et al., 1989). The chromatin of the Hpa II tiny fragment (HTF) islands is very highly acetylated, at least in H4 (Tazi & Bird, 1990). This indicates that the most highly acetylated histone, H4, is associated with a defined region of certain active genes. It is, however, not known whether a gene has to be actively transcribing to carry a high level of acetylated histones or whether its association with HTF is sufficient to cause acetylation.

There is ample evidence that acetylation of histones, especially of H4, H3, and H2B, is involved in the activation of transcription (see Csordas, 1990, for review). The relationship between acetylation of core histones and transcriptionally active chromatin has also been shown by fractionating chromatin with an antibody against acetylated histones and then probing the DNA of this chromatin with an active gene (Hebbes, Thorne & Crane-Robinson, 1988).

Waterborg and Mathews (1984) have studied the role of acetylation of histones in DNA synthesis since histone synthesis is linked to DNA synthesis and occurs during the S phase. Acetylation specific to the S phase occurs in all histones, but involves mainly mono- and diacetylation of H3 and H4. Chambers and Shaw (1984) found high levels of diacetylation during rapid cell division in sea urchin embryos.

In summary, acetylation of nucleosomal histones, especially of H3 and H4, may relax the chromatin structure at specific sites and thereby make it accessible for the enzymes and transcription factors involved in transcription.

Phosphorylation

This is an energy-dependent postsynthetic modification that occurs both in the cytoplasm and in the nucleus (Oliver et al., 1972; Ord & Stocken, 1975; Ruiz-Carrillo et al., 1975). H1 histones are more phosphorylated than other histones (Balhorn, Chalkley & Granner, 1972). The side chains of serine and threonine residues are the sites that are phosphorylated by specific cAMP-dependent and independent protein kinases. These phosphates, which are stable in acid, are removed by alkaline phosphatase. Lysyl, histidyl, and arginyl residues are also phosphorylated to some degree. Both phosphorylating and dephosphorylating enzymes are present in the NHC fraction. The phosphorylation reaction of histones is

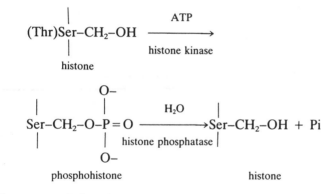

Two types of phosphate linkages are found in chromosomal proteins (Chen et al., 1977). One involves the acid-stable P–O linkage that is present in seryl and threonyl residues. The other involves the acid-labile P–N linkage that is present in lysyl, histidyl, and arginyl residues. The phosphorylation–dephosphorylation reaction exhibits a rapid turnover. The sites of H1 phosphorylation are different at different stages of the cell cycle. Ser-37 is phosphorylated at G1, Ser-114 at S and G2, and Ser-180 at M (Kurochkin et al., 1978). This may be due to multiple forms of H1 kinases having different specificities. Rapidly growing cells have a histone kinase that catalyzes the phosphorylation of threonyl residues and not Ser-37 and Ser-105 (Langan & Hohmann, 1974). Rattle et al. (1977) found that phosphorylation of either or both Ser-37 and Ser-105 decreases the strength of binding of H1 to DNA. This may have functional significance with respect to the role of H1 in chromatin condensation. Phosphorylation at different sites may cause decondensation of chromatin in various ways and expose different segments of DNA.

A high rate of phosphorylation is seen during mitosis of CHO cells, as well as HeLa cells (Lake & Salzman, 1972; Lake, 1973). The sites of phosphorylation in H1 at mitosis (H1m) are different from those at interphase (H1i) (Lake, 1973; Hohmann, Tobey & Gurley, 1976). It has therefore, been postulated that phosphorylation of H1 may be necessary for condensation of interphase chromatin into chromosomes. This is consistent with the finding that when H1 has three phosphate groups, it binds more strongly to DNA than dephosphorylated H1 (Knippers, Otto & Bohme, 1978). A high degree of phosphorylation of H1 and H3 is reported to be necessary for chromatin condensation during mitosis, and dephosphorylation imposes a restriction on this process at interphase. At prometaphase and anaphase when chromatin is aggregated, all H1 histones are superphosphorylated and have 3 to 6 phosphates. Also, all H3 histones are phosphorylated, which may be due to six- to ten-fold increase in specific ATP-histone phosphotransferase in the mitotic cells (Lake & Salzman, 1972; Lake, 1973). It is, however, intriguing to note that a high degree of phosphorylation, which is expected to dissociate H1 and H3 from DNA due to an increase in their negative charges, causes greater condensation.

Phosphorylation of H1 is higher in the liver of the developing rat, but is negligible in the adult liver. It greatly increases after partial hepatectomy (Balhorn, Reike & Chalkley, 1971; Balhorn et al., 1972). It has been suggested that histone phosphorylation is necessary for DNA replication, which is followed by cell division. Ord and Stocken (1975) have shown that the transcription of isolated nucleosomes of the rat liver increases after they are phosphorylated. The nucleosomal histones – H2A, H2B, H3, and H4 – are also phosphorylated in certain organisms at specific sites, though to a small degree.

Similar to H1, H5 histone also undergoes phosphorylation. Sung and Freedlender (1978) have shown that H5 is phosphorylated soon after its synthesis in avian erythrocytes. Subsequently, it is dephosphorylated as the cells mature. Dephosphorylation of H5 corresponds to genomic inactivation and chromosome condensation, which is the opposite of what is reported for other histones. The phosphorylated serine residues are found in that region of H5 that is highly basic and is involved in binding to DNA. Fifty percent of the phosphate is found in the 1–28 segment, and the rest in the 100–200 segment. Marushige and Marushige (1978) have shown that during spermatogenesis in several mammalian species, the protamines undergo phosphorylation–dephosphorylation, which seems to be necessary for the packaging of DNA.

Phosphorylation sites are generally located in the nonglobular tails of histone molecules. Since it is the folded, globular regions of histones that are believed to be involved in histone–histone interactions in the nucleosomes, histone phosphorylation may have a minor role in modulating the conformation of nucleosomes. Studies on histone phosphorylation in CHO cells (Gurley et al., 1978) give some insight into this role. In CHO cells, H1, H2A, and H3 incorporate ^{32}P when the cells are incubated with the labeled chemical. H1 goes through a complex series of phosphorylation and dephosphorylation during the cell cycle. When CHO cells arrested in early G1 are released, phosphorylation occurs at one site in the C-terminal region. During S phase, two more sites in this region are phosphorylated. As the cells prepare to enter mitosis, two additional phosphorylations occur at the N-terminal region, one in serine and another in threonine. Another threonine in the C-terminal region is also phosphorylated. This has been termed super-phosphorylation, the implication of which is not clear. Actually, less is known about the phosphorylation than the acetylation of histones.

The role of phosporylation of H1 at specific sites has been studied to determine the role of specific domains of H1 in binding to DNA and chromatin conformation (see Lennox & Cohen, 1988 for review). H1 is usually phosphorylated at two sites, but during mitosis, most H1 subtypes are phosphorylated at five out of six sites, mostly in the C-terminal domain. The extra phosphorylation correlates with chromosome condensation. It is intriguing to speculate on how H1 phosphorylation, which increases net negative charge, promotes chromatin condensation, unless the phosphates form specific salt linkages with basic side chains. It has been proposed that phosphorylation relieves constraints through loosening of protein–DNA interactions, allowing some other mechanisms to condense the chromatin, possibly at the level of the nuclear scaffold (Lennox & Cohen, 1988).

The studies of Hill et al. (1991) on H1 and H2B histones of sea urchin sperm suggest that the regions of H1 and H2B capable of being phosphorylated have a predominantly cross-linking role in the chromatin of the sperm rather than a role in the formation and stabilization of condensed chromatin filaments. This supports the earlier suggestion that the bulk of chromatin condensation occurs while H1 and H2B are phosphorylated, and that dephosphorylation provides DNA binding arms that interact with exposed "linker" DNA in adjacent filaments (Poccia, Simpson, & Green, 1987; Hill, Packman & Thomas, 1990). Hill et al. (1991) suggest that separate domains of H1 and H2B act independently and may have distinct functional roles. Ser-Pro–X-basic motifs are clustered in the N-terminal domains of H1 and H2B. When serine residues are phosphorylated, these otherwise DNA-binding domains are released, though the DNA binding of the remaining molecule is unaffected. Thus phosphorylation–dephosphorylation of the N-terminal and distal end of the C-terminal tail of H1, and/or the N-terminal tail of H2B controls intermolecular interactions between adjacent chromatin filaments, and thus plays a role in chromatin packing in the sperm nucleus. The tails of H1 are thought to interact with the linker DNA to achieve chromatin condensation (Allan et al., 1986), whereas the central globular domain of H1 interacts with the nucleosome at the dyad.

Poly(ADP)-ribosylation

Sugimura (1973) reported the presence of a chromatin-bound poly(ADP-ribose) polymerase that covalently attaches (ADP-ribose) moieties to the glutamyl side chain of H1 histone and to a lesser extent to H2B. It requires NAD^+ as the substrate. The reaction is

$$\underset{\text{histone}}{\overset{|}{\underset{|}{Glu}}-(CH_2)_2-COO^-} \xrightarrow[\text{+ NAD}]{\text{poly(ADP–R)polymerase}} \underset{\text{ADP–ribose histone}}{Glu-(CH_2)_2-\overset{\overset{\textstyle O}{\|}}{C}-O-(ADP–ribose)_n}$$

The enzyme is inhibited by nicotinamide, cytokinins, and methylxanthine (Levi, Jacobson & Jacobson, 1978). ADP-ribose linkage is alkali labile (Hayaishi, 1976). Poly(ADP-R) glycohydrolase degrades the polymer (Miwa et al., 1974) by cleaving the ribose–ribose linkage. Poly(ADP-ribose) polymerase catalyzes the attachments of ADP-ribose moieties of NAD^+ to the glutamyl residues of histones to form a polymer of poly(ADP-ribose). The enzyme has a M.W. of ~113 kd and has 1,014 amino acid residues. It has three proteolytically separable domains – an N-terminal region for binding to DNA, the central region for automodification, and the C-terminal for binding to its substrate, NAD^+ (Kameshita et al., 1984; Ueda & Hayaishi, 1985). The N-terminal region has two zinc fingers by which it binds to DNA. The enzyme is involved in DNA repair, DNA replication, RNA synthesis, and cell differentiation. The gene for the enzyme does not have the typical TATA and CCAAT boxes, which are also absent in housekeeping genes. It has SP1 and AP-1 binding sites in its promoter. The gene, which stretches over 43 kbp in humans and has 23 exons, is located in chromosome 1. Its transcription is enhanced by cyclic AMP (see de Murcia, Menissier-de Murcia & Schreiber, 1991, for review).

ADP-ribosylated histones dissociate from chromatin more easily than other modified histones. This may be because the introduction of ADP-ribose would not only vastly decrease the net positive charge on histones, but also the large size of the polymer may distort the chromatin conformation. The level of the enzyme increases three- to fourfold at the G1 phase and during differentiation of erythroleukemic mouse cells. In HeLa cells, polyADP-ribosylation is highest at the G1 phase and lowest at S phase. ADP-ribosylation of histones may dissociate them from DNA and allow it to be replicated at the S phase. This is supported by the finding that DNA synthesis in nuclei isolated from the liver of chick embryo increases after ADP-ribosylation (Tanigawa et al., 1978). Kanai et al. (1981) found that maximum ADP-ribosylation occurs at the G2 phase of the cell cycle. The level of polymerase activity is also the highest at this phase. Polyamines, spermine, spermidine, and putrescine stimulate ADP-ribosylation of nuclear proteins in the order mentioned (Tanigawa et al, 1978).

Although poly(ADP-ribose) chains are relatively short, longer chains also occur. Frequently, they are branched, with branch points occurring at intervals of 20 to 30 residues. Ogata et al. (1980) proposed that –COOH group of Glu-2 and Glu-116 in H1, and Glu-2 on H2B act as sites for ADP-ribosylation.

Though several functions have been assigned to ADP-ribosylation, the only function for which sufficient experimental support exists is DNA repair (Man & Shall, 1982; Kreimeyer et al., 1984). Ohghushi, Yoshihara, and Kamiya (1980), and Benjamin and Gill (1980) observed that the polymerase is stimulated by nicked DNA. In support of this, Ohashi et al. (1983) found that DNA ligase is stimulated by ADP-ribosylation of chromatin.

Methylation

Another posttranslational covalent modification that is seen in histones is methylation (Allfrey et al., 1964). Whereas acetylation, phosphorylation, and ADP-ribosylation decrease the net positive charge on histones and thereby cause their dissociation from DNA, methylation may, on the other hand, interact with some of the positive charges and stabilize the binding of histones to DNA. The histone methyltransferase III present in the NHC protein fraction methylates the $-NH_2$ group of lysyl residues using S-adenosyl methionine (SAM) as the $-CH_3$ group donor. The protons of the ϵ-NH_2 group of the lysine side chain are replaced by $-CH_3$ groups to give mono-, di-, and trimethylated derivatives (Paik & Kim, 1980). The reaction is

$$\underset{\text{histone}}{\overset{\text{(His)}}{\underset{\text{(Arg)}}{\text{Lys}}}\text{--}(CH_2)_4\text{--}NH_2} \xrightarrow[\quad]{\overset{\text{methyltransferase}}{+ \text{ SAM}}} \underset{\text{trimethyllysinehistone}}{\text{Lys--}(CH_2)_4\text{--}\overset{+}{N}\text{--}(CH_3)_3} + \underset{\text{homocysteine}}{\text{HS--R}}$$

Methylation occurs after histones bind to DNA. Methyl groups on histones do not turn over, unlike acetyl and phosphoryl groups (Byvoet et al., 1972). NHC proteins are methylated by a different enzyme. Up to three $-CH_3$ groups may be incorporated sequentially on the $-NH_2$ group of the lysyl residue. So lysyl residues may exist as mono-, di-, or trimethyllysines. Methylation occurs mostly in H3 and H4 histones, and H3 exhibits greater methylation than H4. All lysyl residues are not methylated. These sites on histones can be methylated after their isolation. Methyllysines are located near acetylated lysines or lysines capable of acetylation.

Methylation of H3 and H4 histones occurs only at their N-terminals. H3 is methylated at Lys-9 and Lys-27, and H4 at Lys-20 in the calf thymus (Duerre & Chakrabarty, 1977). Methylation sites are highly conserved. Methylation of HeLa cells occurs mainly during the S phase (Lee, Paik, & Borun, 1973). Though methylation is seen to occur throughout the cell cycle in tissue culture, maximum methylation occurs between the late S and G2 phases prior to mitosis (Tidwell, Allfrey & Mirsky, 1968; Thomas, Lange & Hempel, 1975). So methylation of histones may be necessary for the preparation of chromatin for mitosis. A demethylase that removes $-CH_3$ groups was reported in nuclei and mitochondria of rat liver (Paik & Kim, 1973); it is also rich in kidney cells.

There is a gradual shift toward the more methylated forms of H3 and H4 histones with increasing age. Neither H3 and H4 histones nor their methyl groups turn over in the brain of adult rats (Duerre & Lee, 1974). When labeled lysine and methionine are administered to young rats, significant amounts are incorporated into histones of the brain. In adult rats, only trace amounts are incorporated (Honda, Dixon & Candido, 1975). When brain nuclei of adult rats are incubated with SAM, no methyl groups are incorporated into H3 and H4. These studies show that methylation of histones is completed before adulthood. Methylated histones may have several functions. Methylated lysyl residues of histones, particularly trimethyl residues, could have a higher pK and increase the basicity of histones, which would strengthen the binding of histones, especially of H3 and H4, to DNA and cause compaction of chromatin and prevent transcription. However, such changes may occur only in specific regions of the chromatin, rendering them inactive and leaving other regions of the DNA for transcription in postmitotic cells. This may be one of the mechanisms that renders a large portion of the chromatin inactive, leaving only specific regions with the required genes for transcription. How this regional difference in chromatin is brought about by methylation of histones needs further studies. If methylation of histones occurs universally in the entire genome, it may prevent replication and drive the cells to an inactive state.

The four types of covalent modifications of histones are shown in Figure 3-2. It is seen that such modifications may have profound effects on the chromatin structure and function. The major characteristics of these modifications are: (1) Phosphorylation and ADP-ribosylation occur mainly in H1 histone, and acetylation and methylation occur in nucleosomal histones. (2) Phosphorylation, acetylation, and ADP-ribosylation are reversible processes; they decrease the net positive charge of histones and make them more easily dissociable from DNA, whereas methylation stabilizes the positive charge on histones and strengthens their binding to DNA. (3) These modifications may occur independent of each other as they are catalyzed by different enzymes that are located in the NHC protein fraction. (4) Phosphorylation appears to be a general phenomenon and is less specific; it occurs largely in dividing cells and throughout the cell cycle; and it appears to be necessary for DNA replication and mitosis. The three other types of modification may have specific roles. Acetylation occurs largely in metabolically active cells and appears to be required for transcription. Methylation is an irreversible process and may be involved in repression of gene activity and differentiation: (5) The four types of modification are specifically modulated by specific endogenous effectors, including hormones whose levels may change with age. Changes in modifications may alter

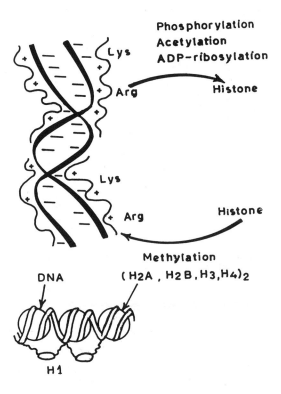

Figure 3.2. Chromatin structure showing association of histones with DNA and covalent modifications of histones that alter their association with DNA. (Kanungo, 1980)

chromatin conformation at specific sites, render these sites active, and influence expression of genes in these regions.

Ubiquitination

Besides the four types of modifications that are seen in histones (acetylation, phosphorylation, methylation, and ADP-ribosylation), H2A specifically undergoes ubiquitination to give rise to a distinct variant, UH2A. Initially, this was thought to be a different protein and was named A-24. Later, it was shown that it consisted of a ubiquitin molecule that was covalently linked by its C-terminal to the ϵ-NH_2 group of the Lys-119 of H2A. It is represented below

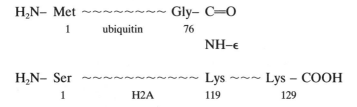

H₂N– Met ~~~~~~~~ Gly– C=O
1   ubiquitin   76

NH–ε

H₂N– Ser ~~~~~~~~~~~ Lys ~~~ Lys – COOH
1   H2A   119   129

The presence of ubiquitinated H2A in the chromatin is intriguing because ubiquitin has been shown to be involved in the degradation of proteins (Wilkinson, Urban, & Haas, 1980; Ciechanover, Finley & Varshavsky, 1984). Proteins that are marked for degradation are first ubiquitinated by an ATP-dependent reaction. Then the protein is degraded by proteolysis in a stepwise manner by six enzymes (see Hershko, 1988, for review). UH2A and probably UH2B are present in nucleosomes. UH2A binds to UH2B to form a dimer and cross-links with H1 as does H2A (Bonner & Stedman, 1979). UH2A can replace H2A in reconstituted core particles. The particles are "normal," as seen from MNase and DNase I digestion.

It has been shown that the ubiquitin moiety is lost from H2A at the onset of metaphase, and reappears in early G1. This suggests that ubiquitination may hinder the chromatin condensation that occurs at metaphase. It is likely that the ubiquitin moiety, since it lies on the outer surface of core histones, may hinder chromatin condensation. If this is true, it may have a role in transcription. Indeed, Goldknopf et al. (1980) found that UH2A is converted to H2A when transcription ceases during erythropoiesis. The heat shock (Hsp 70) and copia genes of *Drosophila* contain one UH2A per two nucleosomes during transcription, but in an untranscribed gene there is one UH2A per 25 nucleosomes (Levinger & Varshavsky, 1982). The nucleosomes of the active DHFR gene contain high levels of UH2A, particularly at their promoter regions. These data suggest that high levels of UH2A may disrupt the higher-order structure of chromatin and enable it to attain an active state for transcription. It is important to note that ubiquitinated H2A turns over more rapidly than the more stable H2A. It has been suggested that rapidly transcribed genes lose their nucleosomal structure due to proteolysis after ubiquitination of H2A (Levinger & Varshavsky, 1982), although there is no experimental evidence to support this. In any case, ubiquitination of H2A appears to have a significant role in chromatin structure and function, and needs further investigation.

Synthesis and turnover of histones

Histone genes are transcribed by RNA polymerase II. The mRNAs have no 3' poly-A^+ tail, but are capped at their 5' end by $m^7G(5')pppN^m$ or $m^7G(5')ppp$ N^mp^N. Their transcription is tightly coupled to DNA synthesis, and their synthesis occurs at the beginning of the S phase. The histone mRNAs are translocated to the cytoplasm where they are translated. The mRNAs have a total life of 10–12 hours, which is equal to the S period in humans. When DNA synthesis is inhibited by cytosine arabinoside or hydroxyurea, transcription of histone mRNAs also stops, the existing mRNAs are degraded, and histone synthesis stops. Also, if histone synthesis is terminated, DNA synthesis stops (Weintraub, 1972). The cell has a mechanism to turn histone genes on or off in conjunction with DNA synthesis. Furthermore, the stoichiometry of the histones synthesized is H1:H2A:H2B:H3:H4 = 0.5:1:1:1:1. This shows that the four nucleosomal histone genes are transcriptionally linked and are transcribed in a coordinated manner. Transcription of H1 may not be linked to the other histone genes even though its gene is located in the same tandem repeat. In *Drosophila*, in which the arrangement of histone genes is different from that of sea urchin, H1 gene is separated from H3 gene by 1,200 bp of DNA. Hence it may be controlled by a separate promoter (Lifton et al., 1978). Furthermore, H1 synthesis in G1 is threefold greater than that of other histones.

There are several exceptions to the coupling of histone synthesis with that of DNA. In *Xenopus laevis,* no such synchrony is seen in the early cleavage stage of embryogenesis, while in the sea urchin, histone synthesis begins at G1 and extends up to G2. At the onset of differentiation, however, histone and DNA synthesis become synchronized (Arceci & Gross, 1977). In HeLa cells, transcription of histone mRNA occurs throughout the cell cycle, but its translation occurs only at the S phase. Hence histone synthesis may be regulated at two steps, transcription and translation.

Leffak, Grainger and Weintraub (1977) found in in vitro chick myoblast cultures that when the cells divide, the pre-existing nucleosomal histones remain in one of the daughter cells, and the newly synthesized histones go to the other cell. So the new histones do not mix with the old ones, and their assembly is conservative. Moreover, these histones remain segregated for three to four generations. Hancock (1978) also showed that nucleosomal histones are conserved for many generations.

This is of significance as the newly synthesized histones become associated with the newly made DNA. It has also been suggested that some NHC proteins may be segregated likewise during cell division. Conservation of nucleosomes and NHC proteins with consequent transcriptional specificity may be a mechanism by which differentiation of a cell is achieved and retained. H1, however, has a turnover of 15% per cell generation. It also undergoes phosphorylation late in G2 phase of the cell cycle that corresponds to chromosome condensation. This may have implications for mitosis which follows.

Nonhistone chromosomal proteins

The proteins, other than histones, that are associated with the chromosomal DNA are nonhistone chromosomal (NHC) proteins, discovered by Mirsky and Pollister (1946). They can be removed from chromatin by 2 M NaCl–5 M urea and are implicated in the structure of chromatin, gene expression/repression, replication, and transcription. Nuclei contain from 0.3 to 0.8 gram of NHC proteins per gram of DNA, unlike histones which are present in approximately equal proportion with DNA. Their isoelectric points vary from 3.7 to 9.0. They are highly heterogeneous in size, their M.W. ranging from ~8,000 to several hundred thousands. Their $T_{1/2}$ values are much shorter than those of histones. Though NHC proteins are present in all cell types, tissues differ in their NHC protein pattern, both qualitatively and quantitatively. They are tissue and species specific. High-resolution techniques show that there are several hundred types of NHC proteins in each type of tissue. Nearly 1,500 NHC proteins have been detected in glia cells using isoelectric focusing and microdisk electrophoresis. They are synthesized throughout the cell cycle, unlike histones which are synthesized mainly at the S phase.

If calf thymus chromatin is treated by 0.3 M NaCl and and the proteins derived are resolved by SDS gel electrophoresis, two groups of NHC proteins are obtained: high mobility group (HMG) proteins of M.W. <30,000, and low mobility group proteins of >30,000 (Johns, 1964).

HMG proteins include four highly charged proteins: HMG1, HMG2, HMG14, and HMG17. They constitute 3% of DNA by weight, and have 25% basic and 30% acidic residues. They are present in all types of tissues, where they are associated with nucleosomes. HMG proteins can be isolated from chromatin by extraction with 0.35 M NaCl and 2% trichloroacetic acid. HMG 14 and HMG 17 have M.W.s of ~10,000 and

HMG 1 and HMG 2 have M.W.s of ~28,000. They are soluble in 0.35 M NaCl and 5% perchloric acid. They are present in mammals, birds, and probably in animals of lower phyla. They are not conserved as are the core histones. In HeLa cells, HMG 1 and HMG 2 are synthesized in the early G1 phase. Johns (1982) has given a detailed description of the HMG proteins.

The primary structures of HMG 14 and HMG 17 of certain animals have been determined. There is considerable homology in the structures of calf and chicken HMG 14 and HMG 17. They have a large number of basic amino acid residues at their N-terminal region as in histones, but have acidic amino acids at the C-terminal domain unlike histones. HMG 1 and HMG 2 of calf thymus contain a long stretch of aspartic and glutamic acid residues at their C-terminal regions (Pentecost, Wright & Dixon, 1985). The residues from 1–92 have a strong homology with those of 98–176.

HMG proteins also undergo postsynthetic modifications like the histones. Acetylation, phosphorylation, and ADP-ribosylation have been reported for all the HMG proteins (Reeves & Chang, 1983), whereas methylation has been reported for HMG 1 and HMG 2 (Boffa et al., 1979). The four HMG proteins also undergo glycosylation (Reeves & Chang, 1983). These authors found that glycosylated HMG proteins bind to DNA more strongly than unmodified forms.

HMG 14 and HMG 17 are rich in proline and hydrophilic residues, and are low in hydrophobic residues, which are the reasons they have little secondary and tertiary structures (Bradbury, 1982). HMG 1 and HMG 2, on the other hand, have a substantial degree (30%–50%) of α-helix. It is the regions 1–92 and 98–176 that have sequence homology and have this secondary structure.

Bradbury and his colleagues (Bradbury, 1982) have shown that the N-terminal half of HMG 14 and HMG 17 have binding sites for DNA. The C-terminal half remains as a random coil. Recent studies have shown that these two HMG proteins are present in the nucleosome, and more specifically in the nucleosomes of the active chromatin. HMG 1 and HMG 2, on the other hand, are present in the internucleosomal region. They preferentially bind to single-stranded DNA and appear to act as helix-destabilizing proteins (Yoshida & Shimura, 1984). Binding of these proteins to supercoiled DNA leads to unwinding. HMG 1 interacts with H2A·H2B dimer and H3·H4 tetramer (Bernues et al., 1983). Both HMG 1 and HMG 2 interact with the lysine-rich H1 histone.

A group of HMG proteins other than the ones just described bind strongly to α-satellite DNA and show a specificity not found in other proteins (Levinger & Varshavsky, 1982; Solomon, Strauss & Varshavsky, 1986). The remaining NHC proteins include (1) transcription-regulating proteins that bind to specific sites on DNA; (2) hormone-receptor proteins that can be identified by labeling with radioactive hormones; (3) chromatin-bound enzymes; and (4) chromosomal scaffold proteins that bind very tightly to DNA and can be isolated after removal of all other proteins. They are implicated in maintaining chromosome integrity and higher-order structure.

The NHC proteins also include several enzymes. Particularly, three types of acetyltransferases have been identified which have different specificities for histone acetylation. Deacetylase is chromatin bound. The enzymes for other types of modifications of histones (phosphorylation, methylation, and ADP-ribosylation) are also present in the chromatin. Both DNA polymerase and RNA polymerases are chromatin bound, as is the DNA repair enzyme. Surprisingly, DNases and histone proteases are present in chromatin. There must be some mechanism by which the activities of these enzymes are kept in check so that the chromatin is not degraded randomly. Some of the other important enzymes bound to chromatin are DNA ligase, histone phosphatase, cAMP phosphodiesterase, and topoisomerase.

There are several transcription-regulating proteins in this fraction. They include the trans-acting factor, SP1, that binds to GGGCGG sequence in the 5' flanking promoter region of several genes. It facilitates expression of genes. Another important protein (NG-1) having a M.W. of 52–66 kd that binds to the CCAAT sequence in the promoter region of several genes is also present in this fraction. Proteins that bind to promoter regions of globin genes and render them DNase I sensitive are also present. Another group of proteins found in the chromatin are hormone receptors, which on binding with the respective hormones, are able to bind to specific cis-acting sequences in the promoter regions of specific genes and in turn regulate their transcription. It is significant that the protein receptors for steroid hormones, T3 and retinoic acid, belong to one family. They have an N-terminal domain for binding to the specific cis-acting element in the promoter region of genes that they activate, and a C-terminal domain that binds to the hormone (ligand) that brings about a conformational change in the protein to enable it to bind to the DNA. It is of interest that even though T3, steroid hormones (17β-estradiol, glucocorticoid, testosterone), and retinoic acid have different structures, their receptors belong to one family, and their M.W.s are about 50 kd. T3 receptor is a component of NHC proteins (Jump & Oppenheimer, 1983). The estradiol receptor has been shown to be bound to chromatin (Welshon et al., 1984).

Structure of chromatin

The chemical composition of chromatin has been known for several years. However, the functions of its components, the way they are organized, the mechanism by which chromatin undergoes condensation during mitosis and elongation thereafter, the way certain genes are expressed in certain types of cells and repressed in others, the way certain genes are expressed at certain periods of the life span and not in others, and the way the expression of genes varies during the life span and under the influence of inducers and repressors are beginning to be understood only recently. It is necessary to understand the structure and organization of chromatin in order to understand the mechanism of gene expression and its regulation.

Olins and Olins (1974) treated interphase nuclei of rat thymus and liver and chicken erythrocytes with a hypotonic solution and examined the nuclei by electron microscopy after positive staining. The chromatin appeared as an exquisite string of beads having a diameter of ~ 100 Å and linked with each other by a DNA strand of a diameter of 15 Å (Fig. 3.3). At the same time van Holde et al. (1974) showed that digestion of chromatin by staphylococcal or micrococcal nuclease (MNase), which cleaves both the strands of DNA, yields particles that are ~100 Å in diameter and contain 200 bp of DNA. Noll (1974a) also digested chromatin by MNase and confirmed the repeating structure of chromatin (Fig. 3.3c). Noll (1974a) then digested chromatin by pancreatic DNase (DNase I) which makes single-strand cuts in double-stranded DNA, and resolved the fragments by gel electrophoresis. The pattern in Figure 3.4a shows a ladder of DNA fragments which are 10 bp and their multiples. On the basis of the presence of 200-bp DNA in the subunits of chromatin and its digestion by DNase I into 10-bp fragments, Crick and Klug (1975) proposed that the DNA helix is wound round a histone core and is periodically "kinked." DNase I cuts the DNA where it is maximally exposed (Fig. 3.4b). Also, Kornberg and Thomas (1974), using biochemical and X-ray diffraction techniques, showed that the particles obtained by MNase digestion contain a 200-bp-DNA repeat in chromatin and two molecules each of H2A, H2B, H3, and H4 histones

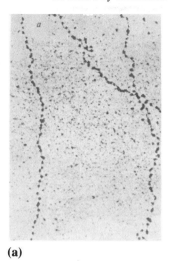

(a)

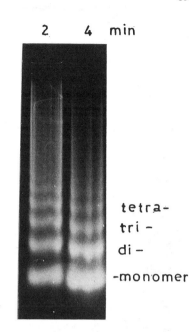

(c)

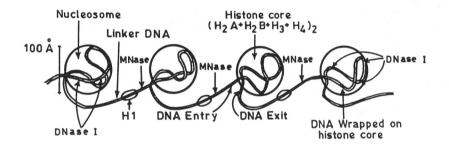

(b)   **NUCLEOSOMES - BEADS ON A STRING**

Figure 3.3. (a) Electron micrograph of chromatin. (b) Generalized structure of chromatin. Cutting sites for MNase (200-bp intervals) and DNase I (10-bp intervals) are shown. (c) Digestion of nuclear DNA by MNase showing cuts at intervals of 200 bp. Rat liver nuclei were digested for 2 and 4 minutes by MNase, nucleotide fragments were resolved in 1.8% agarose gel in Tris–acetate–EDTA buffer and stained by ethidium bromide. Bands represent dsDNA fragments. Band at bottom (200 bp) represents a mononucleosome. Its multiples (400, 600, 800) represent di-, tri-, and tetranucleosomes.

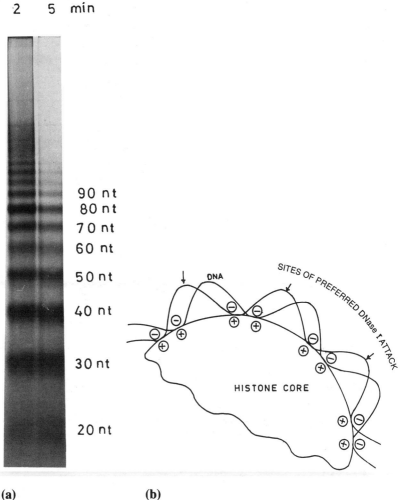

(a)   (b)

Figure 3.4. (a) Digestion of nuclear DNA by DNase I showing cuts at intervals of 10 nucleotides (nt). Nuclei of rat brain were digested by DNase I for 2 and 5 minutes, nucleotide fragments were resolved in 12% polyacrylamide gel containing 8 M urea, and stained by "Stains-All." Ladder of bands is visible from 20 nt on. Bands represent ssDNA. Band at 10 nt is not visible as these fragments are not effectively precipitated by ethanol. Band at 80 nt is intense as the DNA on the nucleosome has a sharp bend here and is digested preferably. Bands at 10-nt intervals indicate kinks at these sites. (b) Explanation for the pattern of digestion of DNA by DNase I. Enzyme cuts preferentially where DNA is maximally exposed on the nucleosome surface. (Adapted from van Holde, 1989) that form an octamer. This particle was called a nucleosome by Gross-Bellard and Chambon (1975). Reconstitution of the four types of histones and DNA yields such particles. H1 histone, therefore, is not present in these particles. It has an internucleosomal location. Therefore, it was proposed that the basic structure of chromatin consists of repeating units of octamers of four types of histones and 200 bp of DNA (Kornberg, 1974). When the nucleosomes containing 200-bp DNA were further digested by MNase (Söllner-Webb & Felsenfeld, 1975), stable core particles were obtained which contained about 140-bp DNA and an octamer of nucleosomal histones (H2A, H2B, H3, and H4). H1 histone was not present but was released along with the oligomeric DNA that was released from the nucleosomes. On the basis of these data, it was suggested that H1 was associated with the spacer or linker DNA that links two successive nucleosomes. Studies on various species of organisms including yeast, invertebrates, and vertebrates showed that whereas the length of the DNA in the core particle was more or less 146 bp, that of the nucleosome varied from about 160 to 260 bp (Mirzabekov et al., 1978; Prunell et al., 1979). Also, reconstitution of core particles from core histones and DNA showed a DNA length of 146 bp associated with the octamer of histones (Simpson & Kunzler, 1979). The above studies established that (1) the basic structure of chromatin consists of nucleosomes linked by spacer DNA; (2) each nucleosome core contains an octamer of histones (H2A, H2B, H3, H4)$_2$ around which 146 bp of DNA is wound; and (3) H1 histone is associated with the spacer DNA (Fig 3.3).

Finch et al. (1977) carried out detailed X-ray and electron microscopic studies and showed that the core particle is a wedge-shaped flat disk, with dimensions of 5.7 × 11 × 11 nm. They proposed that 140 bp of DNA is coiled as 1.75 turns; the diameter of the coil is 9 nm and the pitch is 2.8 nm. This corresponds to nearly 80 bp per superhelical turn of the B-form of DNA. Histones are partly buried in the major groove of DNA and leave the minor groove exposed. The length of 140 bp of DNA is six to seven times longer than the dimensions of the core. So the DNA is compacted six to seven times, which is achieved by its binding to the basic regions of the eight histone molecules of the octamer (Sussman & Trifonov, 1978). This protects the core DNA against MNase. DNase I, however, cleaves this DNA at 10-bp intervals, since there are kinks at 10-bp intervals in the DNA chain that are exposed on the outside of the core. The four core histones are electrostatically bound through their positively charged groups of the side chains of basic amino acid residues to the negatively charged phosphate groups of the DNA. They are arranged as follows:

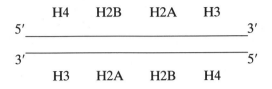

H3 and H4 histones interact with the two ends of the DNA segment. When these histones come in contact with the double-stranded DNA, they form characteristic beaded structures as seen by electron microscopy (Germond et al., 1975). Also, reconstitution of core histones with 140bp DNA produces particles that have the same sedimentation coefficient as that of nucleosomes obtained from chromatin (Tatchell & van Holde, 1976). H3 and H4 histones alone can produce nucleosome-core-like structures with DNA, which are resistant to trypsin and give X-ray diffraction patterns similar to those of native nucleosomes (Moss et al., 1977). Klevan et al. (1978) showed that when H3 and H4 are added to DNA, they bind to 140 bp of DNA which forms a 1.5 superhelical turn around the tetramer and forms a cylindrical structure of $45 \times 8 \times 8$ nm. If H2A and H2B are added, the cylinder condenses and assumes the appearance of the native nucleosome. This is in agreement with the original suggestion of Kornberg (1974) that H3 and H4 are essential for nucleosome structure. These two histones are more conserved, have a higher proportion of β-pleated sheet, and interact with each other more than with other histones.

The degree of binding of various histones to DNA is in the following order: H3 and H4 > H2A > H2B > H1 (Papanov et al., 1978). Cross-linking studies show that H3–H4, H2A–H2B, and H2B–H4 are linked (Chung, Hill & Doty, 1978). At first 2(H3) and 2(H4) form a tetramer and then bind to 140 bp of DNA to form a kernel or basic core. Then, two H2A and two H2B associate to complete the nucleosome. The C-terminal domains of the four core histones are involved in interaction with DNA since the removal of the N-terminal domain does not affect the nucleosome structure (Whitlock & Stein, 1978). H2A and H2B form dimers by interacting through their central apolar regions, leaving the N- and C-terminal ends free. H3 and H4 form dimers through their central apolar and C-terminals, leaving the basic N-terminal regions of the nucleosomal histones available for interaction with the acidic groups of DNA. Each histone is bound to about 10 bp of DNA as seen from

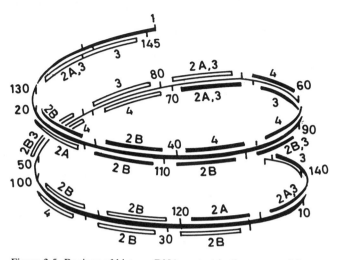

Figure 3.5. Regions of histone–DNA contact in the core particle, as proposed by Shick et al. (1980).

cross-linking of histones to the 5' end of DNA (Mirzabekov et al., 1978; Shick et al., 1980). (See Fig. 3.5.) It is significant that the assembly of nucleosomes is directed by NHC proteins. For example, NHC protein fraction purified from the eggs of *Xenopus laevis* converts pure DNA into nucleosomes in the presence of histones in a cell-free system.

The basic structure of chromatin is a linear array of beads or nucleosomes of 100 Å diameter; this stringlike structure is called the nucleofilament. The nucleosomes are then coiled to give a higher-order structure having a diameter of 200–300 Å with a 100 Å pitch. The coiling of nucleosomes is brought about by H1 histones, which bind to the linker DNA between adjacent nucleosomes. When H1 is added to H1-depleted chromatin, an increase in affinity for H1 is seen up to but not beyond octanucleosome formation. The next higher-order structure of chromatin after the string of beads is a coiled unit of octanucleosomes initiated by H1, or H5 in the case of nucleated erythrocytes.

Two significant features of both yeast and mammalian nucleosomes are a core length of DNA of 146 bp, and highly conserved H3 and H4 histones that form a tetramer and assemble the DNA. DNA (146 bp) will form 14 turns in the B-form with 10.5 bp per turn. If there are 10 bp per turn, then there are 14.5 turns in the DNA surrounding the nucleosome core. The conservation of this length of DNA during evolution may be due to conservation of the core histones, because any mutation in these histones would affect the length of the DNA. Any change in the length of the core DNA would also affect the length of the linker DNA, and the organization of the chromatin would be disrupted. Therefore, mutations in the core histones have been suppressed just as mutations in the genetic code are strongly suppressed.

Noll's (1974a) observation that digestion of chromatin DNA by DNase I yields 10 bp DNA fragments and their multiples suggested that the entire core DNA is on the surface of the histone octamer of nucleosomes. This was supported later by neutron scattering studies of nucleosomes in solution. The radius of gyration of histone core is about 30–35 Å, and that of DNA is 40–50 Å, indicating that DNA is on the surface of the core (Pardon et al., 1975; Braddock, Baldwin & Bradbury, 1981).

The association of histone octamers with DNA appears to be a spontaneous process. DNA and histones are dissociated by high salt (>2 M) concentrations. If, then, the ionic strength is lowered gradually, they reassociate. No other factor appears to be necessary for this assembly. How is the DNA bound to the histone core? Is the binding localized at certain regions, or is it uniform throughout? To answer these questions, Simpson and Whitlock (1976) labeled the 5' end of the core particle DNA with ^{32}P using T4 polynucleotide kinase. The DNA was then cleaved by DNase I, and fragments were resolved by electrophoresis. Lutter (1978) used the same technique after labeling the 3' end of the DNA. These studies revealed that all sites are not cleaved with equal probability. The sites at 30, 60, 80, and 110 bases from the 5' end are cleaved slowly. This is also the case if the 3' end is labeled. Also, DNase II and MNase cleave slowly at 30, 60, and 110 bases. There is resistance to cleavage at sites 30 and 110, 0 and 80, 60, and 140, each of these sites being separated by about 83 bp. On the basis of these data, it was suggested that one turn of DNA on the nucleosome corresponds to about 83 bp (Lutter, 1978).

McGhee and Felsenfeld (1979, 1980) found that almost all the residues of core DNA react with dimethylsulfate. They also analyzed the effect of ionic strength on the interaction of phosphate groups with the basic residues of core histones. These studies have led to the conclusion that DNA may not be buried among the histones, but may lie on the surface and make contact with the side chains of histone residues. This is shown in Figure 3.4b (van Holde, 1989).

The DNA regions that are in contact with specific histones have been examined by cleaving the nucleosomes by nucleases and determining which regions of DNA are associated with which histones. Studies carried out by Klevan et al. (1978) have shown: (1) the $(H3:H4)_2$ tetramer is more strongly bound to DNA than other histones. This tetramer associates first with the DNA when the ionic strength is lowered, and dissociates last when the ionic strength is increased. (2) A single $(H3:H4)_2$ tetramer protects about 70 to 80 bp of DNA. If a longer DNA is used, then it binds to H2A:H2B dimer, provided it is already primed by the $(H3:H4)_2$ tetramer. Once the DNA is bound to a $(H3:H4)_2$ tetramer, it will bind only to a $(H2A:H2B)_2$ tetramer, not another $(H3:H4)_2$ tetramer. The $(H2A:H2B)_2$ tetramer seems to be bound comparatively weakly since 0.25 M NaCl + 4 M urea removes it (Sibbet & Carpenter, 1983).

Hydrodynamic and electron microscopic studies have shown that the diameter of the core particle is 100 Å (Olins et al. 1976). It is a disk about 50 Å thick with the DNA wrapped in two coils or helix of about 110 Å outer diameter around a disklike protein core of 60–70 Å diameter (Braddock et al., 1981). Klug and his group (Finch et al., 1977) carried out X-ray diffraction studies at 25 Å resolution on crystals of nucleosome core particles. This was followed by neutron diffraction at 25 Å resolution (Finch et al., 1980) and 16 Å resolution (Bentley et al., 1984), and X-ray diffraction at 7 Å resolution (Richmond et al., 1984). The conclusions from these studies are that DNA is wrapped around histones in a left-handed double helix. It is of B-type and is not uniformly bent. There are four sharp bends, which correspond to the positions observed for protection of the DNA from DNase I cleavage. Klug et al. (1980) postulated that the $(H3)_2$ dimer forms the organizing center of the nucleosome, and that the $(H4)_2$ dimer makes substantial contacts with the DNA. Thus the $(H3:H4)_2$ tetramer is able to protect about 70–80 bp of DNA as suggested by earlier studies. Later, both physical (for review see van Holde 1989) and biochemical (Svaren & Chalkley, 1990) evidence was discovered to support the conclusion that H3 and H4 form a tetramer $(H3:H4)_2$, which binds to DNA and directs the subsequent association of H2A and H2B. Hayes, Clark and Wolfee (1991) have shown that the $(H3:H4)_2$ tetramer binds to 120 bp of DNA. The tails of these histones do not contribute to DNA binding, i.e., the central regions of these histones actually bind to DNA.

DNA–histone interactions

Although there are over 200 lysyl and arginyl residues in each histone octamer, only a few of them interact with the negatively charged phosphate groups of the DNA. The N- and C-terminals are specifically rich in these residues. However, digestion of core particles by trypsin that removes these residues from the tails does not affect the core structure. Also, histone cores from which the tails containing arginyl and lysyl residues are removed can form core particles with DNA. Furthermore, treatment of histone cores with chymotrypsin does not affect nucleosome core formation by DNA. These studies show that the central globular domains of the core histones interact with the DNA to confer additional stability. The studies of Ichimura, Mita, and Zama (1982) show that 14 arginine residues in the globular regions are involved in DNA–histone interactions. The remaining 84 lysyl and arginyl residues in this region form salt bridges with the aspartic and glutamic acid residues. Arginyl residues are preferred as they can participate in both H-bond and electrostatic interactions, and thus bind more strongly with the DNA, as follows:

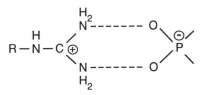

Higher-order structure of chromatin

The first-order structure of chromatin is the beads-on-a-string formation or a string of nucleosomes. It has two components: the core particle and the linker region that varies among chromatin from different sources and conditions.

Studies of several workers (Varshavsky, Bakayev & Georgiev, 1976; Noll & Kornberg, 1977; Albright et al., 1980) have shown that when chromatin is digested by MNase and fragments are resolved by various methods, two major fractions are obtained, one with a longer DNA fragment which is associated with H1 histone, and another with shorter DNA fragments from which H1 is absent. It was shown that the nucleosome that has 160-bp DNA has the H1. When it is digested to 140 bp, H1 is lost. Simpson (1978) isolated and characterized a nucleoprotein particle (called a chromatosome) containing about 160 bp of DNA, a histone octamer, and one molecule of H1. Later it was found that chromatosome has 166–168 bp of DNA, which corresponds to two superhelical turns around the core. It was shown by end-labeling the DNA followed by DNase I digestion that about 10 bp of DNA are present at either end of core DNA of 146 bp. H1 stabilizes the particle. The ends of the 166-bp DNA fragment of the chromatosome make contact with the core histones, possibly H3. The 10-bp DNA fragments at either end are also associated with H1. Cross-linking studies show that H1 is linked to H2A and H3.

Studies of Allan et al. (1980) have shown that it is the globular central domain of lysine-rich H1 that is responsible for stabilization of the chromatosome as is the case for core histones. The stoichiometry of the occurrence of H1 histone in the chromatin gives a value of one H1 per nucleosome in the regions of the chromatin that do not have actively transcribing genes. In the regions that have active genes, 20%–40% less H1 is present (Kamakaka & Thomas, 1990). This would partially open the chromatin and make the DNA accessible to RNA polymerase for transcription. For replication too, a similar partial dissociation of H1 may occur to allow DNA polymerase to bind to DNA and replicate.

Laybourn and Kadonaga (1991) studied the role of H1 on the transcription of reconstituted chromatin. In chromatin containing only nucleosomal cores with an average density of one nucleosome per 200 bp DNA as is normally the case, transcription is 20%–50% of that of naked DNA. This repression occurs as nucleosomes are located at the RNA transcription start site, and cannot be counteracted by transcription stimulatory factors, Sp1 and GAL4–VP6. When H1 is added to chromatin at 0.5–1.0 molecule per nucleosome (200-bp DNA), transcription is reduced to 1%–4% of that observed with chromatin containing only nucleosomal cores. This clearly shows that, besides being important for the higher-order structure of chromatin, H1 acts as a gene repressor.

Before one considers how the primary structure of chromatin, that is, the beads on a string, is further condensed, it is necessary to know whether the nucleosomes (1) are arranged or positioned on the DNA randomly, (2) are associated with sequence-specific DNA regions, or (3) are spaced from each other at uniform distances separated by defined lengths of linker DNA. Regular spacing of nucleosomes along the length of DNA would mean that they are "phased." This has been studied by digesting chromatin with MNase, separating the DNA fragments, and studying their lengths. However, MNase has a preference for cleaving at AT-rich regions and, therefore, does not give meaningful data. Methidiumpropyl-EDTA·Fe(II) on the other hand, selectively cleaves DNA in the linker region and has little sequence specificity (Cartwright et al., 1983).

Wu (1980) used an indirect end-labeling method for locating positioned nucleosomes in chromatin DNA. Chromatin was cleaved by MNase, DNA fragments were isolated, cleaved by a specific restriction endonuclease, and resolved by gel electrophoresis. After transfer to filter paper, the fragments were hybridized to the labeled probe that abuts the restriction site. Then fragments that begin at the site and extend to the internucleosomal site that has been cut by MNase were identified. Zhang, Fittler, and Honz (1983) used another method to examine the phasing of nucleosomes in repetitive DNA sequences. This required a restriction site in the repetitive region of the DNA. The chromatin was digested by MNase and the extended DNA sequences beyond the core particle were trimmed by exonuclease III, which produces single-stranded DNA (ssDNA) ends. The ends were removed by S1 nuclease and the DNA fragments were cleaved by a restriction enzyme and resolved by gel electrophoresis.

Although the length of the DNA in a single eukaryotic cell (2N) is about 2 meters, the DNA is in such a compact form that it is contained in a nucleus that is only a few microns in diameter. In the first-order structure of the chromatin, the DNA is wrapped around nucleosomes and this condenses it approximately sixfold. It has to undergo a great deal more compaction to be accommodated within the nucleus. The interphase chromatin represents an intermediate state of condensation, and the metaphase chromosome has the DNA in the most highly compacted state. Several techniques, especially transmission electron microscopy (EM), have given useful information on the way the DNA is compacted. Thoma and Koller (1981), and Woodcock, Frado and Rattner (1984) digested rat liver nuclei by MNase and fixed the chromatin in glutaraldehyde in buffers of varying ionic strength and studied it by EM. At very low ionic strength the chromatin is seen as a fiber containing nucleosomes of 100 Å width. The entry and exit points of the DNA in the nucleosome appear to be near each other. If H1 histone is removed by treating chromatin with a reagent of low ionic strength, most of the nucleosomal structure is lost.

At higher ionic strength, the chromatin appears as a ribbon of 250 Å width. At still higher ionic strength, the fiber has a width of about 300 Å (McGhee et al., 1983), which is believed to be due to the solenoid structure of the string of beads (Fig. 3.6; Finch & Klug, 1976). Each turn contains six nucleosomes and has a pitch of 110 Å. Another view

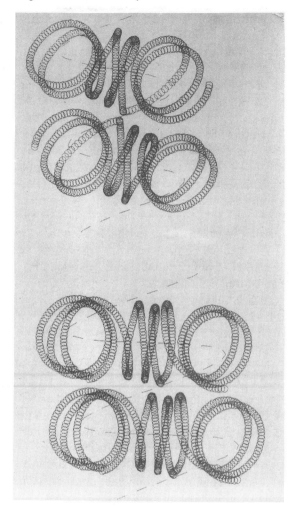

Figure 3.6. Fiber model (300 Å) of chromatin as proposed by McGhee et al. (1983). The chromatosomes are proposed to lie radially with flat faces tilted 26° + 6° from the solenoid axis. The linker DNA between nucleosomes is supercoiled about the helix (dashed lines) that passes through the chromatosome centers. For clarity, only the three nucleosomes on the front surface of the fiber are shown, and histone cores have been omitted. The model is for sea urchin chromatin, which has a linker DNA length of 77 bp and a tilt angle of 20°.

is that it is due to linear aggregation of "superbeads," each of which is an aggregation of several nucleosomes. Using the photochemical dichroism technique, Mitra, Sen and Crothers (1984) found that the flat faces of core particles lie parallel to the axis of the 300 Å fiber. Thus, the second-order structure of chromatin appears to be a solenoid having 6–8 nucleosomes per turn, a pitch of 110 Å, and diameter of 300 Å. There is, however, no general consensus of opinion over the exact second-order structure of chromatin. The maintenance of the second-order structure requires a high concentration of monovalent cations or a low concentration of divalent cations, which reduce the electrostatic repulsion between DNA chains. This structure also requires H1 histone (Watanabe, 1984). The fibers condense into irregular aggregates without H1 and in the presence of high salt. It is suggested that H1 may set the regular spacing of nucleosomes, which helps in the formation of a solenoid.

Allan et al. (1982) removed H1 from chromatin and then added H1. The chromatin refolded and formed the 300 Å fiber. H1 was then removed from chromatin, which in turn was digested by trypsin to cleave the tails of core histones. If H1 histone was then added, the 300 Å fiber was not formed. This clearly showed not only the importance of H1 in the solenoid formation, but also the involvement of the tails of core histones. Also, removal of H2A and H2B relaxes the solenoid even if H1 is present.

The formation of the solenoid condenses the DNA by another six- to sevenfold. That is, the DNA may be compacted by a total of about 50-fold. The 2-meter DNA still cannot be accommodated in an interphase nucleus that is 6 to 8 μ in diameter. When interphase chromosomes of *Drosophila* are gently isolated and digested with low concentration of DNase I to produce single-strand nicks, the sedimentation coefficients of chromosomes decrease gradually and reach a plateau. This indicates that chromosomal DNA is arranged in a large number of independently supercoiled domains or loops, each of which becomes relaxed by a single nick. The average loop size was estimated to be about 85,000 bp. Igo-Kemenes and Zachau (1977) digested rat liver nuclei by MNase and restriction enzymes and found the largest size of the DNA loop to be about 75 kbp. The average size of the loop was calculated to be about 35 kbp. It was also found that the loop size varies between 35 and 80 kbp among species. It is likely that different domains may have different sets of active or inactive genes. This may provide a mechanism for controlling the expression of genes.

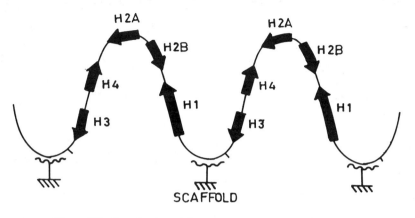

Figure 3.7. Organization of the genes of *Drosophila* histones and their mode of attachment to the interphase scaffold. (Mirkovitch et al., 1984)

Interphase chromosomes

The individual chromosomes remain distinct at interphase and are highly extended. Based on electron microscopic studies, Comings (1968) first observed that the chromosomes are connected to the nuclear envelope and are projected as loops into the nucleoplasm. Lebkowski and Laemmli (1982) and Jackson, McCready, and Cook (1984) observed that DNA loops extend into the proteinaceous scaffold of the nuclear matrix. The proteins in the scaffold are lamins – A, B, and C. SC1, a major protein component of the metaphase scaffold, is also present. Thus it appears that DNA is attached to two types of sites – some are on the peripheral laminar structures, and others are on the matrix fibers within the nucleus. Mirkovitch, Mirault, and Laemmli (1984) found that histone gene clusters of *Drosophila* are attached to the scaffold through A-T rich regions present between H1-H3 spacers. A-T sequences between heat-shock genes are also attached to the matrix protein. The arrangement and attachment of histone genes in *Drosophila* are shown in Figure 3.7.

Metaphase chromosomes

Wray, Elgin, and Wray (1980) carried out EM studies on metaphase chromosomes isolated from nuclei without osmotic shock. The intact chromosome isolated in high ionic media shows ~500 Å fibers which extend radially from a central core. In media of low ionic strength, loops of 300 Å fibers are seen. If divalent cations are removed, loops of 100 Å fibers are observed. When histone-depleted chromosomes are examined by EM, large loops of DNA appear to emerge from a scaffold structure (Paulson & Laemmli, 1977). These loops have dimensions similar to those of domains in the range of 40 to 80 kbp, and appear to arise and terminate in the scaffold at the same point.

Later, Lewis, and Laemmli (1982) found that the scaffold has two major proteins, SC1 of Mr = 170 kD, and SC2 of Mr = 135 kD. SC1 appears to be eukaryotic topoisomerase II (Gasser et al., 1986), which regulates supercoiling of individual domains. SC1 and SC2 comprise 40% of the scaffold protein of the metaphase.

Chromatin structure and transcription

Although the basic mechanism of transcription of prokaryotic and eukaryotic DNA is very similar, there are several differences that influence the rate and modulation of their transcription. The major differences are shown in Table 3.3.

In multicellular organisms, differentiation of cells into various types has been possible because of the complex nature of chromatin. The permanent repression of certain genes in certain cell types, the expression of certain genes only at certain times when appropriate signals are received (inducible genes), and the expression of a few genes at all times (housekeeping genes) from among thousands of genes in the entire genome have been possible due to several types of control mechanisms.

Several questions arise from our present knowledge concerning gene expression and repression. How does a cell know that a specific gene is to be transcribed? Why is a gene transcribed in a specific cell type? Are there signals for transcription and repression? Why is a gene in a particular cell transcribed at a specific life stage of the organism and then permanently switched off? There are also genes that switch on and off, but can be switched on again after an appropriate signal is received. Is there a marker in this region of the chromatin to indicate that the gene may be transcribed again? Why does a gene that was silent in the early phase of development become active later? What role does the nucleosome have in these processes? These questions are beginning to be answered by the use of genetic engineering technology. We are beginning to understand how a gene becomes active (or expressed) and how a gene becomes inactive (or not expressed).

The sequence of DNA representing a protein-coding gene is far

Table 3.3. *Major differences between prokaryotic and eukaryotic DNA*

Prokaryotic DNA	Eukaryotic DNA
1. One RNA polymerase for transcribing all types of RNA.	1. Three types of RNA polymerases I – rRNAs II – mRNAs III – tRNA and 5S RNA
2. Polycistronic mRNAs transcribed from two or more tandemly arranged genes that are linked.	2. mRNAs are monocistronic and transcribed from single genes.
3. Genes are not split.	3. Genes split into exons and introns. Entire gene is transcribed and sequences corresponding to introns are excised. Exons are spliced together to form mRNA.
4. Control of expression is comparatively simple.	4. Control of expression complex due to tissue specificity of gene expression. Positive and negative control mechanisms and response to modulators are complex and diverse. All genes are not transcribed in all cells; at most ~5% genes are transcribed in any cell type from the entire set of ~50,000–100,000 genes, due to differentiation of cell types. Hence certain regions of chromatin are transcribed (active) and others are not transcribed (inactive).

longer than 146 bp. Therefore, the gene must extend through several nucleosomes and internucleosomal regions. Do the nucleosomes open up, dissociate, and reassociate again to permit the RNA polymerase to pass through and transcribe the gene? Or is there any specific region of the gene (e.g., the promoter or transcription start site) earmarked in the internucleosomal region such that RNA polymerase II can more easily bind to and begin transcription? Are the nucleosomes present selectively in the introns or in exons? In the context of the higher-order structure of interphase chromatin, how does an inducer pass through the complex organization to reach the specific site in the promoter? If the chromatin DNA coding for a gene is depleted of all the histones, and the gene is cloned into a plasmid along with its promoter, it can be transcribed by RNA polymerase II. This transcription can be modulated through cis-acting elements in its promoter. Is it necessary for the DNA sequences to be present in such a complex organization for stimulation of transcription? Has this complex organization evolved for keeping far more genes inactive than active? Even if it were so, it does not appear to be an all-or-none situation, because an inactive gene in one type of cell is active in another, and in certain abnormal situations inactive genes do become active.

Active and inactive chromatin

The presence of an active gene in one tissue and its inactivity in another, or its activity at one life stage in a tissue and its inactivity in another or vice versa show that the chromatin structure plays an important role in keeping the gene in a functional/nonfunctional state. Except for the direct modification of DNA (such as cytosine methylation that may promote binding of specific trans-acting factors), no other DNA modifications are known to keep a gene functional or nonfunctional. However, there are several mechanisms whereby chromosomal proteins may play a role in keeping a gene active or inactive. Histones may dissociate from specific regions to make the gene accessible to RNA polymerase, or may bind to it more firmly to make it inaccessible. Trans-acting inhibitory factors may bind to the region and lock it by associating this region with another. Such association/dissociation of proteins may be achieved by modifications such as phosphorylation, acetylation, ADP-ribosylation and methylation of histones/NHC proteins/trans-acting factors. These may increase the accessibility of specific sites on the DNA, making them hypersensitive or insensitive. Such modifications may permit a RNA polymerase to pass through the DNA faster or slower, and thus regulate the rate of expression of a gene. The problem is greatly compounded because a cell has only a few active genes, most of which are not located at a specific region of the chromatin, but are dispersed among a great number of inactive genes. Not only is this specific site of active chromatin passed on from one cell generation to another in a tissue, but it is also inherited. How does the RNA polymerase detect these active sites and transcribe the genes? Surely, certain features must be imprinted in these sites of the chromatin to serve as markers for RNA polymerase. Such identification marks, if any, may not be so difficult to locate for histone and ribosomal genes that are greatly amplified and tandemly located in one chromosome, but for single-copy unique genes it would be difficult.

The endonuclease, DNase I, has been useful in identifying active and inactive genes (Weintraub & Groudine, 1976; Wu, 1980). These workers showed that DNase I preferentially cleaves the chromatin at specific hypersensitive sites (DH-site). These sites are also sensitive to MNase. By indirect end-labeling it is possible to locate these sites in the chromatin after mild digestion by MNase.

Protein-coding genes

Protein-coding genes occur generally as a single copy per haploid genome and are transcribed by RNA polymerase II. Though these genes are present in all cell types, they show tissue- and stage-specific expression. For example, the insulin gene is expressed only in the β cells of the islets of Langerhans, and the glucagon gene is expressed in α cells. For example, the globin gene is expressed in reticulocytes, immunoglobulin genes in B lymphocytes, and casein in the mammary gland. In other cell types, these genes are not expressed. On the other hand, genes for glycolytic enzymes, tRNAs, and rRNAs are expressed in all cell types. Genes for Krebs cycle enzymes are expressed in all cells having mitochondria. Genes for histone and NHC proteins are expressed in dividing cells. Certain genes are expressed in a stage-specific manner. For example, α-fetoprotein gene is expressed in the hepatocytes of fetal mammalian liver. After birth it is switched off, and the albumin gene becomes expressed and continues to be expressed throughout the remaining part of the life span. Surely, the chromatin structure, in particular that of nucleosome, must be different not only in the active and permanently inactive genes, but also in the genes that are active and then become inactive or vice versa, as in the case of α-fetoprotein and albumin genes.

One general observation is that genes that are active have sites that are more sensitive to digestion by DNase I(DH-site) and MNase. Furthermore, the genes that are active and become inactive later, still retain the same DH-sites, as if retaining some memory. The genes that are inactive but become active in response to an inducer like a steroid hormone also are more nuclease sensitive than the permanently inactive genes. Some of these characteristics are given in Table 3.4.

Table 3.4. *Nuclease sensitivity of genes in chromatin*

Gene	Organism	Observation
Albumin	Rat	DH site in liver, not kidney
α-Fetoprotein	Rat	DH site in liver, not kidney
Amylase	Rat	MNase sensitive
Dihydrofolate reductase	Mouse	Nucleosomal structure different
Fibronectin	Rat, chick, humans	3 DH sites in 5' region, liver
Globin	Chick, humans	DH sites
Gly–3P–DH	Mouse	DH sites upstream
Histone H5	Chicken	S1 nuclease sites
Ig (μ chain)	Human	New DH site
Ig genes (L_κ)	Mouse	New DH site
Interferon	Mouse	New DH site
Lysozyme	Chicken	DH site
Preproinsulin	Rat	DH site
Vitellogenin	Chicken	DH site

Source: van Holde, 1989.

One of the genes whose chromatin structure has been studied in detail is globin. In the human, α-globin gene and its variants are located in chromosome 16. The β-globin gene and its variants ε, γ, and δ are located in chromosome 11. They show well-defined switching beginning in fetal life. Two α variant genes $ρ_1$ and $ρ_2$ are expressed in the fetus, and $α_1$ and $α_2$ are expressed after birth. The β chain variant, ε, is expressed in the early fetus, followed by γ; β is expressed after birth along with a minor variant δ. In the chick, no globin gene is expressed in the day-1 embryo. On day 2, an α variant, U, is expressed, and two α variants are expressed after hatching. The β variants, ρ and ε, are expressed in the embryo; βH is expressed at hatching, and β in the adult.

Weintraub and Groudine (1976) first showed in the chick that globin genes have nucleosomes, whose sensitivity to DNase I changes during development. At day 1, all globin genes are DNase I insensitive and the DNA is highly methylated. From days 2 to 6, α and β genes become DNase I sensitive. Certain DH-sites also appear (Weintraub et al., 1981), and certain sites show S1 nuclease sensitivity.

Figure 3.8 shows the sequence of DNase I sensitivity of the α gene variants of the globin gene in the chick beginning from the embryonic stage. Seven DH-sites are present in the 5-day embryo. A DH-site is present in the 5' flanking region of the U gene which is expressed.

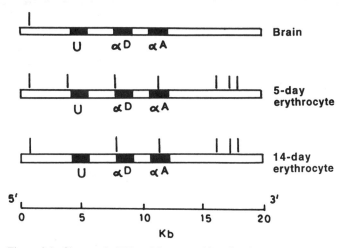

Figure 3.8. Changes in DNase I hypersensitive sites in erythrocyte chromatin during development in the chicken. The α-globin domain is shown including the U gene and the adult α^D and α^A genes. Only one DH-site is detected in the brain. Six new sites appear in erythrocytes from the 5-day embryo. In 14-day embryo, when only the adult genes are expressed, site 5' to the U gene has disappeared. (Adapted from Weintraub et al., 1981)

However, though DH-sites also appear in the α^D and α^A genes, they are not expressed. Only after hatching, when U gene expression is switched off, are α^D and α^A genes expressed. From day 14 on, DNase I sensitivity continues in all the genes that were expressed in the embryo earlier but not expressed after day 14. The most sensitive DH-sites are seen in the adult globin genes. The embryonic genes while being expressed were undermethylated, but became methylated from day 14. Also, the DH-sites in the 5' flanking regions of the embryonic genes disappear and new DH-sites appear at the 5' flanking regions of adult genes (Weintraub et al., 1981; Wood & Felsenfeld, 1982).

These studies suggest that significant changes must occur in the chromatin structure of these genes, particularly at their 5' flanking regions for expression. Emerson and Felsenfeld (1984) and Jackson and Felsenfeld (1985) found that a 60-kD protein is required to confer DNase I sensitivity at the 5' flanking region of the β^A gene. Certain proteins have been shown to bind to specific regions of the 5' flanking sequences of the gene (Nickol & Felsenfeld, 1983; Plumb et al., 1986). The region -175 to -200 has 16 guanine residues and is S1 nuclease sensitive.

Three stages of the chromatin conformation of a gene may thus be deciphered (van Holde, 1989): (1) the *incompetent* stage, which does not have DNase I sensitivity at any stage; (2) the *competent* stage, which is DNase I sensitive, but is not expressed; and (3) the *active* stage, which is DNase I sensitive and is expressed. These stages suggest the requirement for certain proteins that confer on the gene a specific state. A competent gene which is not expressed indicates that some additional factor is needed to confer on it the ability to be expressed. It also means that the competent state is propagated through one or more cell generations (Groudine & Weintraub, 1982).

The types of changes that occur in the chromatin of the globin gene when it passes from the competent to active state has been studied using murine erythroleukemia cells (Friend cells). These cells transcribe the globin gene at a very low level, but when induced by dimethyl sulfoxide (DMSO), transcription is enhanced to over tenfold (Sheffery, Rifkind & Marks, 1982; Salditt-Georgieff et al., 1984). Induction increases the overall DNase I sensitivity of the chromatin, new DH-sites appear (Smith & Yu, 1984), and MNase sensitivity also increases (Yu & Smith, 1985). These changes are blocked by dexamethasone, which inhibits differentiation. No changes in DNA methylation occur during the transition from the competent to the active state (Sheffery et al., 1982). In media of low ionic strength, all regions become DNase I sensitive. Also active β-globin genes sediment slower than nonexpressed genes in sucrose gradient (Kimura et al., 1983). These observations imply that the higher-order structure of the chromatin containing the gene must relax to become active.

Heat-shock genes of *Drosophila* respond to an external stimulus, heat. *Drosophila* kept at 25°C show several puffs in their chromosomes, indicating expression of genes at these loci. If the temperature is raised to 35°C, nine new puffs form at 33B, 63BC, 64F, 67B, 70A, 87A, 87C, 93D, and 95D. Most of the transcripts of these loci are translated into heat-shock proteins (*hsp*), which are mostly nuclear and are not associated with nucleosomes (Levinger & Varshavsky, 1982). These proteins are named on the basis of their M.W., for example, the 87A locus codes for a 70-kD protein, which is called *hsp* 70. The chromatin at the 87A gene becomes more DNase I sensitive when *Drosophila* are subjected to heat shock. The noninduced chromatin has a distinct nucleosomal pattern as elucidated by MNase digestion. However, this organization is totally lost after induction. If the insect is brought back to 25°C, the original nucleosomal pattern is regained. The gene also is MNase resistant in the uninduced state, and a distinct nucleosomal pattern exists

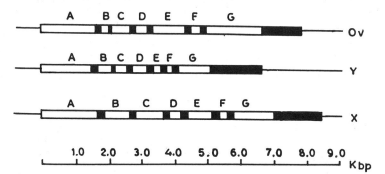

Figure 3.9. Organization of X, Y, and ovalbumin (Ov) genes in ovalbumin cluster in chicken. Dark bands represent exons; introns are designated by letters. The order of the three genes is X-Y-Ov in the domain. They are transcribed from the left to the right (Colbert et al, 1980; © 1980 American Chemical Society)

in the coding region of the *hsp* 70 gene. This pattern is lost on heat activation (Levy & Noll, 1981). These studies show that the nucleosomal arrangement becomes disrupted before the gene is expressed.

Craine and Kornberg (1981) found that a protein factor binds to the 5' flanking region to activate the gene. Parker and Topol (1984) and Wu (1984) have identified two protein factors, one binding to the TATA region and another upstream from TATA. Their binding may disrupt the chromatin that is required for transcription. It is suggested that synthesis of new protein may not be necessary for activation of the gene because the activation is so rapid.

Karpov, Preobrazhenskaya and Mirzabekov (1984) carried out histone-DNA cross-linking after heat shock and found that H1 histone is dissociated first, which is followed by dissociation of the core histones. Such studies have also been carried out on yeast and on fibroblast cells in in vitro culture. Most significantly, identical *hsp* are produced in both the systems, and the proteins are highly homologous. Thus these genes have been conserved during evolution, and perform a basic function when organisms are exposed to heat. Their actual function and mechanism of their action remain to be elucidated.

Studies on the genes for ovalbumin and two other related proteins, X and Y, have thrown much light on their chromatin structure. They are induced by 17β-estradiol and are expressed in the tubular cells of the oviduct of birds. These genes are tandemly arranged and occupy ~100 kbp domain (Fig. 3.9). Their transcription has a ratio of 100:10:1.

The function of the products of the X and Y genes is not known. Their transcription is stimulated on estradiol administration, and ceases when it is withdrawn. The ovalbumin gene contains seven introns. The active gene is more sensitive to DNase I and MNase than the inactive gene. MNase sensitivity decreases upon withdrawal of the hormone, but sensitivity to DNase I remains, indicating that the gene continues to be competent even though not expressed. Thus DNase I and MNase sensitivities indicate two different conditions of the chromatin. The former may exist before, during, and after the gene is transcribed. The latter may occur before transcription and may cease after transcription.

Chambon et al. (1984) have suggested that a repressor protein associated with the upstream region of the ovalbumin gene is responsible for repression of the gene in competent oviduct cells in the absence of estradiol. On administration of estradiol the hormone-receptor complex dissociates this repressor and relieves the repression. Robinson, Nelkin and Vogelstein (1982) reported earlier that the competent DNase I-sensitive domain of the gene is associated with the nuclear matrix. Such an association is not seen in liver cells. Stumph et al. (1982) found that a repetitive DNA family is located at the end of the DH-site domain, which indicates that this sequence may be involved in the attachment to nuclear matrix. Hence during cell differentiation, a group of genes may be segregated for their later selection for expression. Transcription may occur later when appropriate inducers such as a hormone and other trans-acting factors make further selection of genes, depending on their specificities, and stimulate their expression. The presence of DH-sites in the upstream domain has been reported for several competent or active genes.

The discovery of a phosphatase inhibitor, *p*-chloromercuribenzenesulfonic acid (CMBS), which enhances DNase I sensitivity of active chromatin by tenfold and that of inactive chromatin by threefold, may be of value in studying the role of DNase I in chromatin conformation (Feng, Irving & Villeponteau 1991). It may produce this effect by enhancing phosphorylation of H2A.

Several biochemical studies have shown that the nucleosomes along the active genes are low in H1 histone content, are enriched in acetylated and ubiquitinated histones, and are complexed with HMG proteins, especially HMG14 and HMG17 (Singer & Berg, 1991). These changes may cause irregular spacing of nucleosomes in the chromatin regions containing active genes and render the DNA more DNase I sensitive. How does a competent gene remain in such a state as to be brought into activity when needed? It is likely that its chromatin, possibly the region in the gene promoter has certain DNA sequences and proteins that recognize the appropriate inducer. This is followed by uncoiling of the region and dissociation of H1 so that the gene is brought to an active state. Such proteins may be tissue- and stage-specific. It may be useful to look for such proteins that confer competence for activity to genes.

Transfer RNA and ribosomal RNA genes

Unlike the unique genes, tRNA and rRNA genes are present in multiple copies and are aggregated together. Hence the nucleosomal organization in these genes may be different from those of unique genes. Transfer RNA genes are small (about 75 bp long), occur in multiple copies, and are transcribed by RNA polymerase III. In the frog, *Xenopus laevis,* blocks of eight tRNA genes are separated by interspersed sequences, occupying over 3 kbp in total length. This is repeated nearly 100 times per haploid genome. 5S rRNA genes, which are also transcribed by RNA polymerase III, occur in clusters. The sequences of these genes that are needed for initiating transcription lie within the gene (Hofstetter, Kressman & Birnsteil, 1981). This was studied by subjecting chromatin to mild digestion by MNase, isolating the tetranucleosomes, with subsequent resolution by gel electrophoresis, Southern transfer, and hybridization to tRNA genes. It was found that the sequence for initiating transcription was within the nucleosome (Wittig & Wittig, 1982), and that the TTCGA sequence was important for transcription. It appears that the polymerase binds at this sequence and opens the nucleosome for transcription.

Bryan, Hofstetter, and Birnsteil (1983) compared the chromatin of tRNA genes of mature chick erythrocytes in which they are not transcribed with that of liver in which they are actively transcribed. After mild MNase digestion, hybridization was carried out using a tRNA gene probe. In both cases, the regular oligonucleosome ladder was observed. However, the repeat length in erythrocytes was 198 ± 5 bp and in the liver it was 185 bp. The longer repeat in erythrocytes and the phasing of nucleosomes may promote a higher-order structure containing those genes that are not transcribable. In the liver, phasing is different or absent, and transcription is promoted.

5S rRNA genes are separate from the other three rRNA genes, 18, 5.8, and 28S. The 5S rRNA gene is transcribed by RNA polymerase

III, whereas the other rRNA genes are transcribed by RNA polymerase I. In *Xenopus,* the 5S rRNA gene is highly reiterated with as many as 100,000 copies. Each copy of the gene is 120 bp long and is flanked by a nontranscribed region. It also has an internal control sequence near the middle of the gene. The transcription factor, TFIIIA, binds to this region for transcription (Smith, Jackson & Brown, 1984; Bieker, Martin & Roeder, 1985). It has two classes of 5S genes – those that are transcribed in the oocyte and those transcribed in somatic cells. In mature erythrocytes, somatic rRNA genes are inactive. In erythrocytes, there appears to be regular phasing of nucleosomes containing the 5S genes (Young & Carroll, 1983). In the cells in which it is transcribed, the initiation and termination sites and control region appear to lie in the linker DNA (Reynolds et al., 1982). Similar observations for 5S rRNA genes have been made in other organisms.

The chromatin structures of tRNA and 5S rRNA genes may be unique and may not be similar to those of other genes. 5S rRNA and tRNA genes are shorter than the DNA sequence contained within a nucleosome core particle, and RNA polymerase III may have different requirements. The rRNA genes (5.8S, 18S, and 28S) and the protein-coding genes are longer and are transcribed by different polymerases. The three rRNA genes are transcribed by RNA polymerase I as a ~45S RNA precursor in the nucleolus which is processed into the three rRNAs. In higher eukaryotes, rRNA genes are clustered and arranged in long tandem arrays and are flanked by nontranscribed spacer regions. Ribosomal DNA transcriptional units during transcription show a nonbeaded structure.

The chromatin containing inactive rDNA is compact with nucleosomes. Actively transcribing regions do not appear to have nucleosomes, or if they do, they exist in a modified unfolded form. It has been proposed that two NHC proteins of M.W. 30 and 32 kd bind near the nucleosome dyad axis, opening the structure that is called a "lexosome" (Prior et al., 1983). This permits reading of the rRNA genes.

Age-related changes in chromatin structure and function

Transcription of RNA is the first and primary event in the sequence of steps required for the transfer of information from DNA to protein. Since the DNA is complexed with histones and NHC proteins to form chromatin, the rate of transcription depends on the structural organization of the chromatin. This includes (1) conformation of DNA; (2) accessibility of the transcription initiation site of the gene for RNA polymerase and other transcription factors; (3) accessibility of cis-acting elements of the gene for the trans-acting regulatory factors; (4) the extent of association of histones and NHC proteins to DNA; and (5) modifications of histones and NHC proteins that alter their association with DNA.

Changes in chromatin, and hence in transcription, occur during differentiation. When actively dividing myoblasts transcribing several types of mRNA, fuse to form nondividing differentiated myotubes, the types and amounts of mRNA synthesized are greatly reduced (Yaffe & Fuchs, 1967). The few types of mRNA that are synthesized are transcribed in vastly larger amounts. Thus differentiation results in the synthesis of more and more of lesser and lesser types of mRNA. So both qualitative and quantitative changes in the chromatin occur during and after differentiation. Also, the transcription of rRNA whose genes are reiterated, decreases. So there exist control mechanisms that operate at both unique and reiterated genes. Consistent with this observation is the finding that the fusion of myoblasts is accompanied by a great increase in creatine phosphokinase (CPK) activity that shows higher transcription of its mRNA (Shainberg, Yagil & Yaffe, 1971; Morris, Piper, & Cole, 1976). In the chick, the activity of CPK increases 20-fold during differentiation. Cardiac muscle cells also stop dividing soon after birth, and this is accompanied by a decrease in transcription. Its DNA becomes less susceptible to DNase I and its melting point increases (Limas & Limas, 1978).

On the other hand, during embryogenesis when cells are dividing rapidly, transcriptional activity is high, and the types of mRNA synthesized are also enormous. Surely, the chromatin structure in the actively dividing and in the differentiated states must be very different to enable the cell to transcribe the required types and requisite amounts of mRNA.

Various studies have been carried out to determine what changes take place in chromatin during development, differentiation, and aging. The melting temperature (T_m) of chromatin of the liver and thymus of rats has been measured by various workers, and all have noted an increase in the T_m in old age (von Hahn & Fritz, 1966; O'Meara & Herrmann, 1972; Berdyshev, 1976). The melting profile of the chromatin shows an increase in hyperchromicity and T_m in old age. This may be due to increasing covalent links between DNA and chromosomal proteins with increasing age (von Hahn, 1963, 1970). This is consistent with the finding that the amount of proteins that can be extracted from chromatin by salt decreases with age (O'Meara & Herrmann, 1972).

Another finding is that there is an increase in single-strand breaks or nicks in the DNA in old age as the incorporation of ^{3}H-thymidine into DNA increases with age (Samis, Falzone & Wulff, 1964; Price, Modak & Makinodan, 1971; Chetsanga et al., 1975). Chetsanga et al. (1977) have further shown that DNA of the liver of 20-month-old mice is more sensitive to nuclease S1 than that of 1- to 15-month-old mice. Furthermore, alkaline sucrose gradient sedimentation of DNA in the brain of old mice shows four bands while that of young mice shows one band. This indicates that in old age the DNA gets nicked, and such DNA may form covalent links with chromosomal proteins.

The histone content per cell in each tissue remains fairly constant throughout the life span (Carter & Chae, 1975). However, Medvedev, Medvedeva, and Huschtscha (1977) and Medvedev, Medvedeva, and Robson (1978) have shown that though qualitative and quantitative changes occur in the nucleosomal histones, subfractions of H1 histone change in the liver and spleen of rats and mice. Especially, the methionine-containing H1 increases in old age. This change certainly would affect the chromatin structure and function but how it does so remains to be elucidated. Furthermore, the NHC protein content of the chromatin of the rat is reported to decrease with age (Dingman & Sporn, 1964; Kurtz, Russell & Sinex, 1974).

Kanungo and his co-workers carried out in vitro studies on age-related covalent modifications of histones, and alterations in these modifications by various endogenous effectors using slices of rat cerebral cortex. After incubation of the slices for 1 hour with ^{32}Pi or ^{14}C-acetate or ^{14}C-methionine, the chromatin was isolated, and the histones and NHC proteins were purified and resolved by polyacrylamide gel electrophoresis. The histone bands were removed and used for counting the amount of label (Kanungo, 1980). It was found that phosphorylation of histones of the cerebral cortex decreases with age of the rat (Kanungo & Thakur, 1977; Kanungo, 1980). Particularly, phosphorylation of H1 and H4 histones, which is high in young rats, decreases sharply in older animals. Furthermore, calcium inhibits phosphorylation of H1 and H4, and this effect decreases with age. 17β-Estradiol stimulates phosphorylation of histones, particularly in the adult, but this effect is not seen in old animals (Kanungo & Thakur, 1979a). The decrease in phosphorylation of histones may increase their binding to DNA due to a decrease in the number of negative charges. Similar studies using cerebral slices and ^{14}C-acetate have shown that acetylation of histones, especially of H3 is lower in old age (Thakur et al., 1978). Epinephrine stimulates H1 acetylation, and 17β-estradiol that of H3. These modulating effects decrease in old age. Calcium has no significant effect on acetylation of histones.

The effects of butyrate and 17β-estradiol on acetylation and their role on chromatin transcription were then studied (Kanungo & Thakur, 1979b). Butyrate or 17β-estradiol added prior to the addition ^{14}C-acetate, stimulated acetylation of histones in the cerebral slices of immature rats. This effect was greatly reduced in adult rats and was not observed in senescent rats (110 weeks). When nuclei were isolated from these slices and used for transcription by incubating them with ^{3}H-UTP and other nucleotide triphosphates, the highest stimulation of transcription was seen in the immature animals, and this effect was greatly reduced in the adult. The two modulators had no effect on transcription in senescent rats. These studies show not only a direct correlation between acetylation of histones and transcription of chromatin, but also a decrease in the modulating effects of butyrate and estradiol in old age. This may be due to increasing condensation or compaction of the chromatin in the cells of the cerebral cortex, which has largely postmitotic neurons that do not divide.

Studies on the methylation of histones in rat brain show that there is a decrease in the methylation of H3 and H4 histones with increasing age (Kanungo & Thakur, 1979a; Thakur & Kanungo, 1981). Methylation is a stable covalent modification. Hence even if a decrease in $-CH_3$ incorporation occurs with age, in effect, the number of $-CH_3$ groups in histones increases since those incorporated in earlier ages are not removed. Methylation of histones, particularly the trimethyllysyl residues of H3 and H4 would strengthen the binding of histones to DNA as they carry positive charge.

The amount of NHC proteins in the chromatin is reported to decrease with age (Kurtz et al., 1974). NHC protein content in metabolically active tissues is higher than in inactive tissues. Polyacrylamide gel electrophoresis of the NHC proteins of rat brain shows not only a qualitative change in these proteins, but also a quantitative change with age (Kanungo & Thakur, 1979c). Furthermore, covalent modifications of NHC proteins and their modulations by various endogenous effectors change with age (Kanungo & Thakur, 1977, 1979a, 1979c; Thakur et al., 1978). Phosphorylation, acetylation, and methylation of NHC proteins of the brain decrease significantly in old age. Calcium stimulates their phosphorylation both in the adult and the old animal. However, the stimulatory effect of 17β-estradiol that occurs in the adult does not occur in the old animal. Polyamines, spermine, spermidine, and epinephrine stimulate phosphorylation of specific NHC proteins in the immature animals. These effects decrease with age (Das & Kanungo, 1980; Bose & Kanungo, 1982). Calcium does not stimulate acetylation of NHC proteins, but both epinephrine and 17β-estradiol stimulate it only in young, but not in adult and old rats. Methylation of NHC proteins is not affected by either calcium or 17β-estradiol. It is likely that phosphorylation of NHC proteins is more important for gene expression than their acetylation and methylation. These studies further suggest that the decrease in covalent modifications of NHC proteins in old age may be due to increasing compaction of the chromatin and the inaccessibility of the sites for modifications of these proteins.

Changes in ADP-ribosylation of chromosomal proteins have been studied in the brains of 3-, 14-, and 30-day-old developing rats (Das & Kanungo, 1986). Although these studies were not done on old rats, they do, however, give insight into the types of changes that occur in this modification in these cells that stop dividing around day 14 after birth. These cells do not synthesize DNA and their chromosomal proteins, especially core histones, rarely turn over for the remaining period of the life span. Nuclei purified from the cerebral hemispheres were incubated with ^{32}P-NAD$^+$ for 35 minutes and histones and NHC proteins then extracted. Histones were resolved by high-resolution SDS polyacrylamide gel electrophoresis and the radioactivity was counted in each band. It was found that ADP-ribosylation of histones is higher than that of NHC proteins. ADP-ribosylation of histones decreased with development. Spermine stimulated, whereas 3-aminobenzamide and benzamide inhibited it. When chromatin was mildly nicked by MNase/EcoRI/Alu I prior to incubation with ^{32}P-NAD$^+$, ADP-ribosylation was stimulated. The stimulation decreased after day 14. These studies show that the chromatin of neurons becomes condensed after the cells stop dividing. This is corroborated by the finding that digestion of chromatin of the brain by DNase I and MNase is the highest in 3-day-old rats and progressively decreases thereafter (Chaturvedi & Kanungo, 1985a, 1985b; Das & Kanungo, 1986). Furthermore, ADP-ribosylation decreases transcription of the chromatin (Das & Kanungo, 1987).

The above studies show clearly that the four types of covalent modifications that are seen in histones and NHC proteins have profound effects on the transcriptional activity of chromatin. Whereas acetylation of histones clearly has a stimulatory effect, methylation and ADP-ribosylation inhibit this activity, and phosphorylation may stimulate or inhibit transcription. These effects are brought about by conformational changes in the chromatin.

That condensation of chromatin does occur with increasing age is evident from the studies on its digestion by DNase I. DNase I digests chromatin DNA at 10-bp intervals and its multiples. When nuclei of young and old brain tissue are digested by DNase I for various time periods, and the DNA fragments are resolved by gel electrophoresis, beautiful ladders of DNA bands are produced (Fig. 3.10; Chaturvedi & Kanungo, 1983, 1985a). It is seen that for any time period of digestion, 2, 5, or 15 minutes, the intensities of the 10-, 20-, and 30-bp bands are greater for the young rat. The longer fragments are more in the old (110 weeks). This clearly shows that DNase I is unable to detect the sites for digestion in the nucleosomes of the old as well as it does in the young animal. Digestion by MNase which cuts the chromatin DNA at linker regions does not show any difference. The kinetics of digestion of DNA by DNase I were studied to measure the amount of acid-soluble DNA produced. At any time point, from 1 to 60 minutes, the amount of acid-soluble DNA is lower in the old animal. The M.W. of DNase I is ~31,000, and that of MNase 16,000. Since MNase is a smaller molecule, it may find its way to the linker region more easily. Besides, its sites of digestion are farther apart (200 bp) in the linker DNA. DNase I is a larger molecule, and its closely spaced digestion sites (10-bp intervals) in the nucleosome may not be as easily accessible as the 200-bp intervals for MNase.

Similar studies were undertaken in the brain of 3-, 14-, and 30-day-old developing rats (Das & Kanungo, 1985). The kinetics of digestion by DNase I show that the rate of digestion is the highest in 3-day-old rats, and it gradually decreases as development proceeds. No change is seen for MNase. This indicates that the condensation of the chromatin of the brain of the rat begins early in life when cell division slows down and cells become postmitotic.

Supakar & Kanungo (1983) found not only that the digestibility of the chromatin by DNase I decreases from the 3- to 30-day-old brain, but also that $ZnCl_2$ stimulates this digestion by three- to fourfold. Zn is a cofactor not only for RNA polymerase II, but is essential for certain proteins like the zinc fingers that regulate transcription. It is likely that $ZnCl_2$ causes chromatin to unwind at specific regions, which are necessary for transcription and are more easily digested by DNase I. Similar

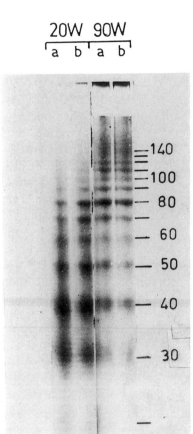

Figure 3.10. Effect of age on digestion of chromosomal DNA by DNase I. Nuclei (a) and chromatin (b) of the brain of 20- and 90-week-old rats were digested by DNase I for 15 minutes, and the DNA fragments were resolved by denaturing polyacrylamide gel electrophoresis. (Chaturvedi & Kanungo, 1985a)

studies on skeletal muscle of 3-, 14-, and 30-day-old developing rats also show that the rate and extent of digestion of its chromatin decrease as development proceeds (Pandey & Kanungo, 1984).

One of the ways to test if the chromatin of the old animal is more condensed than that of the young animal is by nick-translation. This involves nicking of the DNA in nuclei by endonucleases such as DNase I, MNase, and restriction enzymes and then carrying out DNA synthesis by incubating nuclei with the four dNTPs, one of them radiolabeled. The DNA is then purified and its radioactivity determined. If the chromatin is more compact, DNA polymerase will move more slowly from its starting point from the nick. Therefore, the incorporation of nucleotides should be less. This indeed was found by Chaturvedi and Kanungo (1985b) when they studied the chromatin of the brain of 22- to 23- and 118- to 119-week-old rats. Nuclei were digested by DNase I, Eco RI, Msp I, and Hpa II and then nick-translated using ^{3}H-dTTP as the label. The incorporation of the label was nearly 40% lower in the old brain. A progressive decrease in incorporation of the label into the DNA is also seen in the developing rat brain after it is digested by EcoRI (Das & Kanungo, 1986). Nucleotide incorporation in the 30-day-old rat is about 20% of that of the 3-day-old rat.

Summary

All information in a eukaryote is stored in DNA, which is complexed with histones and NHC proteins to constitute the genetic apparatus called chromatin. Histones are basic proteins of low M.W., rich in lysine and/or arginine residues that are clustered at N- and C-terminal regions. Their structures have been conserved during evolution. All tissues have the same types and amounts of histones, and they do not show any change throughout the life span. There are only five major types of histones – H1, H2A, H2B, H3, and H4. Their genes are linked and are reiterated several times in the chromosome. They are synthesized along with DNA in the nucleus during the S phase and are translocated to the nucleus. Histones are required for the structural organization of chromatin. The basic structure of chromatin consists of a string of nucleosomes of ~10 nm in diameter. The nucleosomes have an internal core that consists of an octamer of two molecules each of H2A, H2B, H3, and H4 around which 146 bp of DNA is wrapped in 1.75 turns. Two successive nucleosomes are linked by DNA, called the linker DNA, to which H1 is attached. Whereas core histones are involved in the first order of compaction of DNA, H1 is involved in the next higher order of compaction. It is also involved in the repression of gene activity.

NHC proteins are highly heterogeneous, several hundred in number, rich in acidic amino acids, and tissue and species specific. Metabolically active tissues have more types and greater amounts of NHC proteins. Synthesized throughout the cell cycle, the NHC proteins are involved in the higher-order structure of chromatin and both the positive and negative regulation of gene expression.

Both histones and NHC proteins undergo several types of postsynthetic covalent modifications – acetylation, phosphorylation, ADP-ribosylation, and methylation. Whereas the first three types of modification decrease the net positive charge on histones, methylation increases it. Acetylation of core histones is involved in transcriptional activation. Phosphorylation is involved in chromosome decondensation and condensation during the cell cycle. ADP-ribosylation is implicated in DNA repair.

Thus changes not only in the histones and NHC proteins, but also in the DNA alter the structure of the chromatin, and thereby influence the accessibility of genes to RNA polymerases and transcription factors necessary for transcription. Indeed studies have shown that such covalent modifications of histones as acetylation and phosphorylation influence transcription. It has been shown that the chromatin in postmitotic cells becomes increasingly compact with age and its digestibility by DNase I is lower in old age. Thus much useful information has been derived on the function of chromatin, and hence on the overall activity of genes, by studying the structure, conformation, and chemical changes of chromatin.

4
Eukaryotic genes

With the advent of genetic engineering technology during the last decade, much valuable insight has been gained into the structure, function, and regulation of eukaryotic genes, especially those that code for proteins. Incisive experimental studies have established that (1) most eukaryotic genes are interrupted or split, and (2) they have sequences at their 5' regions that are responsible for the regulation of their transcription. A generalized representation of a eukaryotic gene is given in Figure 4.1.

Exons and introns

In prokaryotes, the entire nucleotide sequence of a gene codes for a messenger RNA (mRNA), transfer RNA (tRNA), or ribosomal RNA (rRNA), as the case may be. In eukaryotes, however, the corresponding genes are considerably longer than their final products. The protein-coding genes are transcribed into pre-mRNAs (hnRNAs) of equal length, but are then processed into shorter and mature mRNAs in the nucleus. The mRNAs are translocated into the cytoplasm where they are translated into proteins. Likewise, pre-tRNAs and pre-rRNAs are transcribed from their corresponding genes and are processed into shorter and mature tRNAs and rRNAs in the nucleus. They are then translocated into the cytoplasm to carry out their functions in the translation of mRNAs.

The shortening of the three types of transcripts is necessary because their corresponding genes contain sequences that are not required in their final products. The RNA polymerase transcribes the entire gene into a pre-RNA, which is as long as the gene. The sequences that are not required in the final product alternate with those that are required. Using a special mechanism, the cell removes the sequences that are not required in the pre-RNA, joins the required sequences in a given order,

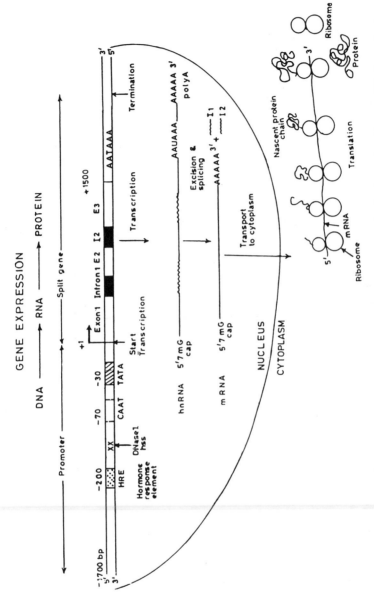

Figure 4.1. Generalized structure of a eukaryotic gene. Gene is split into exons (unfilled space) and introns (filled space). Promoter region located 5′ to the gene has cis-acting elements, CCAAT, hormone-responsive elements, and other entities. RNA polymerase and transcription factors that bind to TATA box initiate transcription. Entire gene including both exons and introns is transcribed into a hnRNA to which is added 7 mG cap at the 5′ end during transcription. A poly-A⁺ tail is added after transcription. The regions in the hnRNA that represent the introns are excised, and the exons are then spliced (ligated) to form the mature mRNA. The mRNA is translocated to the cytoplasm, binds to ribosomes, and is translated into a protein.

Table 4.1. *Split genes and the number of their introns*

Gene	No. of introns	Gene	No. of introns
tRNA	1	Dihydrofolate reductase	5
Actin	1	Cytochrome b	6
Fibroin	1	Ovalbumin	7
β-Globin	2	Ovomucoid	7
Insulin	2	HPRT	8
Growth hormone	2	Cytochrome P-450	8
Ig light chain	2	Albumin	13
myc Oncogene	2	α-Fetoprotein	14
Lysozyme	3	Conalbumin	16
rRNA (large)	3	Fibronectin	30
Ig heavy chain	4	Vitellogenin	33
		Collagen	~50

and produces a mature RNA that is shorter than the pre-RNA. The nucleotide sequence of a protein-coding gene starts with a sequence at the 5' end, which is represented in its mature mRNA. This is followed by a sequence that is not represented in the mRNA. The sequence of the gene that is present in the mRNA is called the exon, and the sequence that is not present in the mRNA is called the intervening sequence or intron (Gilbert, 1978). Consequently, a protein that is translated from the mRNA contains amino acids corresponding to the exons only. Thus the eukaryotic gene is interrupted or split. Exons and introns alternate (Fig. 4.1). The number of exons and introns in a gene depends generally on the length of the protein it codes for (Table 4.1).

The elegant experiments of Chambon and co-workers on the ovalbumin gene (Breathnach, Mandel & Chambon, 1977), and Jeffreys and Flavell (1977) on the β-globin gene showed for the first time that eukaryotic genes are split. This was shown by an R-loop mapping technique developed by Thomas et al. (1976) in which mature mRNA purified from the cytoplasm is hybridized to the DNA containing the gene that codes for it. The DNA is denatured before hybridization. The RNA hybridizes to the strand from which it is transcribed more strongly than the complementary DNA strand. It is seen that the regions of the strand of the DNA that find complementary regions in the mRNA form H-bonds with those regions, and the regions which do not find complementary regions loop out. Breathnach et al. (1977) found that when ovalbumin mRNA is hybridized to its gene, seven R-loops are produced.

Jeffreys and Flavell (1977) showed that β-globin mRNA produces one R-loop. Therefore, the coding sequences of the two genes are represented completely in their mRNAs, which are translated into their respective proteins. This established that the introns that are the noncoding or intervening regions that interrupt the coding regions are not present in the mRNAs, and hence are not represented in the proteins. The exons are spliced or joined together and are represented in the contiguous amino acid sequences of the protein. It was later shown that the β-globin gene has an additional small intron near its 5′ end. Some examples of split genes and the number of introns they contain are given in Table 4.1.

R-loop mapping has shown that the total length of all the introns of a gene far exceeds the total length of all its exons. For example, the chicken ovalbumin gene is 7,700 bp long, out of which 1,158 nucleotides code for the 386 amino acids of the protein. The remaining part is within the seven introns. Also, the introns of the ovalbumin gene vary in length, from 250 bp to 1,600 bp. Exons also vary in length; the shortest is 47 bp and the longest is 1,043 bp. The sizes of the three exons of the β-globin gene of rabbit, mouse, and man are the same, reflecting evolutionary conservation of the gene. However, the two introns differ in length and nucleotide sequence among species, though they occur at the precise locations in the coding regions in the three species. The short intron is about 100 bp long, and the long one is about 600 bp.

The presence of introns and their varying lengths have given rise to speculations about their role. Gilbert (1978) suggested that the introns may play an important role in evolution. Prokaryotes, which multiply rapidly to permit recombination and evolution, do not have introns. Eukaryotes have a much slower rate of reproduction and, therefore, have to resort to additional means for evolution. According to Gilbert (1978, 1985) the presence of introns permits shuffling of coding regions of the pre-mRNA of the gene during splicing, giving rise to different mRNAs and hence different proteins, as for example, growth hormone and ACTH. It is conceivable that the greater the number of introns in a gene, the greater would be the possibility of exon shuffling in the pre-mRNA (intra-pre-mRNA shuffling). Possibly also, coding regions of mRNAs of different genes may shuffle among themselves (inter-pre-mRNA shuffling), giving rise to varieties of mRNAs and, hence varieties of proteins. This may have generated proteins having domains of amino acid sequences that are homologous. Various dehydrogenases and receptor families that have affinities for different substrates may have arisen by this process. Yet another possibility is that recombination between genes may have occurred in the regions of introns, giving rise to new and novel genes and greatly enhancing the types of genes. The immunoglobulin genes are believed to have arisen by this method. Whereas the first two methods of increasing the repertoire of proteins may occur in somatic cells during pre-mRNA processing, recombination occurs during meiosis when new types of genes are generated.

Transcription

Transcription is a process by which information stored in genes is transferred to the functional level. The transfer may be made to a message or messenger RNA (mRNA) that is translated to functional proteins, or directly to functional molecules such as transfer RNAs (tRNAs) and ribosomal RNAs (rRNAs) which aid translation. There are three classes of genes in eukaryotes which differ in their organization, the types of enzymes that transcribe them, and the proteins that modulate their transcription. Class I genes are transcribed by RNA polymerase I. It transcribes 5.8S, 18S, and 28S rRNAs. Class II genes are transcribed by RNA polymerase II (RNA pol II). It transcribes a wide variety of mRNAs that are translated into functional proteins. Class III genes encode tRNAs and 5S rRNAs. They are transcribed by RNA polymerase III. Transcription mediated by RNA pol II has been studied in detail (Saltzman & Weinmann, 1989; Zawel & Reinberg, 1992). The following section deals with the current knowledge about transcription of class II genes.

The rate at which a protein-coding gene is expressed or transcribed largely determines the level of a protein it codes for, though other steps such as the rate of processing of its pre-mRNA in the nucleus, its translocation through the nuclear membrane to the cytoplasm, its translation to protein, and its turnover also influence the level of the protein. Genes are packaged with histones to form chromatin. The chromatin needs to be decondensed and the sequences of the genes made available to RNA pol II for transcription. Also, their control sequences should be made accessible to regulatory proteins for modulation of transcription. The mechanism of chromatin decondensation is not clearly understood. Current information on this process suggests that covalent modifications of histones such as acetylation, phosphorylation, and ADP-ribosylation decrease the net positive charge on the histones and dissociate them from the DNA (see Chapter 3).

RNA polymerase II

This enzyme transcribes genes that have TATAA sequences (elements) at around -30 bp from the transcription start site, and the initiator (Inr) sequence that encompasses the start site. These are minimal promoter elements. At least one of these elements needs to be present for transcription of class II genes. Besides RNA pol II, seven transcription factors (TFIIs) that are needed for transcription have so far been discovered. Certain genes lack the TATA element and instead have several GC-rich sequences (GGGCGG). They are called Sp1-binding sites. Proteins that activate transcription via these elements are modular. They have a DNA-binding domain and an activation domain. They interact with an enhancer located far upstream (1–20 kb) to enhance transcription. Other elements, for example, CCAAT and hormone response elements (HREs) [such as estradiol response element (ERE), glucocorticoid response element (GRE), and cyclic AMP response element (CRE)] are needed for maximal transcription and regulation of respective genes.

RNA pol II contains eight to twelve different polypeptides including two large subunits, one of about 220 kd and another of 140 kd. These subunits as well as three other smaller subunits are similar to those of RNA pol I and RNA pol III. However, four or five subunits are unique to RNA pol II. Yeast RNA pol II is made up of 11 subunits with a total $M_r > 500$ kd (Kolodziej et al., 1990). It is highly sensitive to α-amanitin, a reagent that permits initiation but blocks elongation. The binding site of α-amanitin is in the 220-kd subunit. One significant feature of the largest subunit (220 kd) is that its C-terminal domain (CTD) has a stretch of seven amino acids, −Tyr−Ser−Pro−Thr−Ser−Pro−Ser−, which is tandemly repeated 26 (in yeast) to 52 (in mice) times. This repeat is critical for its activity. Enzymes with repeats of less than 12 to 13 are inactive. Serines, threonines, and tyrosines are phosphorylated in these repeats. TF II H plays a role in this phosphorylation (Lu et al., 1992). If they are not phosphorylated, the enzyme is less active in catalyzing RNA synthesis. Phosphorylation of CTD is believed to control the transition from transcription initiation to elongation. RNA pol I and RNA pol III do not have these repeats. If the repeats are less than ten in yeast, they do not survive. Transcription of RNA by RNA pol II starts accurately from a specific site on the DNA. The enzyme also recognizes a specific termination signal to terminate transcription.

Transcription factors

Several transcription factors are needed for accurately recognizing the site at which transcription should start. To initiate transcription, these factors need to bind to promoters or cis-acting elements in the 5′ flanking region in the immediate vicinity of the start site of the gene. Transcription factors are of three types: (1) those required for initiation of transcription, (2) those required for elongation of the RNA transcript and (3) those required for the formation of an initiation complex but are not required after initiation of transcription. Seven TF IIs as mentioned above have been identified that are required for basal transcription. They are TF II A, B, D, E, F, H, and J. They are also called general transcription factors (GTFs) (Buratowski et al., 1989; Saltzman & Weinmann, 1989; Ha, Lane & Reinberg, 1991; Singer & Berg, 1991; Zawel & Reinberg, 1992).

The first step in transcription is recognition of the TATA element by TF II D (Davison et al., 1983; Reinberg, Horikoshi & Roeder, 1987). It is the only GTF that recognizes the TATA element, binds to it, and then initiates the assembly of other GTFs. TF II A then stabilizes the TF II D–DNA interaction. Then TF II B recognizes and binds to the TF II DA–DNA complex. Nonphosphorylated RNA pol II is then escorted to this complex by TF II F to form the DABF–pol II–DNA complex (Flores et al., 1991). This is followed by the recruitment of TF II E, TF II H, and TF II J to complete the preinitiation complex, DABF–pol II–EHJ–DNA (Carey, 1991; Zawel & Reinberg, 1992; Fig. 4.2). A DNase I protection assay of the initiation complex has revealed that TF II E/F binds downstream of RNA pol II, between +20 and +30, relative to the initiation site (Burtowski et al., 1989).

After the assembly of the initiation complex, an ATP-dependent activation step takes place (Sawadogo & Roeder, 1984; Conaway & Conaway, 1991; Martin, 1991; Wolfe, 1991). Activation coincides with a loss of protein–DNA interactions between positions +20 and +30 of the promoter. This may be due to an ATP-dependent dissociation of TF II E or TF II F (Buratowski et al., 1989). Initiation of transcription, which occurs after the complexes are activated, requires dissociation of the two chains of the DNA to allow base pairing of the elongating transcript with the template strand. In prokaryotic transcription initiation, the unwinding that occurs is called "closed-to-open" complex transition (McClure, 1985). A similar conformational change in the

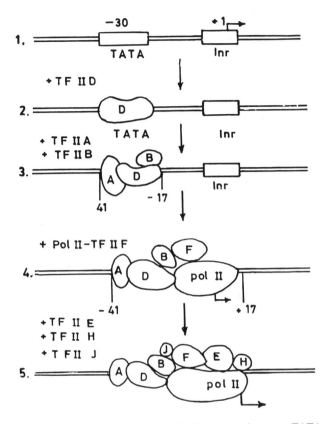

Figure 4.2. Stepwise formation of initiator complex on a TATA-containing promoter. Inr, initiator; pol II, RNA polymerase II. (Zawel & Reinberg, 1992)

eukaryotic promoter DNA has also been demonstrated by Buratowski et al. (1991).

The activity of human TF II D resides in a single polypeptide chain called TATA-binding protein (TBP) which has a Mr of 38 kd. It is one of the subunits of a complex of several subunits that has a Mr of >100 kd. X-ray diffraction studies of Nikolov et al. (1992) have shown that the TBP is saddle shaped, and binds to the minor groove of the TATA consensus sequence through a 180 amino acid domain. TBP is a common factor required by the three classes of RNA polymerases for recognition of the DNA sequence. It is tightly associated with other polypeptides (TAFs) to form the multiprotein complex called TF II D. TBP cannot substitute for the TF II D complex. Truncated proteins having only the

C-terminal domain from yeast, *Drosophila,* and human TBP can bind specifically to the TATA box and interact with the general factors such as TF II A and TF II B. This restores the basal level of transcription to a TF II D–depleted nuclear extract of HeLa cells (Peterson et al., 1990). The integrity of the highly conserved C-terminal domain is essential for function because even small deletions in this region of the yeast TBP disrupt both DNA binding and transcriptional activity in vitro (Horikoshi et al., 1990). Despite the fact that recombinant TF II D proteins from yeast, *Drosophila,* and humans function interchangeably, Gill and Tjian (1991) have shown by complementation assay that its C-terminal conserved domain and the N-terminal divergent domain appear to be involved in species-specific interactions.

TBP genes of yeast (Hahn et al. 1989), *Drosophila* (Hoey et al., 1990), and humans (Kao et al. 1990) have been cloned. A comparison of the sequences of the DNA among different species shows a conserved core sequence of 180 amino acids that is required for binding to TATA and interaction with general transcription factors. Within the 180 amino acid domain are two direct and identical repeats of 66 to 67 amino acids flanking a highly basic segment known as the basic repeat (see Nikolov et al., 1992). It has partial homology to the helix-loop-helix dimerization motif. TF II D may use this motif for interaction with other transcription factors. The N-terminal domain of TF II D differs significantly among species. This region may mediate species-specific interactions.

Human TF II B has been purified and its gene cloned (Ha, Lane & Reinberg, 1991). It is a single polypeptide chain of 33 kd and is coded by a single gene. It has a 76 amino acid repeat in the C-terminal domain and a hydrophobic tail at the N-terminus. Its repeated domain participates in basal transcription by contacting DNA and the general transcription factors, and its hydrophilic tail interacts with co-activators.

Human TF II E has been purified and its gene cloned (Peterson et al., 1991; Ohkuma et al., 1991). It has two subunits, 57 kD (TF II E α) and 34 kD (TF II E β). TF II E joins the pre-initiation complex after RNA pol II and TF II F, and its structure seems to be that of a heterotetramer ($\alpha_2\beta_2$). It is of interest that the N-terminal region of TF II E has three structural motifs – a leucine repeat, a zinc finger, and a helix-turn-helix (HTH) – the first protein known so far that has the three types of motifs. The leucine repeat includes amino acid residues 38 to 66 which may be involved in homomeric/heteromeric interactions. A zinc finger motif of the C_2C_2 type appears next and is followed by an HTH which is believed to bind to the major groove of the DNA. All of these motifs reside within the N-terminal 182 residues. No known motif is present in the C-terminal region. Whereas the N-terminal region is basic, the C-terminal is acidic. TF II E β is highly basic. It is likely that the two subunits form a complex by ionic interactions. TF II E α may also interact with TF II D.

TF II H is important for its role in phosphorylating the serine and threonine residues in the repeating sequences of CTD of the largest subunit of RNA pol II. It is a protein kinase (Lu et al., 1992) and has a Mr of 90 kd (Zawel & Reinberg, 1992).

Processing of pre-mRNA, pre-tRNA, and pre-rRNA

The entire protein-coding gene containing all the exons and introns is transcribed into a pre-mRNA by RNA pol II in the nucleus. The pre-mRNA undergoes modifications at its 5′ and 3′ ends in the nucleus. To the 5′ end is added a 7m5′Gppp5′N cap during the actual transcription. A long poly-A tail containing approximately 200 adenylate residues is added to the 3′ end in the nucleus after transcription. The introns of the pre-mRNA are then excised, and the successive exons are spliced or joined together to form the mature mRNA in the nucleus. Both the excision and splicing have to be accurate because the loss or addition of a single nucleotide to the internal sequence will generate meaningless codons and produce a bizarre protein (Fig. 4.1).

The signal for the removal of introns from the pre-mRNA lies at the exon–intron junction (splice junction) at either end of each intron (Chambon, 1981; Darnell, 1983). Invariably introns begin with –GU– at their 5′ end and end with –AG– at their 3′ end. Also, certain nucleotides are usually present adjacent to these nucleotides:

Exon 1		Intron		Exon 2
5′ _____C A G	G U A A G T..........C A G	G_____3′
A	G	T	

The fact that all introns are flanked by GU at the 5′ end and end with AG at the 3′ end in all genes suggests that just one enzyme may be involved in the excision of introns and splicing of the successive exons from the pre-mRNA.

Thus the formation of a mature mRNA in the nucleus involves (1) 7mG capping of its pre-mRNA at its 5′ end, (2) poly-A tailing at its 3′ end, (3) removal or excision of the regions representing the introns, and

(4) joining or splicing together of the successive exons. The mRNA carrying the 5' cap and the poly-A$^+$ tail is then translocated to the cytoplasm for translation into the corresponding protein. The 5' cap protects it from degradation at the 5' end. The 3' poly-A$^+$ tail not only prevents it from rapid degradation at the 3' end, but also is necessary for its translocation through the nuclear pore to the cytoplasm. These modes of protection at both ends permit the mRNA to engage in several rounds of translation in the polyribosome complex.

Most of the eukaryotic protein-coding genes studied so far have been found to be split except those for histones and interferon. Even the tRNA and rRNA genes are split. The intron segments represented in their transcripts are excised, and the exon segments are spliced together to form mature tRNA and rRNA, which are then translocated to the cytoplasm to take part in the translation process. The processing of the pre-mRNA and the formation of mRNA are shown in Figure 4.1.

Splicing of pre-mRNA

The sequences of all pre-mRNAs studied so far show that their introns begin with −GU− at their 5' end and end with −AG− at the 3' end. Furthermore, nine nucleotides at the 5' splice junction that includes the last three nucleotides of the exon and the first six of the intron are more or less conserved. The 3' splice junction has a pyrimidine-rich sequence at the 3' region of the intron. Mutations in these sequences cause splicing at the wrong sites, which would cause a shift in the reading frame of the codons and abolish or change the informational content of the resulting mRNA. The removal of the intron from pre-mRNA is believed to be carried out by an endonuclease, which may be a component of the small nuclear ribonucleoprotein complex (snRNP) to which the pre-mRNA binds for processing. This endonuclease cleaves the 5' and 3' ends of the intron. The two exons are then ligated by a splicing ligase that requires ATP.

The splicing of a pre-mRNA occurs in two stages. First, cleavage occurs at the 5' splice site, generating a 3'-OH at the 5' exon and a 5' phosphate at the 5' end of the intron to be removed. The 5' phosphate of the intron forms a 2'-5'-phosphodiester bond with the 2'-OH of an adenosine residue in the UACUAAC sequence which is 20–40 nucleotides upstream from the 3' splice site of the intron. Second, cleavage then occurs at the 3'-splice site of the intron, and the 3' end of the upstream exon is believed to be joined to the 5' end of the downstream exon by a RNA ligase. The intron is released as a tailed circle or lariat. These reactions occur in a nucleoprotein complex in the nucleus in which snRNP complexes take part (Sharp, 1985; Watson et al., 1987; Singer & Berg, 1991).

Splicing of pre-rRNA and pre-tRNA

The excision of the intron and the ligation of the exons of pre-rRNAs and pre-tRNAs are carried out by a mechanism similar to that for pre-mRNAs. Cech and his associates discovered a novel mechanism for the removal of introns of pre-rRNAs in the protozoan, *Tetrahymena* (Cech, 1983, 1986, 1987). They found that an intron, 413 nucleotides long, is excised from the pre-rRNA by guanosine, that is, without any enzyme and ATP. The pre-rRNA acts as an enzyme (ribozyme) in this organism in that it both removes the intron and joins the exons.

Structure of eukaryotic genes

With the development of genetic engineering technology it has been possible to isolate a single gene as well as parts of genes and to insert them into carriers or vectors and clone them to determine their sequences and study their functions. The general picture that has emerged from extensive studies is shown in Figure 4.3.

Even though RNA pol II is responsible for the transcription of pre-mRNA from protein-coding genes, the accuracy of its transcription, that is, the nucleotide from which it should start transcription and where it should terminate, and the rate of transcription are controlled or regu-lated by several short sequences located upstream or at the region 5' to the start site of transcription. This region is called the promoter as it generally promotes transcription. The sequences are called cis-acting elements as they are located in the same DNA strand. These regulatory elements may be broadly divided into two categories: (1) those present in genes that are generally expressed constitutively in most tissues (housekeeping genes), and (2) those that are present in specific genes that respond to specific stimuli and are expressed in specific tissues. Some of the better known cis-acting elements that are involved in the regulation of genes are described below.

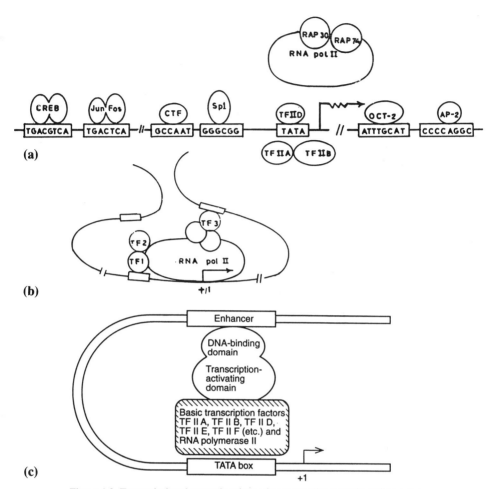

Figure 4.3. Transcriptional control regions of a mammalian protein coding gene. (a) Hypothetical array of cis-acting elements that constitute the promoter and enhancer regions transcribed by RNA polymerase II. RNA pol II, TF II A. TF II B, TF II D, RAP30, and RAP74 belong to the general transcription machinery. CREB, Jun, Fos, CTF, Sp1, OCT-2, and AP-2 are trans-acting factors that activate transcription after binding to specific cis-acting elements. Transcription initiation site is shown by an arrow. It is not implied that all the factors have to bind simultaneously to initiate transcription. (Mitchell & Tjian, 1989; by permission of *Science*, © 1989 AAAS) (b) Possible mechanisms by which cis elements activate transcription. Specific protein–protein interactions between certain transcription factors (TFs) bound to distal elements and factors associated with transcription initiation complexes at RNA start site may be required so that some distally bound factors can participate directly in the initiation process. (Mitchell & Tjian, 1989; by permission of Science, © 1989 AAAS) (c) Another representation of the activation of transcription by trans-acting factor through cis-acting elements acting at some distance in the promoter and by protein–protein interaction. Transcription activator consists of two domains (DNA-binding and transcription activating) and binds at the enhancer site. Arrow represents transcription start site. (Martin, 1991)

DNA sequences that control transcription

Cis-acting elements

These elements (sequences) in the gene act by binding to specific proteins that are synthesized or activated in response to inducing signals. In general, these elements have dyad symmetry since a similar sequence is present at both the 5' and 3' ends of each element. For example, the estrogen responsive element (ERE) which is present at around − 100 bp upstream of estrogen responsive genes, has the following sequence:

5' –A G G T C A N N N T G A C C T– 3'
3' –T C C A G T N N N A C T G G A– 5'

The two halves of the 12-base pair palindrome are separated by three random nucleotides. This suggests that this 15-bp sequence may assume a unique conformation that may bind to a protein in its dimeric form.

Table 4.2 is a list of the response elements of genes showing their various inducers. The important information available from this table is that the same type of cis-acting element is present in several genes but expresses them in a tissue-specific manner. For example, the CCAAT box is present in the promoter regions of the genes for globin and albumin. The globin gene is expressed only in reticulocytes, and the albumin gene is expressed in the liver. The estrogen response element is present in both ovalbumin and vitellogenin genes. The ovalbumin gene is expressed only in the oviduct, and vitellogenin gene is expressed in the liver. The above examples belong to two different classes. The globin and albumin genes are expressed constitutively, but in a tissue-specific manner. The expression of both ovalbumin and vitellogenin genes is dependent on the same hormone, 17β-estradiol, which induces them in a tissue-specific manner.

A CCAAT-binding protein is required for the expression of globin and albumin genes in reticulocytes and liver, respectively. The CCAAT box, however, is also present in the genes for thymidine kinase, c-*myc*, and c-*ras*, which are expressed in other tissues. The CCAAT-binding protein is synthesized in several tissues and remains bound to this sequence. Hence the specificity of the expression of globin gene in reticulocytes and albumin in the liver may be due to one or more additional factors that may be tissue-specific, and interact with the CCAAT-protein complex of the globin gene in reticulocytes such that it activates the transcription initiation complex of the globin gene. Likewise, another

Table 4.2. *Response elements of genes and their inducers and trans-activators*

Consensus sequence	Inducer	Protein factor (trans-acting factor)[a]	Genes containing the sequence[a]
CCAAT box		C/EBP	α,β-Globin, albumin, TK, c-*ras*, c-*myc*
T/GT/ACGTCA	Cyclic AMP	CREB/ATF	Somatostatin, fibronectin, c-*fos*, gonadotropin, hsp 70
GGTACANNN-TGTTCT	Glucocorticoid progesterone	GR, PR	Metallothionein IIA, lysozyme, uteroglobin, tryptophan oxygenase
AGGTCANNN-TGACCT	Estrogen	ER	Ovalbumin, vitellogenin, conalbumin
TCAGGTCAT-GACCTGA	Thyroid hormone retinoic acid	TH, RA	Growth hormone, myosin heavy chain
TGAGTCAG	Phorbol ester	AP1	Metallothionein IIA, α₁-antitrypsin, collagenase
CCCCAGGC (AP2 box)			Metallothionein IIA, collagenase
GGGCGG (Sp 1 box)			Metallothionein IIA, DHFR, type II procollagen
TGCGCCCGCC	Heavy metals	Unknown	Metallothionein
CTNGAATNT-TCTAGA	Heat	Heat-shock transcription factor	hsp 70, hsp 83, hsp 27
ATGCAAAT			Ig gene in B cells

[a] TK, thymidine kinase; DHFR, dihydrofolate reductase; C/EBP, CCAAT-enhancer binding protein; CREB, cAMP responsive element binding protein; ATF, activation transcription factor; GR, glucocorticoid receptor; PR, progesterone receptor; ER, estradiol receptor; TH, thyroid hormone; RA, retinoic acid; AP1, activator protein 1.

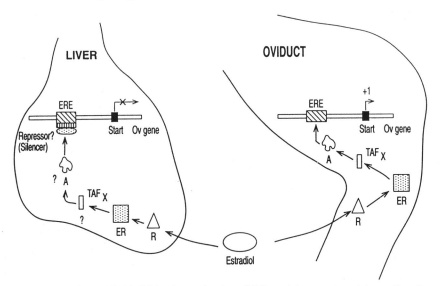

Figure 4.4. Model for the mechanism of differential expression of the ovalbumin gene by estradiol. The gene is expressed in the oviduct, but not in the liver. In oviduct, estradiol (E) binds to a receptor to form estradiol–receptor (ER) complex, which then binds to a specific trans-acting factor (TAFx) to form a complex (A). Complex A then binds to the ERE in the promoter of the gene to stimulate its expression. In the liver, E also binds to an identical R. However, its failure to stimulate expression of the gene may be due to one of the following reasons: (1) the TAFx is absent and so A is not formed, or (2) a repressor is bound to the ERE that prevents the binding of A.

specific factor in the liver may interact with the CCAAT protein and mediate the expression of the albumin gene.

In the case of ovalbumin and vitellogenin genes, the inducer, estradiol (E), is common. The receptor (R) for estradiol both in the oviduct and liver is also the same. The E-R complex that is formed after administration of estradiol, binds to the same estradiol-responsive element (ERE) in the DNA of both the genes. The specificity of the expression of ovalbumin gene in the oviduct and vitellogenin gene in the liver must, therefore, be due to one or more additional factors that are present in the respective tissues (Fig. 4.4). Furthermore, the fact that there is a dose-dependent increase in the expression of the genes shows that these additional factors may be inducible. This interesting aspect can be studied by using the technique of gel retardation of DNA fragments containing the ERE.

There are other similarities among cis-acting elements as well. Not only do both glucocorticoid and progesterone bind to the same DNA sequence in the promoter region, but there is much homology between glucocorticoid/progesterone and estradiol-binding elements. Surprisingly, there is also a similarity between the sequences of the steroid receptors and thyroid hormone (T3)/retinoic acid receptors. T3 and retinoic acid are entirely different molecules, and they bind to different receptors. However, together with their receptors they bind to the same DNA sequence. It is therefore not surprising that the receptors for glucocorticoid, progesterone, estradiol, T3, and retinoic acid belong to one superfamily of receptors, all of which have a conserved central domain that binds to their respective DNA sequences. Their C-terminal domain is for ligand binding, and the N-terminal domain mediates interaction with the transcription complex for stimulation of transcription.

The identification of specific sequences in the promoter and other regions of genes that bind to specific proteins is the subject of very intense research in many laboratories, since it is at the very base of gene regulation. A detailed account of some of the well-studied DNA sequences is given below.

DNA sequences that control transcription are usually grouped into two classes – promoters and enhancers – though they may overlap physically and functionally. For example, many enhancer elements that stimulate transcription from long distances also stimulate transcription when inserted upstream of the TATA box. However, not every promoter element acting alone can enhance transcription over long distances (Serfling et al., 1985). Enhancers and promoters have a modular structure and are composed of an array of sequence motifs. In general, enhancer motifs can act independently, but each additional element usually further increases the transcriptional activity. These sequence motifs are binding sites for transcription factors. In some cases the binding sites for a given transcription factor occur both in the promoter and enhancer regions, as for example, the immunoglobulin (Ig) enhancers and promoters (see Müller, Gerster & Schaffner, 1988, for review). Furthermore, even though some promoters are active in a great variety of cells (as in the case of globin gene promoters), other promoters strongly contribute to the tissue-specific expression of genes such as that for the IgH gene (Müller et al., 1988). Even though the TATA box is involved in the binding of RNA pol II and the transcription factors TFIID, TFIIA, TFIIB, TFIIE, TFIIF, TFIIH and TFIIJ, it is not sufficient for a good response to a remote enhancer. Addition of one or two cis-acting DNA motifs upstream of the TATA box converts this region into a promoter that may be weak by itself but responds strongly to a remote enhancer even if it is located downstream of the gene. Addition of several sequence motifs upstream of the TATA box can result in the buildup of a promoter that does not require the presence of a remote enhancer (Westin et al., 1987).

Generally, cis-acting elements are located within a few hundred base pairs toward the 5' region of the transcription initiation site, but some elements can exert control on transcription from distances of 1–20 kbp. The control regions in the immediate vicinity, generally up to about 200 bp, are called promoters. Elements that control transcription from longer distances and in an orientation-independent manner are called enhancers (Serfling et al., 1985; Singer & Berg, 1991). TATA, CCAAT, and GC boxes, CRE, HRE, Sp1, and AP-1 binding sequences are promoters. They are present in upstream regions of genes transcribed by RNA pol II.

TATA box. One of the sequences in the gene that is very important for initiation of transcription is the TATA box (TATAA), located 25–35 bp upstream of the transcription start site. It binds to TF II D, a 38-kd protein. The box is present in the genes for histones, immunoglobulins, globins, ovalbumin, silk fibroin, and other proteins. Any change in a nucleotide in this box affects transcription. On the other hand, any change in the sequence between the TATA box and transcription start site does not affect transcription. RNA synthesis begins generally at 30 nucleotides downstream from the TATA box. Thus, it is required for specifying the transcription initiation site in certain genes, and in some others it is required for efficiency of transcription.

The sequence AATAAA at the 3' end of a gene acts as the signal for RNA pol II to terminate transcription. The pre-mRNA then dissociates from the DNA and poly-A polymerase adds a poly-A$^+$ tail of approximately 200 adenosine residues to the 3' end of the pre-mRNA. This is a part of the processing and maturation of the pre-mRNA. So far, except for the mRNAs for histones, all other protein-coding mRNAs have been shown to carry a poly-A$^+$ tail. The presence of a poly-A$^+$ tail helps to separate such mRNAs from the rest of RNAs by passing the total RNA in a poly-dT (polythymidylic acid) or poly-rU (polyuridylic acid) column. Cells, of course, do not add a poly-A$^+$ tail to protein-coding pre-mRNAs for that purpose. The presence of the tail enables it to bind to specific proteins in the nucleus for translocation through the nuclear pores to the site of translation in the cytoplasm. It also protects the coding region of the mRNA from degradation at the 3′ end, just as the 5′ cap protects the 5′ end. The 5′ cap, however, consists of a terminal nucleotide, 7mG, in a 5′–5′ linkage with the first nucleotide of the pre-mRNA chain. The cap also aids in mRNA translation by enabling it to bind to the ribosome.

CCAAT. This cis-acting sequence is commonly found between − 70 and − 120 bp upstream from the transcription start site of protein-coding eukaryotic genes. CCAAT is part of a motif that binds to the transcription regulatory factors that activate transcription. DNase I footprinting, methylation protection patterns, gel-mobility shift assays, and binding competition studies have shown that several proteins bind to this motif in mammalian cells. For example, CTF (the transcription factor that binds to CCAAT) binds to the CCAAT of the promoters of thymidine kinase and β-globin genes. HeLa cells contain two CCAAT-binding proteins, CP-1 and CP-2, which are different from CTF and are present in the same cell. CCAAT, which is present in several genes and plays an important role during development, is mainly required for regulating the efficiency of transcription.

Several CCAAT-binding proteins that are active in transcription initiation have been cloned. Chodosh et al. (1988) have identified three such proteins; two of these proteins bind to CCAAT as a dimer. This is the case with yeast HAP2 and HAP3 proteins which bind to CCAAT. Another protein that binds to CCAAT and an enhancer sequence far upstream (a few kilobase pairs) from CCAAT is called C/EBP (CCAAT and enhancer-binding protein). It is heat stable and binds to CCAAT region via its leucine zipper, a domain of the protein in which leucine occurs at every seventh amino acid. The basic region of the protein is also involved in DNA binding. C/EBP was first discovered in the nuclear extract of the rat liver. Although the enhancer sequence in the DNA is approximately 100 bp long, its core sequence needed for the binding of this protein is 5′-TGTGGAAAG-3′.

A C/EBP protein has been purified from the nuclear protein fraction of rat liver. It binds to promoters of albumin, α_1-antitrypsin, and trans-thyretin genes which are expressed only in the liver. Since CCAAT is located at around − 80 bp and the enhancer is located several kilobase pairs farther upstream, it is presumed that binding of this protein to the two motifs bends the DNA and thereby brings certain factors to the vicinity of the transcription complex to stimulate transcription. C/EBP occurs as a stable dimer and has a leucine zipper. Its gene has been cloned. The 30 amino acid domain toward the N-terminal from the leucine repeat is essential for its binding to DNA. C/EBP is present in several types of cells and has been shown to bind to these sequences in several genes (for reviews, see Mitchell & Tjian, 1989; Johnson & McKnight, 1989; Nussinov, 1990; Singer & Berg, 1991). It participates in the arrest of cell growth, followed by their terminal differentiation. It binds to DNA in a sequence-specific manner and functions exclusively in terminally differentiated, growth-arrested cells. In the adult liver, C/EBP is restricted to the nuclei of mature hepatocytes. It is also present in adipose tissue and intestinal cells. In rapidly proliferating cells, its level is negligible. Cultured 3T3-L1 adipoblasts do not express C/EBP protein during proliferative growth, but its transcription is markedly increased when the cells are stimulated to differentiate. Also, premature expression of C/EBP in adipoblasts causes cessation of mitotic growth. Since C/EBP is expressed in several types of tissue, it may have a fundamental role in regulating the balance between cell growth and differentiation in higher animals.

C/EBP also binds to the cAMP-responsive element I (CRE-1) in the liver that confers cAMP responsiveness to the phosphoenolpyruvate carboxykinase (PEPCK) gene. CRE-1 is also present in the kidney, spleen, and brain. It is likely that there are tissue-specific CRE-binding (CREB) proteins. It is also possible that each tissue has a C/EBP-specific protein that mediates its binding to the DNA site.

GC box.   Not all genes are controlled by TATAA and CCAAT cis-acting elements located in their 5'-flanking region. There are certain genes that lack the TATAA- and CCAAT-sequence motifs in their promoter regions, but contain instead multiple GC boxes. These genes include the housekeeping genes such as dihydrofolate reductase (DHFR), adenosine deaminase, hypoxanthine phosphoribosyltransferase (HPRT), hydroxymethylglutaryl ~CoA (HMG ~CoA) reductase, as well as nonhousekeeping genes such as those for malic enzyme, EGF receptor, and NGF receptor. GC boxes are situated at about -200 bp in the 5'-flanking region. They bind to several transcription regulatory factors like Sp1, AP2, and ETF. GC boxes are necessary for efficient promoter activity. In the DHFR promoter, there are four GC boxes within 210 bp 5' from the coding sequence. A GC-box binding factor is necessary for transcription, and a truncated promoter containing only one GC box is transcriptionally inactive. The factors binding to the GC box interact in a position-dependent manner. The GC box has a consensus hexanucleotide sequence, 5'–GGGCGG–3'. Mutations in this sequence abolish the binding of the factors to the GC box. Thus in TATA-less promoters the GC boxes both initiate transcription and promote efficient transcription.

Thus, the general cis-acting elements present in the upstream region within about -200 bp of the transcription start site are the TATA, CCAAT, and GC boxes. They are present in several genes, and the proteins that bind to them are distributed widely. Therefore, they are not likely to be responsible for cell-specific expression of genes, nor do they confer specificity of expression to particular genes. They are mainly responsible for transcription.

Besides the general cis-acting elements – the TATAA, CCAAT, and GC boxes that are required for both initiation and efficiency of transcription – there are several tissue-specific cis-acting sequences in the promoter regions of genes that are involved in regulation of transcription. They have specificity for binding to receptors for steroid and thyroid hormones, cAMP, retinoic acid, Sp1, *fos,* and *jun.*

Hormone-responsive elements. Steroid and thyroid hormone receptors as well as retinoic acid receptor belong to a family of receptors that undergoes conformational changes after binding to their respective ligands. The receptor–ligand complexes then bind to defined cis-acting sequences (Table 4.2) in the promoter regions of genes which are expressed in a tissue-specific manner and are responsive to these hormones (for reviews, see Allan et al., 1991; Wahli & Martinez, 1991; Luisi et al., 1991). The estradiol receptor binding site (estradiol responsive element, ERE) has been shown for vitellogenin and ovalbumin genes. Specific binding of progesterone receptor to PRE has been reported in the genes for ovalbumin, transferrin, ovomucoid, lysozyme, and uteroglobin.

The steroid receptors as well as cAMP and retinoic acid receptors have a cysteine-rich, highly conserved DNA-binding domain in the central region which binds to an 8–10 bp palindromic sequence in the promoter. It has two cysteine-rich zinc fingers by which it binds to DNA. The C-terminal region binds to the ligand and is moderately conserved. There are two well-conserved subregions, which may form nonspecific contacts with the hormone or other trans-acting factors. The N-terminal regions, which modulate transcription, vary extensively both in size and sequence because these regions may have to interact with different types of tissue-specific protein factors for the expression of specific genes. The C-terminal region, besides binding to the ligand (hormone), is required for dimerization of estrogen receptor and inhibition of transcription by glucocorticoid receptor.

The binding of the hormone to the receptor results in activation or transformation of the receptor, allowing it to dimerize and recognize specific binding sites in the flanking regions of the HRE. Mutational studies have shown that three amino acids in the stem of the first zinc finger (P-box) are critical for specifying the half-site sequence in the HRE, and five amino acids in the second finger (D-box) can alter the selection pattern of the half-site spacing (Mader et al., 1989). NMR spectroscopic studies on the GR and ER indicate that an α-helix including the P-box contacts the major groove of the DNA double helix, and the D-box stands as a potential interface of receptor dimerization within the DNA-binding domain (Schwabe, Nieuhaus & Rhodes, 1990).

Consensus sequences have been defined for steroid response elements (SREs) that mediate induction by glucocorticoids (GREs), estrogens (EREs), thyroid hormones (TREs), and retinoic acid (RARE). SREs consist of two closely related motifs. The two copies are arranged either immediately adjacent to each other or with a short space in between, and are either directly repeated (RARE) or are dyad symmetrical (all other SREs). Though no distinct elements have been found for progestins or androgens, both these hormones are capable of induction via the GRE (also called GRE/PREs).

Umesono et al. (1991) found a TRE that consists of a direct repeat, not a palindrome, of the half-sites. Unlike palindromic TREs, direct repeats do not confer a retinoic acid response. The tandem TRE can be converted into a RARE by increasing the spacing between the half-sites by one nucleotide, and the resulting RARE is no longer a TRE. Decreasing the half-site spacing by one nucleotide converts the TRE to a vitamin D_3 response element, while eliminating the response to T3. These results show a correlation with the DNA-binding affinities of T3, RA, and vitamin D_3 receptors. Hence, the tandem repeat SREs have a physiological role, and a physiological code may exist in which half-site spacing plays a critical role in achieving selective hormonal response.

Enhancers.   Grosschedl and Birnsteil (1980) first reported that DNA sequences far upstream of the transcriptional start site of the H2A gene are responsible for its high level of transcription. This sequence itself is unable to act as a promoter, but it increases initiation of transcription greatly from an adjacent promoter element when located in either orientation relative to the start site of transcription. Subsequently, Banerji, Ruscony and Schaffner (1981) and Moreau et al. (1981) showed the presence of DNA sequences that enhance gene expression in other systems. Banerji et al. (1981) showed that a 200-bp DNA segment of SV40, when linked to a test gene, could enhance its transcription more than 100-fold. This activation occurred even from distances as long as 3 kbp. Furthermore, activation also occurred when it was placed at the 3'-end of the gene. The 200-bp sequence of SV40 is located between position − 300 and − 100 bp 5' of the initiation site for the early genes. In later studies, more enhancers were found in other genes. The characteristics of all enhancers are: (1) they activate transcription of a linked gene from its correct initiation site; (2) they activate transcription independently of the orientation of the enhancer; and (3) they exert their enhancing effects from large distances of over 1 kbp from a position either upstream or downstream of the initiation site (Müller, Gerster & Schaffner, 1988; Fig. 4.5).

This property of remote control by enhancers may be involved in the overexpression of certain genes and certain pathological processes. During a viral infection, an enhancer sequence of the virus may be inserted in the vicinity of a cellular gene that controls cell proliferation (a so-called proto-oncogene). This insertion converts the proto-oncogene into an oncogene, leading to its deregulation and overexpression. This, in turn, may lead to uncontrolled cell growth and malignancy (Weber & Schaffner, 1985).

The first cellular enhancer was discovered by Banerji, Olson and Schaffner (1983), Gillies et al. (1983), and Neuberger (1983) in the immunoglobulin (Ig) heavy chain gene. This showed that remote control of gene activity was not specific to viruses, and that the Ig enhancer stimulated transcription in a cell-type-specific manner. Moreover, the Ig enhancer was found to be located within the gene. Later, several cell-type-specific enhancers were discovered (for reviews, see Maniatis et al., 1987; Serfling et al., 1985; Wasylyk, 1988). Cell-type-specific and stage-specific enhancers were also discovered in genes that control embryogenesis in *Drosophila* (Fischer & Maniatis, 1986; Hiromi & Gehring, 1987). This implies that tissue-specific factors present in these cells are necessary for binding to enhancer sequences, and thereby conferring on them a specificity for interaction with the transcription initiation site to enhance transcription.

That enhancers are tissue specific, that is, a gene that is expressed

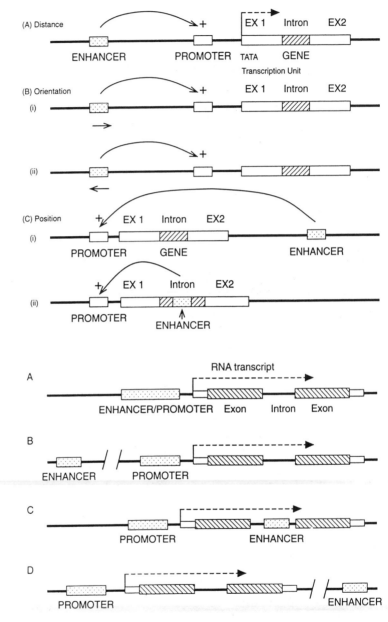

(a)

(b)

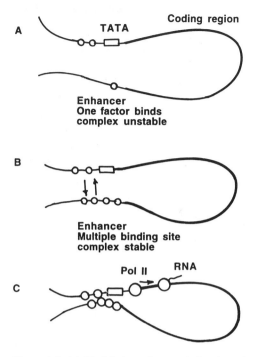

(c)

Figure 4.5. (a) Modulation of transcription by enhancers. (A) An enhancer element that can activate a promoter at a distance: (B) in either orientation relative to the promoter; (C) when (i) positioned upstream, (ii) positioned downstream, or (iii) within an intron. (Adapted from Latchman, 1990). (b) Another representation of enhancer activation of transcription. (Müller, Gerster & Schaffner, 1988) (c) Looping model of enhancer mechanism. (A) Remote enhancer with a single binding site for a transcription factor cannot easily form a functional complex. (B) Greater number of factor-binding sites in enhancer greatly facilitates formation of a functional complex by protein–protein contacts, including the TATA box factors and/or RNA polymerase II. (C) Transcription is initiated. (Müller, Gerster & Schaffner, 1988)

only in a specific tissue has an enhancer that can activate the promoter of the gene only in that tissue and not in other tissues, was demonstrated by Gillies et al. (1983). In B lymphocytes in which the genes for light and heavy Ig protein chains are expressed, an enhancer is present in the intron between the joining and constant regions of the gene. If this sequence is linked to the β-globin gene which is then placed in B lymphocytes, it increases the activity of its promoter dramatically, and β-globin gene is transcribed even though these cells do not normally transcribe this gene. However, if this enhancer is linked to another gene which is then placed in fibroblast cells, it does not activate its promoter. It is clear that the enhancer of Ig gene acts in B lymphocytes because of certain factors present in these cells and not in other cells. Such tissue-specificity of enhancers has been shown for α-fetoprotein, albumin, and α_1-antitrypsin genes which are expressed in the liver, insulin gene in the β cells of the islets of Langerhans of the pancreas, and prolactin and growth hormone genes of the pituitary. Rutter and his associates (Karlsson et al., 1987) showed that an enhancer element located at 250 bp upstream of the transcription start site of the insulin gene activates the promoter of the insulin gene in the pancreas. Furthermore, mutations in these sequences abolish the tissue-specific stimulatory effects of enhancers.

In an elegant study, Hanahan (1985) linked the enhancer and promoter of the insulin gene to the gene encoding the T-antigen of the eukaryotic virus SV40. This construct was introduced into a fertilized mouse egg, which was then placed in the oviduct of a female mouse. The transgenic mouse that was born expressed the T-antigen gene only in the β cells of the islets of Langerhans and not in other tissues. Thus the enhancer of the insulin gene can stimulate the expression of another gene provided it is in β cells. This shows not only tissue-specificity of enhancers but also that tissue-specific enhancer binding protein (EBP) is required for the expression of a gene.

Significantly, a sequence in the enhancer of Ig heavy chain gene in B lymphocytes is not different from a sequence in its promoter. The octamer motif −ATGCAAAT− is present in both. So the enhancer and promoter bind to the same transcription factors, which may act together to activate transcription (Sen & Baltimore, 1986). Enhancers may stimulate gene expression, as in the case of promoters, by displacing the nucleosomes and altering the chromatin structure at the transcription start site. Binding of enhancer-specific proteins may displace nucleosomes in either direction. Alternatively, the binding of enhancer-specific protein may bend the chromatin toward the transcription complex so that it interacts directly with the complex. Indeed such bending of DNA has been shown for progesterone-responsive genes (Fig. 4.6; Theveny et al., 1987). It was shown by EM studies that when progesterone together with the progesterone receptor (P-R complex) binds to the progesterone-responsive element (PRE), the intervening DNA between the PRE and transcription start site bends so that there is direct interaction between the P-R and transcription initiation complex. With which component of the initiation complex the P-R interacts is not known, however, but it is clear that the role of both enhancers and promoters such

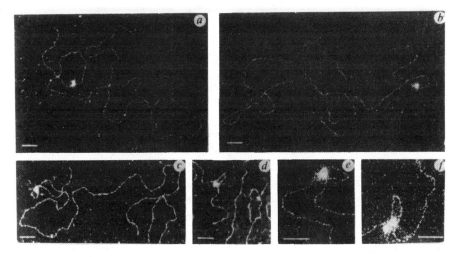

Figure 4.6. Mechanism of action of cis-acting elements in gene activation. Progesterone receptor binds to DNA upstream from the uteroglobin gene. Receptor oligomers shown bound to *(a)* site I (− 2427 to − 2376) or to *(b)* site II (− 2709 to − 2620). *(c)* Two receptor oligomers shown attached to both binding sites to make a DNA loop. *(d)* Two DNA-bound receptor oligomers inducing a DNA loop shown beside a single oligomer bound to site I. *(e)*, *(f)* DNA loops shown at higher magnification. (Theveny et al., 1987; by permission of *Nature*, © 1987 Macmillan Magazine Ltd.)

as HRE, CRE, and CCAAT is to interact with the transcription initiation complex at the start site. This is possible only if the intervening sequence is sufficiently long to permit bending.

It has been shown that such bending is possible if the intervening DNA sequence is more than 30 bp, possibly due to the steric constraints of the bending of a shorter sequence. This raises the question of synergism between an enhancer and a promoter. Each sequence alone does not activate transcription to any significant extent as shown for Ig genes. Only when the enhancer and promoter of the gene are present does maximal expression of the gene occur. It has been postulated (Dynan & Tjian, 1985) that the enhancer-specific factor, on binding to the enhancer, makes contact with the promoter-bound transcriptional apparatus either by looping of the intervening DNA or by sliding along the DNA. There is no evidence so far for the latter possibility. The occurrence of similar octamer motifs both in the enhancer and promoter of certain genes raises the possibility that a subunit protein binds to the enhancer and causes the DNA to bend towards the transcription start site where the promoter is already positioned along with a similar sub-unit. The enhancer and promoter bound subunits dimerize and stabilize. This stimulates transcription and hence the synergism (Ptashne, 1986, 1989).

Not all enhancers activate their promoters. Only in a tissue that has the appropriate enhancer-binding protein will the enhancer activate its promoter. If the protein is absent, it will have no effect. Also, in certain cases the enhancer inhibits the promoter action by binding to a factor present in several tissues. In the presence of an inducer, the inhibitory factor is released from the enhancer, and the positive factor then binds to the enhancer which then activates transcription. The positive-acting factor is normally present but is prevented from binding to the enhancer due to the binding of the negative factor. Thus, enhancers may either activate or repress gene expression in a tissue-specific manner and in association with respective promoters. Hence they are important for the regulation of gene expression.

Repeated DNA sequences

Only about 2%–5% of the mammalian genome is transcribed and codes for proteins, tRNAs, and rRNAs. The role of the remaining 95% of the genome is largely unknown. Approximately 25% of this component in humans comprises repeated DNA sequences that are present in many copies. Those that are repeated 100,000 times or more are called highly repetitive; those repeated 1,000–100,000 times are called moderately repetitive, and those that are repeated 10–1,000 times are called low repetitive sequences. The repetitive sequences are about 200–300 bp long and are interspersed throughout the genome, between the unique protein-coding genes and even in the introns of these genes. Those that are present in the introns of genes are transcribed by RNA pol II but are excised from the pre-mRNAs during their processing to form mRNAs. It has been speculated that such a large component of the genome may have important functions to perform, but so far there is no clear evidence for any specific role for these sequences except for the report that small transcripts of the identifier (ID) repeats are present in the rat brain. They are transcribed by RNA polymerase III. However, the function, if any, of ID sequences is not known.

It has been postulated that some of these repeats have promoters for RNA polymerase III, which recognizes tissue-specific factors and initiate transcription of these repeats. This opens up the adjacent chromatin that contains a protein-coding gene which is then transcribed by RNA pol II. It has also been suggested that these repeats are transcribed and play a role in tissue-specific selection of pre-mRNAs as proposed by Britten and Davidson (1969). It is surprising that the role of repetitive DNA sequences has remained elusive despite tremendous strides in genetic engineering technology. This remains a very important area for study and may provide insight into the mechanism of aging.

DNA-binding proteins

Though protein–protein interaction has been known for decades, studies on DNA–protein interaction is a more recent development. DNA footprinting, X-ray studies on DNA-bound proteins, and gel-mobility shift studies have established that proteins bind to specific sequences of DNA as they bind to proteins. Hydrogen bonding and ionic and hydrophobic interactions are involved in DNA-protein binding as for protein–protein binding. So far four main types of structural motifs or domains are known through which proteins bind to DNA. They are helix-turn-helix, helix-loop-helix, zinc finger, and leucine repeats.

Helix-turn-helix (HTH)

One class of proteins that binds to specific DNA sequences has a HTH motif which has two α-helices separated by a β turn (Fig. 4.7). They bind to DNA as dimers. The distance between the two helices is matched to the distance separating successive major grooves on a single face of β-DNA. The helices are aligned in an antiparallel conformation so that their orientations (amino to carboxyl dipoles) are matched relative to the DNA sequence on each half of their dyad-symmetric recognition sites. The C-terminal helix (called helix 3) from one subunit is juxtaposed to the helix 3 of the other subunit. Helix 3 recognizes a stretch of six bases in the major groove of the DNA sequence through H-bonding and van der Waals interactions. The β turn that occurs between two helices begins generally with glycine and is invariably followed by a residue bearing a hydrophobic side chain. The geometry of the major groove changes after the contacts are made. Helix 2 stabilizes the binding by making contact with the sugar–phosphate backbone and possibly with the bases flanking the recognition sequence.

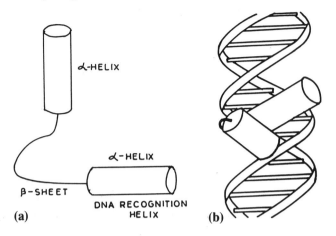

Figure 4.7. (a) A helix-turn-helix motif. (b) Binding of a helix-turn-helix motif to DNA with the recognition helix in the major groove of DNA. (Latchman, 1990)

Although proteins containing HTH were first discovered in the Cro and CI proteins of bacteriophage λ, the catabolite activator protein (CAP) of *E. coli,* and MAT 1 and MAT 2 of *Saccharomyces cerevisiae,* later such motifs were discovered in the products of *Drosophila* homeotic genes – *antennapedia (Antp), fushi tarazu* (ftz), *engrailed,* and *ultra-bithorax.* In still later studies, the mouse and human proteins – OCT1, OCT2 and PIT 1 – were found to possess HTH motifs.

The homeodomain is a conserved polypeptide motif present in the proteins coded by *Drosophila* homeotic genes (McGinnis et al., 1984; Scott & Wiener, 1984). This DNA-binding domain has been found in several transcription factors (Rosenfeld, 1991; Murtha et al., 1991; Free-mont, Lane & Sanderson, 1991; Laughon, 1991). The genes that code for these proteins are called homeotic genes since they have a 180-bp homeobox that specifies the 60 amino acid homeodomain within the protein. The homeodomain forms an HTH. It has a high content of basic amino acids (30% of residues being lysine and arginine). The importance of this homeodomain for DNA binding is seen from the fact that point mutations and frame-shift mutations in the homeodomain completely impair the DNA-binding capacity of the protein. However, mutations introduced elsewhere in the protein do not cause loss of DNA binding. They bind to the DNA sequence as monomers, unlike those of prokaryotes that bind as dimers (Harrison, 1991; Singer & Berg, 1991). An important characteristic of some homeoproteins is that they can bind to different-but-related DNA sequences, though with different binding affinities. So differential-binding specificity may enable one homeoprotein to modulate transcription of more than one kind of gene. After binding to DNA, these proteins influence transcription from their target promoters, though the mechanism is not known. The acidic nature of their C-terminal, and clustered prolines and glutamines in their N-terminal, may influence transcription.

Studies on the three-dimensional structures of two homeodomain–DNA complexes have shown that homeodomains utilize an HTH fold to make contact with the major groove of DNA (Otting et al., 1990; Kissinger et al., 1990). In addition, homeodomains utilize an N-terminal arm to contact DNA in the minor groove, which is sequence specific and contributes substantially to the high affinity of homeodomains for DNA. The homeodomain contains three helices, with helices 2 and 3 forming HTH. Helix 1 lies across helix 3 parallel to helix 2 (Fig. 4.7). The entire structure is held together by a pocket of hydrophobic amino acids at the interface between the three helices. The homeodomain binds to DNA as a monomer. The equilibrium binding constants for high-affinity sites are about 10^{-9}M for *Antp* and $(2–60) \times 10^{-11}$M for *ftz* (see Laughon, 1991, for review).

Three to four side chains of helix 3 make base-specific contacts with three to four bases in the major groove. The amino acids are Ile (47), Gln (50), and Asn (51). Mutation of Asn (51) to either Ala or Gln reduces its binding 25- to 92-fold. Conservation of Asn (51) suggests that all homeodomains recognize adenine at position 3. The minor groove contains less information than the major groove for base-pair recognition by proteins (Dickerson, 1983). The side chains of amino acids in the N-terminal recognize bases in the minor groove. This is a universal feature of homeodomains. Deletion of the N-terminal of *ftz* homeodomains reduces its binding by 130-fold.

Thus, homeodomain-containing proteins are transcription factors that regulate the expression of genes. Only a few genes are known to be regulated by these proteins in *Drosophila,* but examples of such proteins are more prevalent for mammals (Rosenfeld 1991). Moreover, these proteins act in combination with other transcription factors (Jiang, Hoey & Levine, 1991).

Helix-loop-helix (HLH)

Many genes that have specific functions during development code for proteins that act as transcriptional regulators (Garrell & Campuzano, 1991). These proteins contain HLH domains. This motif was first found in murine transcription factors, E12/E47, which bind to enhancers of Ig genes. Its tertiary structure has two amphipathic α-helices connected to a random-coiled loop of variable length. Adjacent to the HLH is a highly conserved basic domain that binds to DNA (Benezra et al., 1990). The HLH domain mediates the homo–heterodimerization of these proteins that is independent of, but necessary for, DNA binding. It may also play a role in selecting the DNA-binding site.

HLH-containing transcription factors play a role in sex determination in *Drosophila* and myogenesis in mice. In mammals, myogenesis is under the control of MyoD family of HLH muscle-specific transcriptional regulators (MyoD, myogenin, MRF-4). MyoD was the first such gene to be identified. If any of these genes is transfected to mammalian cell lines they undergo differentiation (Davis, Weintraub & Lassar, 1987; Olson, 1990). Deletion mutagenesis shows that the full basic region and the HLH domain are necessary and sufficient for myogenesis. The HLH region is necessary for dimerization, and the basic region binds to DNA. The MyoD-activation domain maps within the N-terminal 53 residues. The activation function of the basic region requires a second factor (Weintraub et al., 1991). It has been suggested that the basic region undergoes a conformational change upon binding to DNA. On this basis, a "scissors grip" model (Fig. 4.8) for the dimeric protein has been proposed (Vinson, Sigler & McKnight, 1989; Garrell & Campuzano, 1991).

In *Drosophila*, groups of ectodermal cells go through a "proneural" state before they become neural cells. This is achieved by a class of proneural genes that code for HLH type of proteins such as *achaete (ac), scute (sc),* and *asense (ase).* Functionally related, each gene promotes a subset of neural elements at specific positions.

Zinc finger

Miller, MacLachlan, and Klug (1985) and Brown, Sander, and Argos (1985) first reported the presence in *Xenopus laevis* of a zinc-binding domain in the transcription factor IIIA (TF III A) that transcribes 5S rRNA. Zinc ions are necessary for its site-specific binding to DNA. TF III A includes nine tandemly arrayed sequences that include two cysteine and two histidine residues separated by 12 to 14 amino acid residues. In addition to cysteine and histidine, two aromatic amino acids and one leucine are conserved as well. The amino acid residues between cysteine and histidine loop out so as to facilitate direct inter-

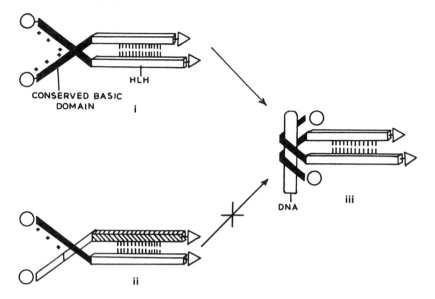

Figure 4.8. Model of helix-loop-helix heterodimers. Two basic regions are needed to bind to DNA (i). The black strip is the conserved basic region adjacent to the HLH domain (blank strip). An altered basic region is shown by a white strip (ii). Circles and triangles represent nonconversed N- and C-terminal regions. Heterodimers are formed by hydrophobic interactions (vertical lines) between HLH domains. To bind stably to DNA (vertical blank rod), both basic regions of the heterodimer probably undergo conformational change (iii). Heterodimers having only one functional basic region (ii) do not bind to DNA effectively. This is based on the scissors-grip model for DNA binding by leucine-zipper proteins. (Garrell & Campuzano, 1991)

action with DNA. TF III A contains seven to eleven atoms of zinc per protein molecule (Miller et al., 1985). Each of the short sequences binds to a zinc ion by the invariant cysteine and histidine to form a structural domain called the "zinc finger." This motif was identified by studies using X-ray adsorption spectroscopy of the zinc sites in the TF III A–5S-rRNA complex. These peptides fold in the presence of metal ions such as zinc and cobalt, but not in their absence. Two-dimensional NMR studies on a single zinc finger show that it consists of a two-stranded antiparallel β-sheet that includes two cysteine residues, a turn, and then a helix that includes two histidine residues. The N- and C-terminals of this structure are well separated from each other. It has been suggested that the zinc finger protein wraps around the DNA helix to which it binds, with the helical portion lying in the major groove of the DNA.

Three base pairs of DNA are included in one zinc-binding domain (Berg, 1990; Schwabe & Rhodes, 1991; Kaptein, 1991).

The presence of zinc fingers has been reported for glucocorticoid receptor (GC) proteins, in which histidine is replaced by cysteine. Zinc is coordinated to four sulfur atoms from cysteine at a distance of 2.32 Å (Yamamoto, 1988; Luisi et al., 1991). That zinc is essential for binding to DNA was shown by gel mobility shift and DNase I footprinting assays. It is of interest that the GC receptor and TF III A proteins are not related, except that both bind to DNA and their binding to DNA is mediated by zinc fingers.

Zinc fingers are small peptide domains consisting of about 30 amino acid residues in which the zinc atom is coordinated by various combinations of cysteine and histidine residues. Three classes of zinc fingers have been discovered so far, based on the number and location of the zinc and the way it is bound to cysteine and histidine. One type of zinc finger includes the first protein discovered with this motif – TF III A – that is responsible for transcription of 5S rRNA, and the mammalian nuclear transcription factor Sp1. It is of the CC/HH (C_2H_2) type in which a loop consisting of about 12 amino acid residues is formed by the coordination of a zinc atom with pairs of cysteine and histidine (Fig. 4.9a). TF III A has nine such fingers. In the second type, the two histidines are replaced by CC, forming the CC/CC (C_2C_2) variety. This type includes steroid, retinoic acid, and T_3 receptors, and yeast GAL4, each of which has two zinc fingers. In the receptors, the C-terminal domain binds to the ligand (hormone), and the central region binds to a specific DNA sequence in the promoter region of the respective gene. It is the first finger toward the N-terminal that carries the specificity. The receptors bind as dimers to the DNA. The N-terminal region apparently takes part in trans-activation of the gene (Fig. 4.9b). The third type of zinc finger is of the CC/HC type, containing shorter central loops (only four residues). These zinc fingers include retroviral proteins coded by *qaq* genes (Fig. 4.9c).

NMR spectroscopy and circular dichroism studies have revealed that the zinc atom is buried in the interior of the molecule surrounded by α-helices and other types of helical regions connected by β-sheets. Mutations in the finger regions prevent their binding to the target DNA sequences. Truncated regions of these proteins containing only the fingers bind to the target DNA. Interestingly, replacing the finger of the estrogen receptor by the finger of the glucocorticoid receptor creates a hybrid receptor that binds to the glucocorticoid-responsive element

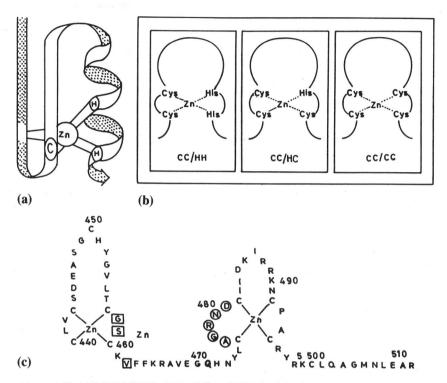

(a)   **(b)**

(c)

Figure 4.9. (a) Model of a zinc finger. (b) Types of zinc fingers classified according to the ligands used for Zn^{2+}. (c) Amino acid sequence of the two zinc-finger DNA-binding domains (DBD) of the rat GC receptor. Boxed residues are essential for discriminating between binding to GRE and ERE. Circled residues are involved in protein–protein interactions in the complex of dimeric DBD with GRE.

(GRE) but not to the estrogen-responsive element (ERE). All the specificity associated with binding to the target gene resides in the first finger (see Kaptein, 1991, for review).

Leucine zipper

The products of several oncogenes, *fos, myc,* and *jun;* a yeast regulatory protein, GCN 4; and a mammalian DNA-binding protein (C/EBP) that regulates transcription and is present in nuclei of the liver have leucine zipper motifs. They have structural similarities and are capable of sequence-specific recognition of DNA. A 35 amino acid region covering roughly half of the C/EBP DNA-binding domain contains a heptad repeat of leucine residues. It is also rich in oppositely charged amino acids (acidic and basic) juxtaposed so as to facilitate interhelical ion pairing. Because of this, it has a stable helical domain. These proteins have structural similarities. The heptad array of leucine in these regulatory proteins is similar to the heptad repeat common to proteins that adopt a coiled-coil quarternary structure, for example, keratin, lamins, and the tail of myosin heavy chain. This suggested that they may dimerize via a leucine-repeat helix. It was hypothesized that leucines extending from the helix of one polypeptide would interdigitate with those of the analogous helix of a second polypeptide to form an interlock termed "leucine zipper" (Johnson & McKnight, 1989; McKnight, 1991). When the leucines of C/EBP are changed to isoleucine or valine, C/EBP fails to dimerize and loses its sequence-specific recognition of DNA. Besides the leucine-zipper region, a 30-amino acid region immediately toward the N-terminal side of the zipper is also essential for binding to DNA. This sequence has basic amino acid residues. Dimerization of the protein facilitated by the leucine zipper positions the basic regions from two subunits, which is critical for recognition of the specific DNA sequence. Their binding sites on DNA exist as rotationally symmetric dyads.

The zipper domain is a 30–40 amino acid residue segment. It is present at the C-terminal half of the protein. The N-terminal region is basic, and is also called bCC because dimerization occurs by means of a coiled-coil mechanism. The normal repeat of an α-helical coiled-coil is about 100 residues. So the α-helices in the (bZIP) are expected to wrap around each other by about one-third of a turn (see Harrison, 1991, for review). These proteins recognize DNA sites 9–10 bp long. Contacts are made exclusively in the major groove over 10 bp (Fig. 4.10).

There is a periodic repetition of leucines in a long α-helical segment. Leucines appear on the same side of the helix every other turn; each seventh amino acid residue is a leucine, and there may be four to six such leucines. There is an unusually large number of negatively charged residues on the opposite side of the helix, as well as a highly basic amino acid sequence adjacent to the helically stacked leucines. The large number of negatively charged residues have the potential to form ion pairs. Also, the stacking of hydrophobic residues contributes to the stability of the long α-helix.

These proteins form dimers, present either as homodimers or heterodimers. Interactions between cognate DNA sequences and amino acids in and around the basic domains probably account for the specificity and relative affinities of DNA binding. Dimers are essential for

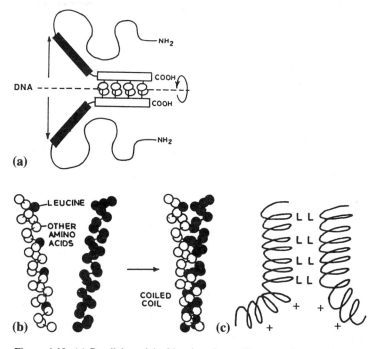

Figure 4.10. (a) Parallel model of leucine zipper. Two proteins with leucines
are arranged parallel with opposing leucines that overlap. The DNA-binding
regions of the two proteins are thus placed in a suitable position for contacting
dyad-symmetric motifs. (b) and (c) Proteins combining to form a coiled-coil.
(McKnight, 1991; © 1991 Scientific American, Inc.)

efficient binding. Impairing the zipper region prevents binding to DNA
even though the basic region may be intact. Alterations in the basic
region prevent DNA binding, but do not affect dimerization. The C/
EBP and CREB present in mammalian nucleus and GCN4 bind to DNA
as homodimers. However, the AP1-*jun* family of transcription factors
bind to DNA after forming heterodimers with *fos* protein. Though *fos*
protein has a basic region and leucine zipper (bZIP) it does not form
homodimer and does not bind to DNA alone. Leucine-zipper proteins
probably bind to DNA cooperatively, each binding to a half-site (Singer
& Berg, 1991; O'Shea et al., 1991; Pu & Struhl, 1991).

Despite the presence of a leucine zipper in *fos* protein (FOS), it does
not efficiently interact with DNA. However, FOS can interact with DNA
when complexed with *jun* protein (JUN) which also has a leucine zipper.
The domain of JUN required for interaction with FOS has been localized

to the leucine repeat. Likewise, the domain of FOS required for interaction with JUN has been localized to its leucine repeat. This FOS–
JUN heterodimer is formed via a leucine zipper.

Modifications of bases in DNA

Methylation

Three to ten percent of cytosines of the vertebrate genome is
methylated in a sequence-specific and tissue-specific manner (Ehrlich
& Wang, 1981). DNA methylation has been implicated in the control
of transcription (Felsenfeld & McGhee, 1983; Doerfler, 1983; Yisraeli
& Szyf, 1984), replication, transposition, DNA repair, chromosome
configuration (Sano & Sager, 1982; Gama-Sosa et al., 1983; Keshet,
Lieman-Hurwitz & Cedar, 1986), and inheritance of specific patterns
of gene activity (Razin & Riggs, 1980; Holliday, 1987; Cedar, 1988).
Tissue-specific differences in the methylation of certain genes (Doerfler, 1983), highly repeated satellite DNA sequences (Sano & Sager,
1982; Gama-Sosa et al., 1983), and the whole genome (Gama-Sosa et
al., 1983), however, often do not correlate with normal transcriptional
activity. Whether methylation is a primary signal for gene expression
(Razin & Riggs, 1980) or is a maintenance signal for patterns established by other mechanisms (Bird, 1986) is not clear. The possibility
of specific trans-acting factors that recognize DNA sequences containing 5-methylcytosine (5mC) in the genome and modulate various
functions of DNA is being increasingly recognized (Razin et al., 1985;
Silva & White, 1988). Trans-acting factors may also modify the meth-
ylases or interfere with them (Lyon, Buonocore & Miller, 1987).

DNA bases may undergo modifications such as methylation of cytosine in CpG doublets and in –CCGG– sequences. Such modifications
may alter the conformation of DNA and attract specific proteins to bind
to it. If such modifications occur in genes or in their flanking regions,
they may have a regulatory role in their expression. A vast amount of
information is available on the role of cytosine methylation in chromatin
structure and gene expression (Razin, Cedar, & Riggs, 1984; Yisraeli
et al., 1988; Boyes & Bird, 1991; Singer & Berg, 1991). Because the
–CCGG– methylation pattern is transmitted by somatic cell divisions,
it has been suggested that it may serve as a marker for the expression
of specific genes. There is good correlation between hypomethylation

of −CCGG− sequences and gene expression, though some genes do not show such correlation.

The methylation state of −CCGG− sequences has been studied by two restriction endonucleases, Msp I and Hpa II. Msp I cuts both methylated ($C^{\downarrow}$mCGG) and unmethylated ($C^{\downarrow}$CGG) sequences at the sites shown. Hpa II cuts only the unmethylated sequence. Though nearly 70% of −CCGG− sequences are methylated in the genome, only 30%–40% of such sequences are methylated in regions that have DNase I-hypersensitive (DH) sites. This is corroborated by the finding that the DNA of active genes is generally hypomethylated (Razin & Riggs, 1980; Razin et al., 1984; Boyes & Bird, 1991; Singer & Berg, 1991). The −CCGG− sequences are distributed at various sites of a gene such as the 5′ flanking region, the 3′ flanking region, exons, and introns as well as some distance from the gene. No general pattern of hypomethylation is seen in any specific region of active genes. The ovalbumin gene in the chicken oviduct is hypomethylated at its 3′ flanking region (Stumph et al., 1982); the active human DHFR gene is hypomethylated at its 5′ flanking region (Shimada & Nienhuis, 1984); and the phosphoenolpyruvate decarboxylase gene in rats is hypomethylated both at its 5′ flanking region and within the gene (Benvenisty et al., 1985).

The above findings are supported by several studies (see Razin & Riggs, 1980; Razin et al., 1984). It was shown that there is an inverse correlation between DNA methylation and transcriptional activity, which was confirmed by introduction of methylated genes into cells in vitro (Razin et al., 1984; Yisraeli et al., 1988). That cytosine methylation acts as a repressor of transcription was shown by experiments using the demethylating agent 5-azacytidine (5-AZT) (Jones & Taylor, 1980). When cells were treated with 5-AZT, genes were reactivated (Mohandas, Sparkes & Shapiro, 1981).

Two possible mechanisms by which methylated CpG may act as repressor have been suggested. One mechanism is based on the postulation that essential transcription factors that bind to regions containing mCpG regard it as a mutation, and therefore, do not bind to it. In support of this theory, it has been shown that CRE-binding protein (CREB) does not bind to CRE if CpG is methylated (Iguchi-Ariga & Schaffner, 1989). However, not all transcription factors are methyl sensitive. For example, Sp1 can bind and activate regardless of methylation (Hoeller et al., 1988). Murray and Grosveld (1987) have shown that CpG methylation at any site in the promoter of β-globin gene represses its expression, indicating that repression is not site specific.

The second mechanism is based on the postulation that specific protein(s) bind to a CpG-containing sequence if it is methylated and prevent binding of the essential transcription factor. This theory is corroborated by data suggesting that methylated CpG regions are protected against nucleases and restriction enzymes that act at that region (Antequera, Macleod & Bird, 1989; Levine, Cantoni & Razin, 1991; Bird, 1992). Meehan et al. (1989) have identified such a methyl cytosine-specific protein (MeCP). Boyes and Bird (1991) have further shown that MeCP-1 plays a role in the methylation-mediated repression of transcription both in vivo and in vitro. MeCP-deficient cells show much reduced repression of methylated genes.

Despite the relationship between hypomethylation and gene activity seen in higher organisms, it is unlikely that methylation is a basic mechanism of gene regulation because several eukaryotes have undetectable or negligible methylated cytosines. For example, yeast DNA is not methylated, and methylation in insects is negligible (*Drosophila* has only one 5mC in 12,500 bases). Thus it is likely that several mechanisms, independently or in various combinations, may be involved in gene regulation. These mechanisms are methylation, DNase I hypersensitivity, interaction between cis-acting elements in promoter regions and trans-acting factors, and binding of enhancers to trans-acting factors. Furthermore, certain genes may have these regulatory elements or sites at different locations or in their flanking regions.

Methylation of rat brain DNA has been studied by restricting nuclear DNA by MspI and Hpa II followed by nick-translation (Chaturvedi & Kanungo, 1985b). The rationale behind this study is that if methylation of −CCGG− sequences increases with age, Hpa II will make a lower number of cuts, and there will be lower incorporation of the label into DNA during nick-translation. Indeed it was seen that the incorporation of ^{3}H-dTMP into DNA was about 40% lower in old age. This was followed by studying the methylation of repetitive DNA sequences (RDS) of the brain by digesting it with Hpa II, Msp I, EcoRI + Hpa II, and EcoRI + Msp I, followed by end-labeling. Two prominent RDS of ~5.0 and 0.4 kbp were obtained by restriction digestion. mCpG doublets were found in both the RDS and throughout the genome. Hemimethylated mCpC doublets were also observed. Both mCpG and mCpC doublets were found in greater amounts in the old brain. This age-related increase in DNA methylation occurs both at −CCGG− sites of RDS and in the entire genome (Rath & Kanungo, 1989). Such an increase in DNA methylation may alter chromatin conformation and gene expression in the rat brain as it ages. This is corroborated by the finding that nick-translation in nuclei of the brain of old rats is lower after restriction with EcoRI (Chaturvedi & Kanungo, 1985b).

Other modifications

Even though H-bonding between the two strands of DNA requires that the protons of the bases have relatively stable locations, under certain conditions such as during replication, transcription, and binding of trans-acting factors to DNA, the two strands of DNA separate, and at such times the bases may undergo tautomerism and acquire imino or enol forms. Even though these forms occur rarely, they may serve as potential candidates for various chemical modifications whenever and wherever they occur in the genome. Of such modifications, methylation of cytosine (Wyatt, 1951) and guanine (Park & Ames, 1988) are well established. Enzymatic methylation of cytosine at C-5 by methyltransferase has been implicated in the regulation of gene expression (Felsenfeld & McGhee, 1983; Bird, 1986; Holliday, 1984). Chemical studies have shown that cytidine is acetylated at N-4 giving rise to N^4-acetylcytosine (Singer, 1980).

Kanungo and co-workers have explored the possibility of enzymatic acetylation and phosphorylation of bases, in addition to methylation because of their possible existence as imino and enol forms, even if transiently (Kanungo & Saran, 1991, 1992). Incubation of liver and brain nuclei separately with (^{3}H)S-adenosylmethionine (SAM), (^{3}H)acetyl~CoA, and (^{32}P) γ-ATP for methylation, acetylation, and phosphorylation of the bases, respectively, was carried out. Methylation of DNA bases was found to be inhibited by S-adenosylhomocysteine; acetylation was stimulated by Na-butyrate; and phosphorylated bases were dephosphorylated by alkaline phosphatase.

The mechanisms of the three modifications of the enol form of guanine are shown in Figure 4.11. The proton of the imino group of guanine moves to oxygen of the keto group (C-6) giving rise to the enol form. The resulting −OH group is reactive and may undergo enzymatic substitution in the presence of a donor, for example, SAM, acetyl~CoA, and ATP for methylation, acetylation, and phosphorylation, respectively. The −CH$_3$ or O=C-CH$_3$ or −PO$_4^{2+}$ group may substitute for the

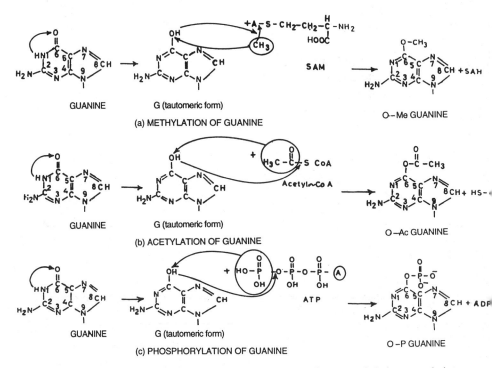

Figure 4.11. Possible enzymatic reactions leading to methylation, acetylation, and phosphorylation of guanine when it exists in tautomeric form in DNA. Thymine and cytosine may also undergo similar modifications. Such modifications in DNA may influence binding of trans-acting factors and thereby alter gene expression. (Kanungo & Saran, 1991)

proton of −OH of guanine, giving rise to the modified base. As the proton involved in base pairing is substituted, base pairing between the two strands may be prevented, thereby hindering formation of a Watson–Crick structure of DNA. Such modifications may expose or mask certain regions of the DNA, making them either accessible or inaccessible to trans-acting factors. If these modifications occur in genes, they may alter their expression. It is speculated that this may be one of the mechanisms by which specific regions of the chromatin and hence genes become permanently inactive or active by binding to trans-acting factors as is seen during differentiation of tissues. It is significant that the degree of these modifications changes in the liver and brain of the rat with increasing age (Kanungo & Saran, 1991).

Mechanism of action of regulatory elements in genes

It is now well established, at least for some of the DNA sequences located in the promoter regions of genes mentioned in Table 4.2, that they bind to specific proteins and stimulate transcription. The activation may be brought about by binding of a specific protein to the promoter region (cis-acting elements). This may displace one or more nucleosomes, which may be the reason why the site becomes hypersensitive to DNase I. The dislocation/shifting of the nucleosome makes the site accessible to other transcription factors, which may then interact directly with the proteins necessary for transcription such as those that bind to the TATA box. This results in the formation of a stable transcription complex, which may enhance the binding of RNA polymerase to the transcription start site or alter its structure so that its activity increases. This is shown in Figure 4.12.

As mentioned above, the N-terminal domain of GR interacts with other transcription factors. In the cases where the factor binds to a DNA sequence situated at a considerable linear distance that may not permit direct interaction with the TATA box factor or transcription complex, the binding of the protein may cause bending of the intervening DNA so that the two factors are brought together for interaction. This is well illustrated by an experiment in which binding of progesterone to the PRE causes bending of the DNA, causing this domain to come near the transcription initiation complex (Fig. 4.6; Theveny et al., 1987). Recently, Kornberg and his group have shown that activator proteins that bind to enhancer stimulate transcription by interacting with a general initiation factor such as TFIID. This interaction is mediated through a molecule distinct from general transcription factors.

Factors that regulate transcription

The gel mobility shift technique has greatly simplified the purification of protein factors that bind to specific sequences in the DNA. The levels of these factors, in general, are too low to be purified by the usual method of protein purification. The gel mobility shift technique utilizes the property of binding of a sequence-specific binding protein to DNA. The DNA is first cleaved by restriction enzymes to obtain small fragments that contain specific cis-acting elements. These fragments are subcloned in plasmids to obtain a large quantity of fragments, which are then removed from the plasmids and incubated with the nu-

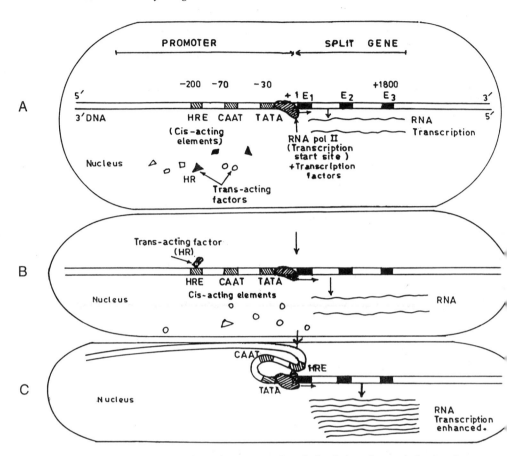

Figure 4.12. Schematic representation of stimulation of transcription in eukaryotes by a trans-acting factor. (A) Binding of RNA polymerase II and the usual TFs form an initiation complex and cause a low degree of transcription. (B) When a steroid hormone or any other effector that has effect on a gene is provided, it first binds to a nuclear/cytoplasmic trans-acting factor to form an HR complex, which then binds to a specific cis-acting element (HRE) at the promoter region of the gene. (C) Binding causes bending of the DNA and brings the HR to the initiation complex, permitting protein–protein interaction. This results in stimulation of transcription.

clear extract that contains trans-acting factors. These factors bind to the corresponding cis-acting elements. Both the control and incubated DNA fragments are subjected to slab gel electrophoresis. If the DNA fragment binds to a trans-acting factor, then its mobility will be less than that of the corresponding fragment that has not bound to the factor. The trans-acting factor can then be dissociated from the DNA fragment, quantified, characterized, and sequenced. Also, its corresponding gene can be searched for in a cDNA library. Using this technique several trans-acting factors binding to cis-acting elements of specific genes have been identified (Kadonaga & Tjian 1986; Kadonaga et al., 1987; Johnson & McKnight 1989; Mitchell & Tjian, 1989).

Regulation of gene expression

Ptashne (1986) proposed a unified model to explain the mechanism by which various regulatory factors modulate gene transcription. A protein bound at a specific site on the DNA influences transcription of a gene hundreds or even thousands of base pairs away. Proteins bound at an enhancer can effect the transcription of a distant gene. Using the example of λ phage repressor that recognizes DNA at a particular site and turns transcription on or off, Ptashne postulated that a similar situation may occur in eukaryotes as well. Even if the regulatory protein binds far away in linear terms, the intervening DNA may bend or loop and bring that protein in contact with the transcription complex to effect a protein–protein interaction. This may be responsible for positive or negative control of expression. Support for this model comes from the finding that the early gene promoter of SV40 has a TATA box and an enhancer farther upstream. Efficient transcription requires efficient binding of proteins to the enhancer. Interestingly, the proteins must be on the same side of the helix in order for the DNA to loop. Protein–protein interaction may have a cooperative effect on binding of additional proteins, thus producing a synergistic effect on transcription.

This model is supported by studies of Tjian and his co-workers. Sp1 transcription factor activates transcription (Courey & Tjian, 1988). Deletion analysis shows that it has four separate regions that contribute to activation, and all lie outside its zinc finger. At least five protein factors are needed to initiate transcription; these include TFIID, TFIIA, TFIIB, TFIIE, and TFIIF, as well as RNA pol II. These factors need to interact to initiate transcription at the correct site. Hence, introduction of one or more factors brought in proximity by DNA looping and their interaction with the initiation complex was suggested as the likely mechanism for activating transcription.

Interaction between DNA and proteins has been visualized by EM studies on progesterone receptor. When purified progesterone receptor is incubated with naked DNA containing the progesterone-responsive gene for uteroglobin, the hormone binds to its cis-acting element (enhancer at -2427 to -2376 or -2709 to -2620) in the promoter region. The DNA then loops and interacts with the transcription initiation complex. Su et al. (1991) have carried out detailed studies on the role of Sp1 in activating transcription in SV40 DNA. Sp1 protein binds to the $-GGGCGG-$ sequence in the promoter by means of its three zinc fingers. One of its two glutamine-rich regions is involved in trans-activation (Courey & Tjian, 1988). Thymidine kinase promoter carries two Sp1 binding sites, which are involved in trans-activation by Sp1 binding. If Sp1 binding sites are inserted in the downstream 3′ region, the activation of transcription is more than additive (synergistic activation) (Courey et al., 1989). This may be due to self-association of DNA bound Sp1 to generate a highly active transcription complex. Su et al. (1991) found that Sp1 binds to the SV40 promoter that carries six GC boxes in a 342-bp DNA fragment. Electron micrographs show that DNA fragments carrying Sp1 self-associate, a reaction that was seen only if glutamine-rich segments of Sp1 were present. How the DNA may loop after binding to the GC boxes is shown in Figure 4.13.

AP-2 is another sequence-specific DNA-binding protein expressed in the neural crest during mammalian development. It is a dimer and binds to the GC-rich region in DNA in this form. The DNA-binding domain, which resides in its C-terminal region, has a specific domain for dimerization. The domain for transcriptional activation resides in the N-terminal region which is rich in prolines. Thus its characteristics are different from those of other transcription activation factors (Williams & Tjian, 1991). AP-2 expression is stimulated by retinoic acid. It is absent in human hepatoma cell line HepG2, but is present in HeLa fibroblast cells. It appears to play a role in the differentiation of neural crest cells.

The present view of the mechanism of action of various transcription activation factors is the one initially proposed by Ptashne (1986) and Ptashne and Gann (1990). Biochemical and EM studies support this view. According to this model the transcription factors after binding to specific cis-acting elements loop and bring the activation domain of the factor in contact with the proteins of the initiation complex. Protein–protein interaction occurs and this somehow stimulates transcription. A model for this mechanism is given in Figure 4.14.

Several questions on the regulation of gene expression have still not been answered. The studies carried out so far have been conducted with naked DNA. If the gene to be transcribed is complexed with histones

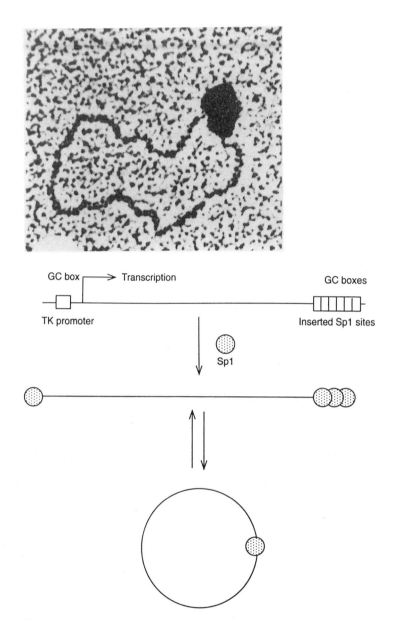

(a)

(b)

Figure 4.13. **(a)** Interaction of Sp1 protein with the engineered TK promoter. Sp1 is the wild-type protein purified from HeLa cells. The upstream-bound Sp1 and the downstream-bound Sp1 associate to loop the intervening DNA. **(b)** Possible Sp1 DNA complexes at the TK promoter. Engineered TK promoter, with GC boxes enlarged for clarity. Transcription starts ~100 bp downstream from the upstream GC box. The six GC boxes downstream are three repeats of sites III and IV in SV40. The distance between the upstream and downstream GC boxes is ~1.8 kbp. (Su et al., 1991; © *Genes and Development*)

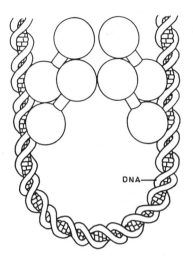

Figure 4.14. Cooperation at a distance operates when binding sites for repressor are moved apart. Intervening DNA shown looping out so dimers can touch each other and thereby help each other to bind. DNA looping may also facilitate interactions between proteins involved in gene regulation in higher organisms. (Ptashne, 1989)

and NHC proteins that are further compacted in several higher orders of structures, how do these factors find the target cis-acting elements? They also need to bind to the same side of the DNA helix to cause looping of the DNA. In the case of factors that bind hundreds or thousands of bases upstream of the transcription start site, the intervening region would contain nucleosomes. Looping of such long segments of chromatin would require much energy. Obviously, looping is an energy-dependent process.

In view of the constraints that trans-acting factors would encounter in effecting looping of the DNA after binding to the targets, it is proposed that the chromatin in a differentiated cell exists in open, flexible and extended coils and loops, which are different in different cell types. Two types of repressors are produced during differentiation – tissue-specific (TSR) and gene-specific (GSR) – to confer various differentiated states on different cells.

Differentiation results in the production of various cell types, which differ in the types of genes being expressed in them, although they may contain certain genes in common (e.g., the housekeeping genes). It is proposed that, concomitant with the process of differentiation of a cell, specific trans-acting factors are produced that are unique to each cell type. Some of these factors are required for the maintenance of the differentiated state of the cell, and some may be involved in the cell-specific response to inducers.

The TSRs have a global role. They bind at junctions of coils in such a way that certain genes in the loops are expressed and some remain in an inactive state. The specificity of the TSRs lies in keeping different segments of the chromatin in a coiled state in different cell types. This inactivates large numbers of genes in various tissues, resulting in differentiated cells in which different sets of genes have been inactivated (incompetent state), leaving a small number of tissue-specific genes in a competent state and some in an active state. The genes in the active state are expressed constitutively. These are generally the housekeeping genes and a few tissue-specific genes such as those for albumin in the liver. An example of cell-specificity is the DH-sites located 50 kbp toward the 5' end and 30 kbp toward the 3' end of the human β-globin gene domain in erythroid cells (Grosveld et al., 1987; Ryan et al, 1989; Talbot & Grosveld, 1991; Jackson, 1991). Long-range sequence organization is critical to genes in their natural chromosomal location. When human α- and β-globin genes are assayed for activity in transgenic mice, their expression is low. However, if constructs are made that contain the DH-sites, the human genes are expressed normally.

The genes that are in a competent state are not expressed because they are repressed by GSRs, which possibly bind to promoter regions of competent genes in differentiated cells. Expression of competent genes requires gene-specific inducers. This induction is mediated by one or more cell-specific trans-acting factors that need to bind to the inducer to confer specificity to the inducer for binding to specific sequences in the gene. In the hepatocytes of egg-laying vertebrates, several genes are expressed in the early developmental stage, but the vitellogenin gene is not expressed until estradiol is produced or administered. This may occur because estradiol, on binding to its receptor, may release the GSR located in the gene. A similar example of this type of activity is seen by the ovalbumin gene in oviduct cells. The interesting question that arises is: why is estradiol unable to induce the vitellogenin gene in the oviduct, but can nevertheless induce ovalbumin gene? The estradiol receptors in the liver and the oviduct are the same, as also are the EREs in the vitellogenin and ovalbumin genes. Hence, the liver and oviduct must have specific trans-acting factors that confer specificity to the E-R complex, and enable it to bind to the ERE and induce the expression of the respective genes.

In B lymphocytes, most of the genes are blocked out by TSRs, but the genes for the enzymes of the glycolytic path and the Krebs cycle and the housekeeping genes remain in an active state. The Ig genes remain in a competent state due to GSRs. Repression of Ig genes is relieved by specific antigens that act as gene-specific inducers and remove the GSR.

Occasionally, however, one or more genes that are in an inactive state and are repressed by TSRs get expressed in certain nontarget tissues when appropriate inducers are given. For example, Gillies et al. (1983) took the enhancer present between the joining and constant regions of the Ig gene, linked it to the β-globin gene and placed it in B lymphocytes. β-Globin gene was expressed even though B lymphocytes are not the target cells for the β-globin gene. This may be due to the absence of the specific TSR in B lymphocytes. Hanahan (1985) linked the enhancer and promoter of the insulin gene to the gene encoding T-antigen of SV40 virus. This construct was introduced into a fertilized mouse egg, which was then placed in the oviduct of a female mouse. The transgenic mouse that resulted expressed the T-antigen gene only in the B cells of the islets of Langerhans and not in other tissues. These studies show that the expression of genes in nontarget tissues may occur due to the release of TSRs brought about by enhancer-mediated inducers. A model for this regulation is depicted in Figure 4.15.

The above model may be tested by analyzing the proteins bound to specific genes before and after differentiation by genomic footprinting. The presence of trans-acting factors that confer specificity for the expression of specific genes may be examined by incubating the genes (chromatin) in homologous and heterologous nuclear extracts.

Conclusions

Tremendous progress has been made in gaining an insight into the structure and function of eukaryotic genes, especially after it was discovered that these genes are split. No doubt, the knowledge that was already available on the prokaryotic genes has helped in the studies of the eukaryotic gene. Moreover, the availability of restriction enzymes, and the techniques of cloning of genes and their fragments; Southern, northern, and South Western hybridizations and the gel mobility shift assay have accelerated the progress in this field during the last decade.

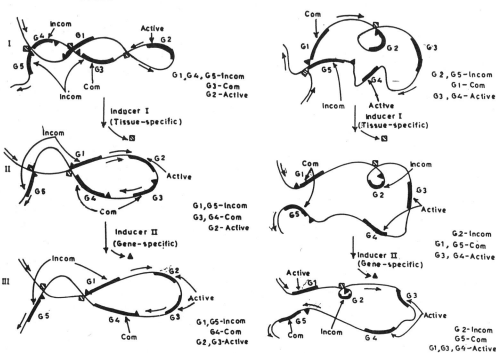

Figure 4.15. Model for tissue-specific repression and induction of genes. Brain and liver are used as models to show the differences in the genes that are expressed in them, beginning from their differentiated states. Incom, incompetent; Com, competent; G1–G6, genes 1–6; hatched boxes, tissue-specific repressor (TSR); filled triangles, gene-specific repressor (GSR). *Step* I – Because of the differences in the differentiated states of brain and liver cells, and the binding of TSRs, the loops and coils of the two chromatins are different, and genes that remain in incompetent/competent/active (expressed) states are different. Gene 2 is active in the brain, while genes 3 and 4 are active in the liver. *Step* II – Tissue-specific inducer for the brain renders the gene 4 competent by removing a TSR. Coiling of the chromatin changes. A separate inducer may render gene 5 competent in the liver by removing other TSRs. *Step* III – Gene-specific inducer, such as a steroid hormone, may remove a GSR from gene 3, not gene 4, and render the former active in the brain. Likewise in the liver, a gene-specific inducer may remove gene 1 and render it active, but not gene 5.

Several discoveries have shown that the structure and regulation of eukaryotic genes are in many ways different from those of prokaryotic genes. One important breakthrough is the discovery that a eukaryotic gene is interrupted by sequences (introns) that are not represented in the mature mRNA or the protein it codes for. The implications of the presence of introns are still not understood. Even though we have tremendous progress in our knowledge of 2%–5% of the eukaryotic genome (the part that is actually expressed), we know practically nothing about the remaining overwhelming amount (~95%) of the genome that includes repetitive DNA sequences. Both introns and the "unknown" fraction of the genome remain fertile fields for thought and experimentation.

The delineation of the processing of pre-mRNA represents significant progress. Although basically the cis-acting elements of prokaryotes and eukaryotes have much in common, the variety and form of the elements in the latter far exceed those of the former. Furthermore, the addition of a large number of receptors and trans-acting factors for increasing the repertoire of regulatory mechanisms became necessary once eukaryotic organisms with different cell types evolved. The varieties of DNA-binding domains in these proteins have aided the fine-tuning of the expression of eukaryotic genes. We are only beginning to understand how these factors locate and bind to specific DNA sequences of genes present in the chromatin complex and interact with their transcription complex.

Cell- and tissue-specific expression of genes is the result of differentiation. Analysis of the process of differentiation, the role of cis-acting elements and trans-acting factors, and the modifications that the bases of DNA undergo is necessary not only for the understanding of the normal functioning of the genes and the organism, but also for exploring the changes that take place in these processes that may lead to aging.

5

Changes in gene expression during aging

Phenotypic changes that occur in an organism after the attainment of adulthood may be traced to the changes in specific proteins/enzymes that are coded by specific genes. For example, changes in the cell membrane that alter its permeability may be due to changes in its lipid components, which are synthesized by specific enzymes. The turnover of collagen decreases with age because the level of collagenase declines. This increases cross-linking between collagen fibrils and in turn the tensile strength of collagen. Wrinkling of the skin in mammals is the result of these changes. During aging, a decrease in the enzyme tyrosinase leads to greying of the hair in mammals. The level of free radicals that damage macromolecules increases with age because superoxide dismutase (SOD) which is coded by a gene decreases. Thus, the functional changes that occur as an organism ages, whether at the organ or cellular or molecular level, are due to specific enzymes that are coded by specific genes. Environmental and intrinsic factors, however, influence the rate and the degree of such changes. Nevertheless, the fact that all individuals of a species have a specific life-span pattern indicates that the primary reason for a functional change is due to one or more genes. Therefore, an understanding of the changes in gene expression during aging may throw light on the basic cause of this process.

Techniques that were developed nearly three decades ago for the study of individual genes in prokaryotes were not suitable for the study of eukaryotic genes because of the complexity of chromatin in which they are housed. It is only since the early 1980s that appropriate techniques have been developed for probing into the expression and regulation of eukaryotic genes. During the last few years, experiments have been conducted on various genes of organisms as they age with the hope that such studies may throw light on the basic cause of aging. However, whether or not the genes being studied are crucial for aging of the organism is not known. The basic question is: Are there one or several genes that cause aging? It is unlikely that one gene is responsible for aging because, if a single gene that caused aging appeared during evolution, it might have been selected out or eliminated as it would be detrimental for the species. It is more plausible that expression of several genes may be involved in the aging process. Neither the number nor the identity of these genes is known at present. This is a challenge to all biologists who study the causes of aging.

During the past few years progress has been made on the types of changes that occur in the DNA, in general, and certain specific genes in particular, during aging in various organisms. The results of these studies on changes in DNA and genes may not tell us if they are the only causes of aging, because whether or not changes in a particular gene or a set of genes lead to aging, or such changes are the consequences of aging, is still to be elucidated. The next question would be to determine why these genes undergo changes in their structure and function. Only when this question is answered will it be feasible to devise therapies to defer the aging process. With the development of genetic engineering techniques to quantitate specific mRNAs, rapid progress has been made in studying the expression of specific genes during aging of various organisms. The alterations in gene expression under various conditions during the life span including aging are described below.

Programmed cell death

Aging of cells followed by their death occurs at two stages of the life span: (1) at the early developmental stage when organs with specific contours are being formed from a large mass of cells, and (2) after adulthood when cell activity decreases, leading to "normal" senescence. Whereas aging of cells followed by death in the early developmental period is rather a rapid process, that after adulthood generally takes a longer time, at least in long-lived animals. Another difference is that the death of cells during early development appears to be precisely programmed, since specific cells are earmarked for death, and the timing of their death is precise. This does not appear to be the case for cells that senesce and die in various organs after adulthood. Unlike brain and heart and skeletal muscle tissues that have highly differentiated and postmitotic cells, the thymus, bone marrow, and epithelium have cells that are premitotic and continue to divide and die more rapidly during the life span than the cells of the above body parts. It is not known whether the triggers for cell senescence during early development and cell senescence during normal aging are the same. Nevertheless, much useful information may be gained from studies on the programmed cell death that occurs during early development.

Programmed cell death is accompanied by a process called *apoptosis,* which is characterized by early collapse of the nucleus, extreme condensation of the chromatin, loss of the nucleolus, and shrinking of the cell. Nuclear collapse is probably due, in most cell types, to fragmentation of the chromatin (Cohen & Duke, 1984; Wyllie, 1980). This is different from accidental death of cells in which the major target organelles are mitochondria, which swell until they are dysfunctional. This leads to death and lysis of cells in a process called *necrosis.*

Owens et al. (1991) studied programmed death of immature thymocytes. Glucocorticoids and γ-radiation trigger programmed death of these cells (apoptosis). This cell death can be prevented by inhibitors of RNA and protein synthesis (Sellins & Cohen, 1987). This indicates that genes normally silent or negatively regulated are activated by inductive stimuli leading to production of proteins that act in the death process. Hence, these researchers looked for the presence of specific "death proteins" that may be synthesized to cause death of cells. The mRNAs that were expressed in thymocytes that were induced to die were used to construct a cDNA library, which was used to determine the sequences that were not present in uninduced thymocytes. By subtractive hybridization, Sellins and Cohen (1987) isolated clones complementary to mRNAs that appeared soon after the cells were exposed to radiation. Owens, Hahn, and Cohen (1991) isolated two clones, RP-8 and RP-2. RP-8 mRNA appeared within 1 hour of irradiation, and RP-2 appeared within 2 hours. The mRNAs were undetectable after 6 hours of irradiation when apoptosis and mRNA degradation occurred (Fig. 5.1). The presence of RP-8 cDNA suggests that it codes for a protein, which was found to have a zinc finger and may have a regulatory role. RP-2 appears to code for a membrane protein. Thus RP-8 and RP-2 appear to be death-associated proteins since apoptosis followed their appearance, and no other new proteins were detectable.

Programmed cell death also occurs during embryogenesis in terrestrial vertebrates whose limbs are initially shaped like flattened paddles. Individual digits are formed from this paddlelike structure by the death of parallel rows of cells; that is, cells at specific locations are marked for death (Saunders, 1966; Hinchliffe, 1981). During the development of the vertebrate brain, many more neurons are produced than are required. Nearly 40%–85% of neurons die during the formation of spe-

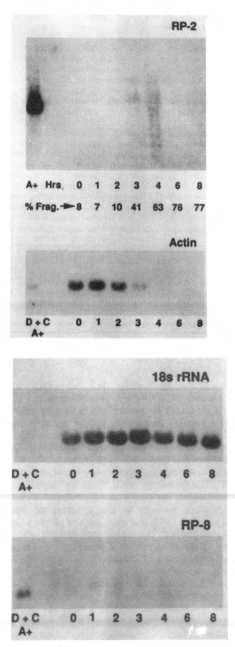

Figure 5.1. Northern hybridization used to study the temporal appearance of RP-2 and RP-8 mRNAs in thymocytes following low-dose γ-irradiation. RP-2 mRNA appears in thymocytes 2 hours following irradiation, but is no longer detectable after 6 hours. Actin mRNA also declines following irradiation and is faintly detectable, if at all, after 6 hours. In comparison, rRNA is relatively stable throughout the period when the death process is going on. (Owens et al., 1991; by permission of the American Society for Microbiology)

cific contours of the brain before birth. This process is analogous to a sculptor who takes a big piece of stone and chisels it to get the desired shape. Inhibition of transcription or translation prevents the death of these (40%–85%) neurons. Hence, expression of genes is required for the death of these cells. It appears, therefore, that programmed cell death is an active process.

During embryonic development of the free-living nematode, *Caenorhabditis elegans,* 671 somatic cells are produced, out of which 113 die prior to hatching (Ellis & Horvitz, 1986). These researchers showed that *ced*-3 and *ced*-4 genes are needed for the initiation of programmed cell death. In *ced*-3 or *ced*-4 mutants, such cells, instead of dying, survive, differentiate, and suffer fates like their wild-type counterparts. *Ced*-3 and *ced*-4 mutants appear generally normal in morphology and function, indicating that programmed cell death is not essential for nematode development. These genes define the first known step of a developmental pathway for programmed cell death and apparently determine which cells will die during development. The degeneration process is initiated in these 113 cells committed for death. They are then engulfed by neighboring cells. It was later shown that 14 genes are involved in the death of these cells (Yuan & Horvitz, 1990).

Holometabolous insects undergo metamorphosis to produce an adult from a pupa. During this process, a large number of cells comprising a significant part of the body degenerate, and giant intersegmental muscle cells (5 mm long and 1 mm in diameter) die during a 36-hour period. The trigger for the degeneration is the lowering of the level of the molting hormone, 20-hydroxy-ecdysone. By lowering the ecdysone level, death of these cells can be accelerated. Conversely, by maintaining its level, cell death can be prevented and blocked indefinitely. Therefore, ecdysone regulates the death of these cells. This has served as a good model to study programmed cell death during insect development.

Schwartz et al. (1990) found that in the insect, *Manduca,* certain unique proteins are synthesized that are needed for cell death. If ecdysterone or actinomycin D is given, these proteins do not appear. These studies show that the polyubiquitin gene is expressed during this period, and its expression is inhibited by pretreatment of the pupa with ecdysone (Fig. 5.2). Polyubiquitin protein is degraded to ubiquitin monomers, which take part in proteolysis of muscle proteins. There is a 10-fold increase in ubiquitin before metamorphosis. Ubiquitin is a polypeptide having 76 amino acid residues and is highly conserved. Its C-terminal glycine binds to the lysine of proteins which are then targeted

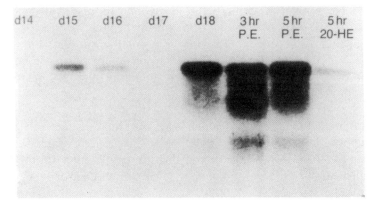

Figure 5.2. Changes in the expression of polyubiquitin gene during interseg-mental muscle development. A northern blot was probed with a clone encoding for polyubiquitin. Transcripts for this gene were transiently expressed during the early period of atrophy (days 15 and 16) and then again on day 18 coincident with the commitment to die. Treatment with 20-hydroxyecdysone repressed polyubiquitin expression on day 18. (Schwartz, 1991)

for proteolysis by an ATP-dependent nonlysosomal proteinase (for re-view, see Hershko, 1988, 1991; Rechsteiner, 1991). The degradation process of proteins is complex and involves eight steps. What is of significance here is the programmed synthesis of a protein to cause proteolysis that leads to cell death.

Programmed cell death during early development is also observed (1) at the proximal end of insect wings that leads to the production of functional wings, (2) the tail of the tadpole of frog, and (3) the pro-nephros and mesonephros of higher vertebrates.

We thus have evidence that in *C. elegans,* 14 genes are responsible for the death of the 113 cells that are necessary for the formation of an adult nematode. However, it is not known how the expression of these genes is triggered, and the nature of the gene products that cause cell death. Similarly, ecdysone clearly has a role in the death of cells during the early development of insects, but how the level of the hormone decreases to initiate metamorphosis is not known. It is well known that ecdysone stimulates puffing at certain gene loci in some insects. Iden-tification of the genes and their products would be useful in understand-ing the mechanism of metamorphosis. The trigger for the expression of the polyubiquitin gene, which precedes metamorphosis and proteolysis in the pupa, is likewise not known. Nevertheless, these studies indicate that in certain cases at least the timing of cell death is dependent on specific genes.

It is evident from the above examples that the death of specific cells during early development is essential for the production of active and reproductively viable organisms. Hence, the death of these cells is necessary for the perpetuation of the species. When cell death occurs during development, particularly before the attainment of reproductive maturity, it cannot mean that the organism is aging while developing, because the death of these cells leads to an active, vigorous, and reproductively viable organism, and not an aged and senile organism. Hence the view that aging begins even at the embryonic stage is not justified because the death of any cell that is necessary for the formation and effective functioning of an organ cannot at the same time cause aging of the organ or the organism. It is evident that natural selection has operated through specific genes in the identification of the cells that should die, and in the timing of their death.

Gene expression in vivo

Unlike the programmed cell death that occurs during early development, cell death also occurs in various organs after the attainment of adulthood, and continues till the death of the organism. In vertebrates such death is of two types: (1) death of cells occurring in premitotic cells of the bone marrow and epithelium, and (2) death of postmitotic cells, such as muscle, cardiac, and nerve cells. The first type of premitotic cells have short life spans. There is a rapid turnover of cells in these tissues throughout the life span of the organism. Bone marrow cells, which divide fast and enter the blood stream, have a relatively short life span in comparison to the life span of the organism. Fibroblast cells and hepatocytes retain the ability to divide throughout the life span, but apparently in vivo this ability decreases with age. This is evident from the fact that fibroblasts isolated from older animals divide for a lesser number of times than those from younger animals (see next section "Gene expression in vitro"). The division of fibroblasts after growth is apparently to replace the dead cells. Liver cells divide rapidly after hepatectomy to replace the cells lost by resection. These two types of cells appear to have a limited life span, but no information is available concerning their life span in vivo, although we know the number of times a human fibroblast divides in vitro, and the duration of the cell cycle that gradually is prolonged until cell division stops. It would be of value to determine the life span of these cells in vivo as an animal ages.

The postmitotic cells comprise those that stop dividing at birth or soon after in higher vertebrates and die at various times during the life span. These are the skeletal muscle cells, the cardiac cells, and neurons, all of which are highly differentiated. It is these cells that begin to die after the attainment of adulthood. Death of neurons occurs in all regions of the brain, though the rate of death apparently varies from region to region. Neurons are not replaced, and their place is taken up by non-neuronal cells. Cardiac and skeletal muscle cells continue to die, and they are replaced by fibroblasts and collagen. The death of these cells is apparently not programmed, because the same cells at the same location do not die in all individuals of a species at a given age after adulthood. The process of death of these cells does not resemble apoptosis or necrosis. There is a gradual decrease in their metabolic activity, expression of their genes, and protein synthesis as is seen from studies on various tissues. These changes lead to a gradual decline of function in various organs and the organism, a decline that leads to aging.

It has been relatively easy to study the changes in specific gene expression during early development as gene changes are rapid, localized, and well timed in all embryos. On the other hand, the changes that occur after adulthood are slow and vary from individual to individual both in location and timing. Moreover, the total number of genes that are being expressed in the embryonic stage is relatively low. In adulthood, the number of genes being expressed in the whole organism is enormous, and to single out from this great number the genes whose expression is changing is extremely difficult and complex. Also, it is necessary to determine which genes are changing in the various organs. Moreover, organs having differentiated cells differ from each other in the types of genes that are expressed in these cells, though certain genes, especially the housekeeping genes, may be expressed in all types of differentiated cells. Is the functional decline or aging in these organs due to the genes that are expressed in all of them, or to the unique genes that confer differentiated status to each of them? If it is the former, a common cause can be found. If it is the latter, the reason for the aging of different organs, especially those having differentiated cells, at different times of the life span, and the reason for different rates of their aging may be found by studying the expression of these unique genes. Then the search for a common set of genes that causes aging of the whole organism may be futile. A more desirable approach to study this problem is to compare the changes in the expression of organ-specific genes as a function of age. It has not been possible so far for researchers to carry out such extensive studies because of the enormous number of genes that are expressed in each tissue and the lack of sufficiently sensitive methods. Hence certain genes for which probes are available have been studied as a function of age with the hope that such studies may give some insight into the types of changes that occur in genes during aging.

Three types of changes may occur in the expression of genes as an organism ages: (1) Those that are being expressed may undergo a gradual decrease in expression due either to a decrease in the regulatory (trans-acting) factors that are needed for their expression or due to structural and conformational changes in the genes or the chromatin. (2) The genes that are being expressed may increase in their expression due to the above reasons which, instead of causing repression, may increase the expression of the genes. (3) The genes that have remained repressed throughout the life span, or were expressed during the developmental period but got repressed on the attainment of reproductive ability, may be reactivated and expressed after the cessation of reproduction or in old age. Examples from each category are available and in the following section we shall describe the changes that occur in the three types of genes as animals age.

Genes for enzymes

Cytochrome P-450s are microsomal enzymes that range in molecular weight from 50 to 60 kDa. They are multisubstrate monooxygenases involved in the biosynthesis and degradation of steroids, fatty acids, prostaglandins, leukotrienes, biogenic amines, pheromones, and phytoalexins. They also metabolize most drugs, chemical carcinogens, mutagens, and several environmental pollutants (Boobis et al., 1985). Mammalian liver contains high levels of these enzymes, and they are also present in extrahepatic tissues. Most cytochrome P-450s have broad but overlapping substrate specificities. At least five cytochrome P-450 gene families, each consisting of two to more than 20 genes and pseudogenes have been reported (Nebert & Gonzalez, 1985). Because of their capacity to metabolize and detoxify various pollutants, they have protective functions for the organism. Their role is analogous to the immune system in that both this system and the enzymes destroy and remove foreign substances that enter the body, though by entirely different mechanisms. Cytochrome P-450s, however, have several synthetic functions. Among the other interesting properties of these enzymes are their inducibility by several xenobiotics, such as phenobarbital (PB) and β-naphthoflavone (β-NF), and their sex differences in induction.

Because of the broad range of functions of cytochrome P-450s, researchers have studied their activities during aging of animals. Several workers have investigated the age-related changes in cytochrome P-450 (see van Bezooijen, 1984, for review), and most have reported an age-related decrease in the cytochrome P-450 content of rat liver, though some have reported an increase (Rikans & Notley, 1981). The age-related decrease is more pronounced in the male than in the female because it is influenced by testosterone in the former. Dilella, Chiang and Steggles (1982) reported that the level of cytochrome P-450 LM 2 mRNA in the liver of old rabbits is lower than that in young rabbits. Also its induction is lower in the old animals. Sun et al. (1986) studied six forms of cytochrome P-450s, forms 1–5 and form B in the liver of rats, and their inducibility by PB and β-NF. Though the activities of these enzymes and their inducibility were studied only in 2- and 52-week-old male rats, the data do show that various fractions of cytochrome P-450s undergo changes with age. These changes would alter the drug-metabolizing capacity of the liver. It is conceivable that the ability of animals to detoxify harmful substances and their sensitivity to different drugs depend, at least partially, on the levels of various cytochrome P-450s which differ in their ability to metabolize drugs. Alterations in the levels of these enzymes in old age may affect these functions.

Horbach and van Bezooijen (1991) measured the levels of cytochrome P-450 II B1 and B2 mRNAs in the liver of 3-, 12-, 24- and 36-month-old male rats by solution hybridization. No significant differences were seen in the mRNA level during aging. However, when the inducibility of the gene was studied by administration of PB, it was seen that its induction in the 36-month-old rat was nearly one-third of that of a 12-week-old rat. They further showed that in the older animals there was a longer lag period for the enzymes to be induced and to reach the maximum level. A similar decrease in induction was seen after administration of isosafrole. No such change was seen with cytochrome P-450 II A1 and A2 mRNAs. The effect of restricted diet on the level and induction of cytochrome P-450 mRNA has also been studied, as feeding restricted diet has been shown to prolong the life span (McCay et al., 1943). Two groups of rats were studied. One was given diet ad libitum, and the other was given 60% of the ad libitum diet. The rats given diet

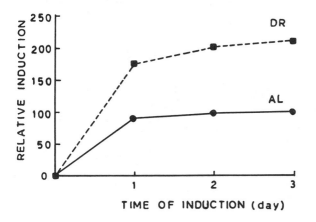

Figure 5.3. Time course of induction of cytochrome P-450 IIB1/ IIB2 mRNAs in 20-month-old rats. AL, ad libitum; DR, restricted diet. (Horbach & van Bezooijen, 1991)

ad libitum lived up to 22–30 months, and those given restricted diet lived up to 36–45 months. Induction of the cytochrome P-450 B II1 and B II2 genes by isosafrole showed that the level of the mRNA was nearly two-fold higher in the rats given restricted diet (Fig. 5.3). The level of the enzyme was also higher in the rats given restricted diet, indicating that the drug-metabolizing capacity may be higher in these rats.

Rath and Kanungo (1988) studied the induction of cytochrome P-450 (b + e) gene in 21- and 120-week-old female rats. The gene is induced in the liver, but not in the brain and kidney. Its induction by PB is far greater in the young female than in the old. In the male rat liver, the level of cytochrome P-450 (b + e) mRNA decreases with increasing age (Kanungo et al. 1992). Also, its inducibility by PB is lower in the old. Young and old male rats were administered PB, nuclei from the liver were digested by MNase, and the purified DNA fragments were subjected to electrophoresis. Southern hybridization with a 1.0-kbp-labeled probe containing 800 bp of the 5' flanking region of the cytochrome P-450 (b + e) gene showed that after PB administration nucleosomes in the promoter region are more dissociated in the young than in the older animals (Fig. 5.4). Hence, the nucleosomal organization may be more compact in the old animals, which is the reason for the lower induction by PB. Furthermore, it was shown that whereas there are three DNase I hypersensitive (DH) sites in the promoter region of the gene in the young rats, there are two in the old animals. Thus, the lower expression

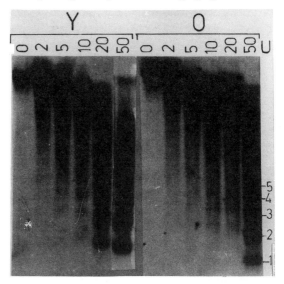

Figure 5.4. Southern hybridization of DNA fragments with 5' cytochrome P-450 gene. Nuclei of the liver of young and old rats that had been administered phenobarbitone were digested with different concentrations of MNase (2, 5, 10, 20, and 50 units/mg DNA). The DNA fragments were resolved in 1.7% agarose gel, transferred to a nytran filter, and hybridized to a ^{32}P-labeled 1.1-kbp fragment of promoter region of the gene. (Singh, 1989)

and induction of the cytochrome P-450 (b + e) gene in old rats are due to conformational changes that occur as a function of age (Kanungo et al., 1992).

Since free radicals have been implicated in the aging process due to their deleterious effects on macromolecules (Harman, 1956), the enzymes that scavenge and detoxify the free radicals have received considerable attention in aging studies. These enzymes are superoxide dismutase (SOD), glutathione peroxidase, and catalase. Vitamins C and E also detoxify free radicals. Studies have focused on the role of SOD and catalase in the aging of several animals. Mammalian SOD is a 32-kDa dimer with two copper and two zinc atoms (Cu–Zn SOD). SOD catalyzes the dismutation of the superoxide anion ($\cdot O_2^-$) to H_2O_2 and O_2. Superoxide radicals are produced as follows:

$$O_2 + e^{\cdot} \longrightarrow \cdot O_2^-$$

$\cdot O_2^-$ is very toxic and harmful to tissues. It inactivates proteins and nucleic acids. It is neutralized by SOD as follows:

$$\cdot O_2^- + \cdot O_2^- + 2H^+ \longrightarrow H_2O_2 + O_2$$

H_2O_2 is less toxic and is scavenged by catalase or peroxidase as follows:

$$2H_2O_2 \longrightarrow 2H_2O + O_2$$

Thus SOD is essential for the neutralization of the superoxide radical. Organisms with mutations in the SOD gene are more susceptible to oxidative stress. A strain of *Drospohila* carrying a null mutation in SOD has a reduced life span (Reveillaud et al., 1991). Mackay and Bewley (1989) have described an acatalasemic *Drosophila* with reduced viability. They also found that a very low level of catalase ($< 2\%$) has a significant life-shortening effect, while restoration of the enzyme to a higher level restores the fly to the wild-type phenotype. Reveillaud et al. (1991) have generated transgenic strains of *Drosophila* that overproduce Cu–Zn SOD. This was achieved by microinjecting *Drosophila* embryos with P-elements containing bovine Cu–Zn SOD cDNA. Adult flies expressed both mammalian and *Drosophila* Cu–Zn SOD. There was a slight but significant increase in the mean life span of the transgenic flies. The level of SOD expressed in adult flies was 1.6-fold higher than that of the normal insects. Thus, free-radical detoxification has a minor but positive effect on the mean life span of *Drosophila*.

Tolmasoff et al. (1980) found that there is a positive correlation between the specific activity of SOD, metabolic rate, and the maximum life span potential in mammalian species. That is, mammals with long life spans have a higher SOD activity per metabolic rate. Semsei, Ma, and Cutler (1989) and Semsei, Rao, and Richardson (1989) found that not only the activities of SOD and catalase decrease in the liver of the rat with increasing age, but also run-on transcription and the levels of mRNAs of the two enzymes are lower in the old animal. If, however, the rats are kept under dietary restriction (60% of normal diet), then both nuclear transcription and the mRNA levels increase. They also found that the average life span of the rat increases by 30% if they are kept on a restricted diet.

Srivastava and Busbee (1992) have reported that both α and β DNA polymerases of the liver from calorie-restricted aged mice show higher fidelity than those of mice fed ad libitum. However, both of these experimental old mice have DNA polymerases that have a lower fidelity than those of weanling mice. Furthermore, UV-initiated unscheduled DNA synthesis is significantly higher in hepatocytes of weanling and calorie-restricted 18-month-old mice than 18-month-old mice fed ad libitum. Thus, it appears that calorie restriction plays a significant role in preventing age-related decline in the activity of DNA polymerases, and thus may postpone the onset of mutation-associated diseases. The level of mRNA of yolk protein of *Drosophila melanogaster* does not change relative to that of alcohol dehydrogenase (ADH) as a function of age. However, in the *tra*-2 mutant (mutation for transformer gene) there is a selective loss of yolk-protein transcripts, but not of ADH transcripts from 5 to 50 days of age (Richardson, 1985; Richardson & Semsei, 1987; Richardson et al., 1987). Katsurada et al. (1982) found that the level of mRNA of malic enzyme of rat liver is lower in old animals. The steady-state levels of mRNAs of tryptophan oxygenase and tyrosine aminotransferase are reported to decrease between 10 to 25 months (Wellinger & Guigoz, 1986).

Genes for other proteins

The level of albumin mRNA of the liver is reported to increase by 121% between 3 to 36 months of age in female rats (Horbach et al., 1984) and by 50% between 6 to 29 months in the male (Richardson et al., 1985). However, Singh, Singh, and Kanungo (1990) have shown by dot-blot hybridization of RNA purified from the liver with a 1.0-kbp cDNA probe and by nuclear run-on transcription that the level of its mRNA and the expression of its gene in the male rat decrease up to 85 weeks. This was found to be due to a decrease in DNase I sensitivity and an increase in cytosine methylation of the gene as studied by Msp I/HpaII digestion followed by Southern hybridization. The levels of α-fetoprotein mRNA in the liver of young (6 months) and old (29 months) rats are, however, the same (Richardson et al., 1985).

Roy and his co-workers have studied the expression of genes that are not only age-specific but also are hormone responsive. α2μ-Globulins are plasma proteins, which are coded by a cluster of genes on chromosome 5 in the rat. The genes are expressed in the liver and code for α2μ-globulins of Mr ~18,500, which are secreted into the blood. Northern hybridization using poly-A$^+$ mRNA and cDNA probe for the gene shows that the gene is expressed from about day 40 in the liver of male rats, and the expression completely ceases after about day 900 (Fig. 5.5; Roy et al., 1983a, 1983b). There is a corresponding decrease in α2μ-globulin synthesis in the liver. The protein has five variants. Variant 2 is the dominant form which is the first to appear at puberty, and variant 4 is the last to disappear at senescence. So the genes in the cluster may

75   750   900

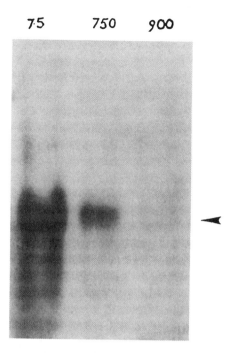

Figure 5.5. Decline in hepatic α2μ-globulin mRNA during aging. Autoradiograph shows the extent of hybridization of a ^{32}P-labeled cloned α2μ-globulin cDNA probe with electrophoretically separated hepatic mRNA (northern blot) obtained from 75-, 750-, and 900-day-old male rats. (Roy et al., 1983b)

be differentially regulated. The age-dependent expression of the gene correlates well with the appearance and disappearance of cytoplasmic androgen-binding protein (AR) in the liver. AR is absent before day 40 and disappears after about 800 days of age. It was shown that the gene becomes sensitive to DNase I after about 20 days of age and retains the sensitivity for the rest of the life span (Roy & Chatterjee, 1985). However, the gene is not expressed after about 900 days of age. Hence DNase I sensitivity is not the only requirement for its expression. Richardson et al. (1987) have shown that if rats are kept on a 40% restricted diet, both the transcription and the level of α2μ-globulin mRNA increase (Fig. 5.6). Thus, the decrease in the protein that occurs with age is due to lower transcription.

Besides the expression of the α2μ-globulin gene, two other genes of the rat liver that code for age-related, senescence marker proteins – SMP-1 and SMP-2 – have also been studied. These genes code for

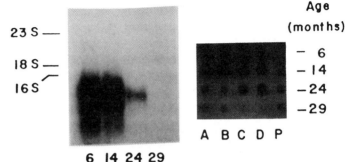

(a)

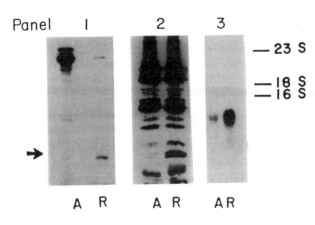

(b)

Figure 5.6. **(a)** Assay of α2μ-globulin mRNA in the livers of rats of various ages. Left panel: hybridization of ³²P-labeled α2μ-globulin cDNA probe to a northern blot. RNA (5 μg) pooled from four rats of each age was applied to the gel. Migration of rRNA standards is shown. Right panel: hybridization of the ³²P-labeled probe to RNA blots from four rats (A, B, C, and D) for each age and from a pooled sample (P). **(b)** Effect of dietary restriction on the expression of α2μ-globulin. Expression of α2μ-globulin was compared to 18-month-old rats fed ad libitum (A), or a restricted diet (R). Panel 1: SDS-polyacrylamide gel electrophoresis of α2μ-globulin (→) obtained from four rats. Panel 2: fluorograph of a SDS-polyacrylamide gel of supernatants from hepatocytes incubated with ³⁵S-methionine. Migration of α2μ-globulin corresponds to the arrow. Panel 3: hybridization of the ³²P-labeled α2μ-globulin cDNA probe to a northern blot. (Richardson et al., 1987; © 1987, The American Society for Biochemistry & Molecular Biology)

proteins of Mr 34 kDa and 31 kDa, respectively. SMP-1 gene is androgen responsive, is expressed from around 40 days after birth, and ceases expression after 750 days. However, SMP-2 is expressed until 40 days after birth and is then repressed. It is again expressed after 750 days. Thus, it is androgen repressible. Its expression is, however, high in young and adult females (Chatterjee et al., 1987). So, whereas the expression of SMP-1 decreases during senescence because of the decrease in androgen level, that of SMP-2 increases. Extending this work, Song et al. (1991) have shown that AR is expressed in hepatocytes. The steady-state level of AR mRNA correlates with the three phases of hepatic androgen sensitivity, that is, insensitivity to androgen during prepuberty, responsiveness during adulthood, and gradual loss of responsiveness during aging. Both prepubertal and senescent rats are relatively insensitive to the induction of $\alpha 2\mu$-globulin. The age-dependent decline in AR mRNA can be delayed by reducing caloric intake by 40%. Thus, it appears that changes in the androgen sensitivity of the liver during aging are due to the age-dependent expression of AR mRNA and the level of AR.

The expression of calbindins (calcium-binding proteins) is regulated by 1,25-dihydroxyvitamin D_3, the active metabolite of vitamin D_3. Calbindins are found in high amounts in the proximal intestine (calbindin-D-9K) and kidney (calbindin-D-28K) and are implicated in calcium transport in these tissues. Absorption of Ca^{2+} declines in the intestine and kidney with increasing age. Armbrecht et al. (1989) have shown by Northern and dot–blot hybridization that the level of calbindin-D-9K mRNA declines from age 2 to 6 months in the intestine, as does the level of the protein (Fig. 5.7). Surprisingly, the level of its mRNA increases between 13 to 24 months, but the level of the protein continues to decline. Apparently, the translation of the mRNA is deficient. In the kidney, the level of the mRNA for calbindin-D-28K declines between 2 to 13 months, then plateaus, as does the level of the protein. No changes are seen in the expression of the gene for calmodulin, which is also known to be involved in calcium transport.

The expression of calbindin-28 kDa gene not only changes during aging but also in neurodegenerative diseases (Iacopino & Christakos, 1990). Both the levels of its mRNA and protein decrease by 60%–80% in different areas of the brain. The decrease also occurs in certain cerebral areas that are the sites of specific diseases, as for example, the substantia nigra in Parkinson's disease, the corpus striatum in Huntington's disease, the nucleus basalis in Alzheimer's disease, and both the

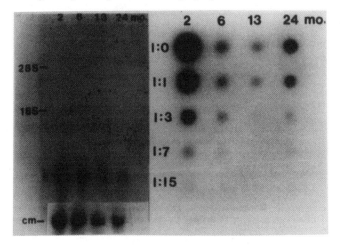

Figure 5.7. Effect of age on calbindin-D mRNA level in rat intestine. Left panel: northern blot of RNA obtained from 2-, 6-, 13-, and 24-month-old male rats subjected to electrophoresis in denaturing agarose gel. Right panel: dot blot of the RNA from same-aged animals. Blots were hybridized with a synthetic cDNA probe for rat intestinal calbindin-D-9K. For comparison, an autoradiograph of the 2,200-bp calmodulin mRNA (cm) is shown at bottom of left panel. (Armbrecht et al., 1989; © 1989 The Endocrine Society)

hippocampus and nucleus raphe dorsalis in Parkinson's, Huntington's, and Alzheimer's disease. No such disease-related decrease occurs in the cerebellum, neocortex, amygdala, or locus ceruleus. The decrease in the expression of calbindin gene and lower levels of calbindin may impair calcium buffering or intraneuronal calcium homeostasis, which may contribute to calcium-mediated cytotoxic events during aging and the pathogenesis of neurodegenerative diseases.

Friedman, Wagner, and Danner (1990) have reported significant changes in the expression of five genes in old age. They prepared cDNA libraries from 3- and 27-month-old whole mice using poly-A$^+$ mRNAs. The mRNAs were screened by hybridization with cDNAs of various tissues from these mice. The functions of these genes are, however, not known. Such approaches are useful in identifying changes in tissue-specific expression of genes.

Gresik et al. (1986) have reported that the level of EGF mRNA decreases by 75% between 12 to 27 months of age in male mice. Whittenmore et al. (1986) found that β-NGF mRNA level increases up to 47 weeks in the brain and then decreases up to 85 weeks.

Astrocytes had received little attention in age-related studies until

Goss, Finch and Morgan (1991) found a significant increase in the level of mRNA for glial fibrillary acidic protein, which is astrocyte specific. There was no change in the level of the mRNA of glutamine synthetase, which is also an astrocyte-specific protein. Astrocytes are active in many brain processes, including neurotransmitter uptake, synthesis and secretion of trophic factors, hormone targets, repair and regeneration of wounds, regulation of synaptic density, and regulation of cerebral blood flow (see Goss, Finch & Morgan, 1991, for other references). Since glial cells outnumber neurons in the brain, more research needs to be done on these cells not only to understand aging, but also to determine their role in brain function.

The role of gonadotropin-releasing hormone (GnRH), and β-endorphin (βE), an opioid peptide derived from pro-opiomelanocortin (POMC), in the reproductive function of male rats has been studied by Gruenewald and Matsumoto (1991) by measuring the expression of POMC. βE has a toxic inhibitory effect on GnRH. POMC mRNA level is found to be significantly lower in old rats. Also, the number of neurons expressing POMC gene decreases with age. Hence βE synthesis also decreases with age.

The above studies on the expression of different genes using various experimental models have made a beginning toward the understanding of the aging process at the genetic level. Merely finding out whether a gene is expressed less or more is not going to provide much insight into the problem. What is of crucial importance is to determine which genes contribute to the aging process, and why their regulation changes. In-depth studies on the promoter regions of these genes are needed to make a breakthrough on this problem.

Genes for acute-phase proteins

Dramatic changes in the concentrations of specific plasma proteins occur following inflammation, which is usually triggered by infection, wounding, or chronic diseases. The response to inflammation takes place primarily in the liver and is known as the "acute-phase reaction" in which hepatic genes are stimulated to produce proteins that enter the plasma. Plasma proteins that increase are called "positive acute-phase reactants," and those that decrease are called "negative acute-phase reactants." In an acute-phase reaction, the levels of the following proteins increase: T-kininogen, fibrinogen, α1-acid glycoprotein, C-reactive protein, hemopexin, ceruloplasmin, and haptoglobin; and the following proteins decrease: albumin, transthyretin, 2-HS-glycoprotein, transferrin, and apolipoprotein A-1. Human genes for α1-acid glycoprotein, serum amyloid-P, and C-reactive protein have been studied in transgenic mice during the acute phase reaction. They are expressed in the same manner as in normal humans even though the corresponding plasma protein genes of the host (i.e., mice) are expressed differently during inflammation. Thus it appears that the human genes that respond to inflammation are recognized by mouse transcriptional factors. However, the response of these genes to the acute phase are different in mice and humans.

Sierra et al. (1989) have taken a step forward toward identifying the genes that may specifically change in expression during aging. They constructed a cDNA library from the mRNA of the liver of an old rat (24 months) and by differential screening with those of young rats (10 months) and old rats (24 months) found that the expression of T-kininogen gene is higher in the old animals (Fig. 5.8). It codes for T-kininogen which is a major acute-phase protein whose level increases in the blood after an acute-phase reaction (inflammation). It is also called cysteine protease inhibitor or thiostatin. Nuclear elongation experiments showed that the increase in the expression of the gene in the liver in old age is controlled at the transcription site. Age-dependent induction operated preferentially at one of the three transcriptional start sites of the gene, as seen from RNase mapping and S1 analysis. In the adult rat, T-kininogen gene is induced after an acute-phase reaction, which is due to stimulation of transcription from two of the transcription start sites. One of these sites is common to both aging and the acute-phase reaction. A specific sequence in one of the transcriptional start sites of the gene has been implicated in its higher expression in old age. A nearby sequence has been reported to be associated with the increase in the level of the protein in response to inflammation. Thus independent changes in genes may occur during aging and inflammation. The question that needs to be answered is what triggers transcription from specific sites.

Sierra et al. (1991) have also studied the expression of T-kininogen gene in young and old rats after subjecting them to an acute-phase reaction by administration of turpentine for 24 hours. No significant difference with age in the expression of the gene or the level of the protein in the serum was seen. On the other hand, the level of another acute-phase protein, γ_2-macroglobulin, significantly decreased in the serum, both during aging and after an acute-phase reaction.

The expression of T-kininogen gene appears to be related to aging,

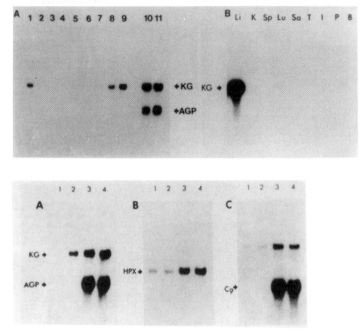

(a)

(b)

Figure 5.8. **(a)** Development and tissue-specific expression of T-kininogen gene by northern blot analysis. Total liver RNA (5 μg) from animals of various ages (A) and from different tissues obtained from 23-month-old animals (B) was electrophoretically fractionated, blotted, and hybridized to radioactively labeled probes. (A) Developmental expression. Lanes 1–9 are for rats of following ages: newborn, 21 days, and 4, 6.5, 10, 13, 18, 21, 23 months. Lane 10: 10-month-old rats, acute-phase induced. Lane 11: 23-month-old rats, acute-phased induced. (B) Tissue specificity. Lanes are for tissue from liver, kidney, spleen, lungs, salivary gland, testis, intestine, pancreas, and brain. **(b)** Northern blot analysis of expression of T-kininogen (KG) and α_1-acid glycoprotein (AGP) genes in young, old, and inflammation-induced animals. (A–C) Lanes 1–4: RNA from young, old, young inflammation-induced, and old inflammation-induced rats. Probes used were KG and AGP for A; hemopexin (HPX) for B; and complement C9 for C. (Sierra, Fey, Guigoz, 1989; by permission of the American Society for Microbiology).

since it is silent up to 18 months in the rat as seen by northern hybridization (Sierra, Fey & Guigoz, 1989), and abruptly increases from about 20 months. The level of T-kininogen in the serum is also higher in the old, while that of albumin does not change with age. These researchers also found that there is no change in the methylation status of the gene that may lead to its higher expression in the old rat. It is believed that this increase may be due to age-related changes in humoral factors, because no differences are seen in its expression when hepatocytes from young and old rats are cultured in vitro. Another gene that was found to increase in expression with age is that for a subunit of cytochrome oxidase. It, however, does not show an abrupt increase in expression as seen for the T-kininogen gene. The role of T-kininogen in the acute-phase reaction is not clear. However, these studies show that the response of certain genes to stress in old age is different from that of young. This may destabilize the physiology of the animal in old age and lead to functional decline.

Positive acute-phase reactants are always elevated in the plasma of older people even when they are healthy and free of infection and trauma (Milman, Graudel & Andersen 1988). Haptoglobin, a plasma protein that binds to free hemoglobin released from ruptured old red blood cells, is a positive acute-phase reactant, which is synthesized in the liver. Its mRNA level increases in the liver by twofold in 16-month-old mice, and by fourfold in 28-month-old mice after inflammation, though their albumin mRNA level shows little change. Ceruloplasmin mRNA also increases like that of haptoglobin. Ceruloplasmin binds to free copper, and protects the brain and heart from copper toxicity. The elevation of ceruloplasmin and haptoglobin mRNA levels indicates that these are related to age and to inflammatory responses.

The transferrin gene that codes for the plasma protein, transferrin, has been well studied. Transferrin binds to ferric iron and transports it to target tissues. Human transferrin gene has been fused to a bacterial chloramphenicol acetyl transferase (CAT) gene and introduced into the mouse genome. Since CAT gene is easily assayed, it served as the reporter gene. The transgenic mice carrying the transferrin-CAT construct were administered intraperitoneally lipopolysaccharide (LPS) which produces an inflammatory response. In humans, transferrin is a negative acute-phase reactant, and decreases during inflammation. However, in rodents it is a positive acute-phase reactant. In the transgenic mice, administration of LPS lowered the CAT activity, but the transferrin level increased. This opposite effect may be due to differences in the regulatory sequences of the mouse and human transferrin genes.

Heat-shock protein genes

The heat-shock response of genes is a useful parameter to study age-related changes in gene expression after exposure to a stress in order to determine whether an old individual withstands the effect of heat stress as well as a young individual. Heat-shock response is an adaptive mechanism that enables cells to survive a variety of environmental stresses that otherwise could be lethal. Heat-shock response was first shown in *Drosophila* (Ritossa, 1962) and was later found to be a universal phenomenon, both in prokaryotes and eukaryotes. Specific proteins (heat-shock proteins, HSP) are induced after exposure to heat. These proteins are also induced by heavy metals, alcohol, oxidative stress, free radicals, and amino acid analogs. These HSP inducers have the common property of causing the accumulation of denatured or damaged proteins within a cell. Finley et al. (1984) suggested that HSPs bind to these proteins after heat shock and prevent their aggregation and consequent cellular damage. Pelham (1986) then proposed that HSPs are involved in the assembly and disassembly of proteins both during normal growth and after heat shock. Thus HSPs prevent cell death due to thermal stress by protecting sensitive proteins from denaturation, and also restore non-heat-shock protein synthesis following recovery from heat shock.

The HSPs are highly conserved. In eukaryotes, induction of HSPs is controlled at the transcriptional level by a positively acting heat-shock factor (HSF) that binds to a heat-shock element (HSE) located in the promoters of genes for HSPs (see Ang et al., 1991, for review). HSPs act as chaperons or aids for attaining and maintaining the tertiary structure of other proteins in the cytosol (Pelham, 1986). One of the major HSPs is a 70 kDa protein (HSP 70), which binds to other unfolded proteins and is present in mitochondria and endoplasmic reticulum. It has a weak ATPase activity. The *hsp* 70 gene is amplified. Whereas in response to a stress the rate of transcription, particularly of rRNA genes, RNA processing and stability, and mRNA translation are significantly reduced, the expression of *hsp* genes is enhanced (Lindquist & Craig, 1988). Also, DNA synthesis and mitosis are arrested.

HSP 60 and HSP 90 are also HSPs and serve as chaperons for maintaining the tertiary structure of cell proteins. It is speculated that under non-heat-shock conditions, a large pool of HSP 70 accumulates. It binds to HSF and degrades it so that *hsp* mRNA is not transcribed. Under heat-shock conditions, HSP 70 binds to other unfolded, misfolded proteins, so that HSF can bind to the *hsp* 70 gene promoter and transcribe it (Hightower, 1991). Because most stresses that induce expression of *hsp* genes disturb protein structure and folding, it is believed that HSPs, especially HSP 70, protect cells from the adverse effects of stress on protein structure and function. This is supported by the fact that inhibition of the expression of *hsp* 70 gene increases the sensitivity of cells to stress.

Several workers have studied the expression of *hsp* genes as a function of age of various organisms to determine whether their increasing inability to tolerate stress is related to *hsps*. Fleming et al. (1988) found that aged *Drosophila* express more HSPs than younger flies. Niedzwiecki, Kongpachith, and Fleming (1991) studied the expression of *hsp* 70 gene during aging of *D. melanogaster*. They subjected the flies to 37°C for 30 minutes and studied the expression of the gene. Its expression increased in flies up to 23–28 days of age. In older flies the expression declined as measured by northern blot. When subjected to 25°C after a 30-minute heat shock, there was a quicker decrease in *hsp* mRNA level in younger flies. Older flies continued to synthesize HSP 70 protein for a longer period after return to 25°C. These researchers also observed that after thermal stress there was more protein–ubiquitin conjugate and increased proteolysis in old flies (Niedzwiecki & Fleming, 1990). Therefore, prolonged *hsp* 70 expression after heat shock may have greater damaging effects in older organisms.

The data on the expression of *hsp* genes in young and old rats after exposure to higher temperature are, however, different. When young and old rats were exposed to 40°C for various durations, the rise in colonic temperature of old rats was slower than in young rats. The expression of *hsp* 70 gene was lower in the brain, lung, and skin of old rats after a specific time period. However, if the colonic temperature of the old rat reached 42°C, it produced the same amount of *hsp* 70 mRNA as that of the young rat at 40°C, as studied by Northern blot hybridization (Blake et al., 1991a).

Since the *hsp* 70 gene is expressed after exposure to stresses other than temperature, Blake et al. (1991b) studied its expression after subjecting rats to mobility restraint stress. Rats were restricted from mobility for 30 minutes to 6 hours, RNA was purified from various tissues, and the expression of *hsp* 70 mRNA was assayed by northern blot hybridization. Following these conditions of stress, the *hsp* 70 mRNA was greatly increased in the adrenal gland. It was also higher in the pituitary, but no significant differences were seen in the brain, muscle, liver, lung, heart, testis, kidney, spleen, and thymus. The expression of *hsp* 70 gene sharply increased within 30 minutes of restraint and gradually declined up to 4 hours and then increased up to 6 hours. The restraint was not continued beyond 6 hours. In comparison, the expression of *hsp* 27 mRNA did not show much increase. Hypophysectomy suppressed *hsp* 70 gene expression after stress, and administration of ACTH induced its expression in the adrenal. Thus, ACTH is a physiologic regulator of the expression of *hsp* 70 gene in the adrenal. Restraint-induced *hsp* 70 expression declines with age. Thus it appears that the hypothalamic – pituitary – adrenal (HPA) axis, which is responsible for maintaining homeostasis under stress, stimulates the expression of *hsp* 70, and that the HPA activity decreases with age. Blake et al. (1991a) have further shown that the cellular heat-shock factor (HSF), which binds to the heat-shock element (HSE) located in the promoter region of the *hsp* 70 gene, is induced in the adrenal after restraint stress. It would be of interest to determine whether such an induction of HSF occurs after heat stress, and if this induction decreases with age.

Since feeding a restricted diet prolongs the life span of rats, Richardson and his co-workers (Wu et al., 1991) studied the expression of *hsp* 70 gene in the hepatocytes of rats kept on a restricted diet that extended their life span by 30%. Hepatocytes were isolated from these rats and incubated for 30 minutes at 42.5°C to induce a heat-shock response and were then returned to 37°C. Synthesis of HSP was measured by incorporation of ^{35}S-methionine into proteins, and *hsp* 70 mRNA was quantitated by northern blot hybridization. HSP 70 protein was found to be elevated by about threefold after exposure to 42.5°C. No detectable synthesis of HSP 70 protein occurred in the hepatocytes of control rats, though the induction of HSP 70 synthesis appeared to be nearly 30% lower in old rats in comparison to that of young rats (Fig. 5.9). Furthermore, the *hsp* 70 mRNA was induced over 100-fold after heat shock. The induction was about twofold greater in the young animals after 60-minute heat shock. It is significant that the induction of both HSP 70 protein and *hsp* 70 mRNA is greater in hepatocytes of 28-month-old rats fed a restricted diet in comparison to those of 28-month-old control rats (Fig. 5.10). Thus dietary restriction enhances induction of *hsp* 70 gene and enables the hepatocytes to respond to heat stress. Also, cells from animals kept on a restricted diet respond more efficiently to thermal stress. It is necessary to find out why restricted diets increase the efficiency of adaption of cells to various types of stress.

Genes for structural proteins

Fibronectins (FNTs) are high molecular weight glycoproteins having two subunits, each of 220 to 240 kDa. They are joined by S–S

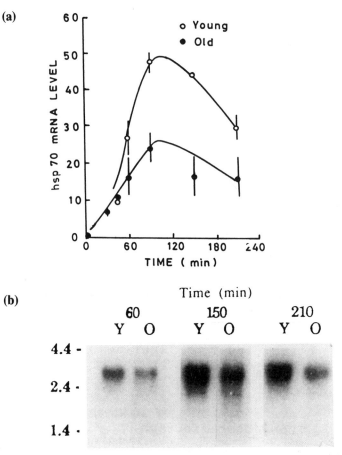

Figure 5.9. **(a)** Time course for induction of *hsp* 70 mRNA levels by hepatocytes isolated from young and old rats. Hepatocytes isolated from 6-month (young) and 28-month (old) rats were subjected to heat shock at 42.5°C for 30 minutes. The hepatocytes were returned to 37°C and RNA was isolated from hepatocytes at the times shown. **(b)** Autoradiograph of the northern blots of RNA pooled from four to six rats. (Wu et al., 1991)

bonds to form the dimer. FNTs are involved in differentiation, migration, and adhesion of cells and also wound healing, hemostasis and tumor metastasis (Hynes, 1985). They are synthesized in the liver and secreted to the plasma. A single FNT gene is present per haploid genome in rodents and mammals (Tamkun et al., 1984). The gene is located in chromosome 2 in the rat, spans ~70 kbp, and has 50 exons. Subunit variants of FNTs are generated by alternative splicing of primary transcripts (Schwarzbauer et al., 1983; Schwarzbauer, 1991). The promoter

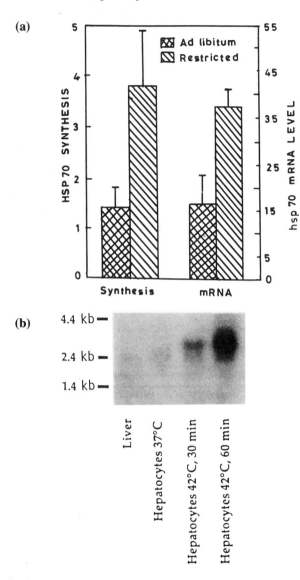

Figure 5.10. **(a)** Induction of HSP 70 synthesis and mRNA levels by heat shock for hepatocytes isolated from 28-month-old rats fed an ad libitum or a calorie-restricted diet. Bars give the mean and standard error of data collected from four animals for each dietary regimen. **(b)** Comparison of *hsp* mRNA levels in liver and hepatocytes before and after heat shock. Northern blot of RNA isolated from liver and hepatocytes incubated at 37° and 42°C. (Wu et al., 1991)

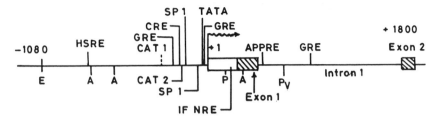

Figure 5.11. Map of rat fibronectin gene showing cis-acting elements and restriction sites in the 5' region (adapted from Patel, 1987). Restriction sites: A,
Ava I; E, Eco RI; P, Pst I; PV, Pvu I. (Hatched box), exon. CRE, cAMP
response element; CAT, CCAAT; GRE glucocorticoid response element;
HSRE, heat shock response element; APPRE, acute phase protein response
element; IFNRE, interferon response element; Sp1, binding site.

region of the FNT gene contains several cis-acting elements including
TATAA, CCAAT, and GGGCGG as well as responsive elements for
cAMP (CRE), glucocorticoid (GRE), and heat shock (HSRE) that take
part in transcriptional regulation (Patel et al., 1987) (Fig. 5.11).

Studies on the expression of the FNT gene of rat liver as a function
of age have been carried out by Singh and Kanungo (1991, 1993),
who show that the level of FNT in the plasma of 125-week-old rats
is significantly lower than that of 20-week-old rats. This finding may
be due to a lower rate of transcription, as seen by run-on nuclear
transcription, and a lower steady-state level of its mRNA as measured
by slot–blot hybridization. Digestion of liver nuclei by DNase I followed by restriction of the DNA and Southern hybridization with a
1.2-kbp probe encompassing the promoter region of the gene show
that there are three DH-sites: one overlapping the CRE, the second
in the TATA region, and the third in the first intron. The DH-sites
in the promoter are more susceptible to DNase I in the young than
in the old. FNT is not synthesized in the brain and transcription of
its gene is negligible. Also its promoter has no DH-sites. There is no
difference in the methylation status of 5'-CCGG-3' sequences of the
promoter region of the gene in the liver as seen by Msp I/Hpa II
digestion of nuclei followed by Southern hybridization with the probe.

Northern blot analysis shows that the FNT transcript is about 8.0 kb.
It is induced by dexamethasone (Singh, 1991). Nuclear run-on transcription assay shows that its transcription in the liver is less than 10%
of that of albumin (Singh & Kanungo, 1991). Of much interest is the
finding that the CRE in the promoter of the FNT gene binds to nuclear

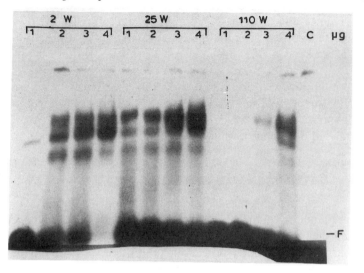

Figure 5.12. Mobility shift assay of 25-mer dsDNA containing the cAMP response element of the fibronectin gene. Nuclear extract (1–4 μg) of the liver of 2-, 25-, and 110-week-old male rats was titrated with a 25-mer dsDNA containing the CRE (–TGACGTCA–) in the promoter region of the FNT gene. (Singh, 1991, unpublished data)

factors whose levels decrease with increasing age. A 25-mer synthetic oligonucleotide containing the CRE

5′–AATTCCCCGTGACGTCACCCGGACA–3′
3′–GGGGCACTGCAGTGGGCCTGTTCGA–5′

was labeled with ^{32}p and incubated with increasing amounts of nuclear liver extract from immature, adult, and old rats. This was followed by a gel-mobility shift assay (Fig. 5.12). The same amount of 25-mer DNA was also labeled and incubated with a given amount of nuclear liver extract from rats at each of the three ages. It was then titrated with increasing amounts of cold 25-mer DNA and resolved by gel-mobility shift assay (Singh, 1991). Both studies showed that three specific trans-acting protein factors are present in the nuclear extract that bind to the CRE, and their levels decrease with increasing age. Furthermore, the binding specificities of the three trans-acting factors undergo differential changes during aging. Thus, the decrease in the expression of the FNT gene during aging is due to both conformational changes that occur in its promoter and a decrease in the levels and affinities of trans-acting nuclear factors that are required for regulation of its transcription.

Wismer et al. (1988) have studied the expression of β-actin and β-tubulin genes in rat brain using cDNA probes for these genes. Northern and dot–blot hybridizations of their mRNAs with the labeled probes do not show any significant change as a function of age. The levels of these mRNAs are different in various regions of the brain, but there is no age-related difference in their levels.

Oncogenes

Several oncogenes have been implicated in tumor growth. Since the frequency of tumors is higher in old age, Kanungo and his co-workers studied the expression of several oncogenes in various tissues of young and old rats by dot–blot hybridization (Rath, Jaiswal & Kanungo, 1989). Many oncogenes are expressed in many tissues of both young and old rats. What is of interest is that the mRNA levels of several oncogenes, such as c-*fos,* c-*myc,* c-Ha-*ras,* c-*mos,* and c-*abl,* are higher in old heart (Fig. 5.13). Such overexpression of oncogenes may make the heart more prone to hypertrophy. The expression of c-*fos* and c-*myc* oncogenes was also reported to be higher in the brain and liver of old rats. Further studies are necessary to find out why these "undesirable" genes are expressed in old age and how they affect the functioning of the heart and brain.

Fujita and Maruyama (1991) measured the levels of mRNAs of the genes for c-*jun* and c-*fos* in the liver of 6- to 7-month- and 24- to 25-month-old male and female rats by northern hybridization. The c-*jun* mRNA was undetectable in the liver of young rats, but its presence was significant in old rats of both the sexes (Fig. 5.14). The level of c-*fos* mRNA was also significantly higher in the old animals. These genes have TPA-responsive elements in their promoter. TPA (12-*O*-tetradecanoylphorbol-13-acetate) is a xenobiotic. Whether or not the elevation of the mRNA levels in old age is due to exposure to TPA is not known. However, since the protein coded by c-*fos* is implicated in cell proliferation, it is significant that its level increases with age. This increase may either be due to elevated transcription of the genes, or a decrease in the degradation of the mRNAs. The rate of transcription may be examined by nuclear run-on transcription. One of the factors that is known to increase degradation of poly-A$^+$ mRNAs is a protein that binds to the ATTTA sequence at its 3'-untranslated region (Meijlink

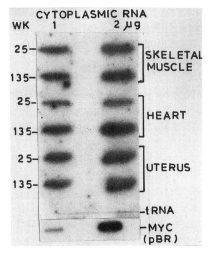

(a)

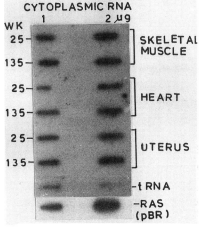

(b)

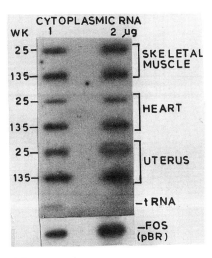

(c)

Figure 5.13. Determination of the levels of mRNAs of **(a)** c-*myc*, **(b)** c-Ha-*ras*, and **(c)** c-*fos* in the skeletal muscle, heart, and uterus of young (25 weeks) and old (135 weeks) female rats by slot–blot hybridization. Cytoplasmic RNA at levels of 1 and 2 μg was used. tRNA was used as negative control. pBR containing c-*myc*, c-*fos*, and c-Ha-*ras* was used with respective blots as a positive control. (Y. K. Jaiswal, 1988, unpublished data)

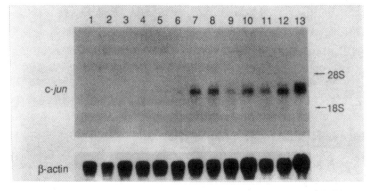

Figure 5.14. Age-associated changes in c-*jun* expression in the rat liver. Lanes 1–3: adult (6–7 months) male; lanes 4–6: adult (6–7 months) female; lanes 7–9: aged (24–25 months) male; lanes 10–12: aged (24–25 months) female; lane 13: lung. (Fujita & Maruyama, 1991)

et al., 1985; Ryseck et al., 1988). If the level of this protein decreases, the mRNA would accumulate and cause the synthesis of the protein.

Transgenes

Transgenic mice have been used to determine what type of changes occur in gene expression during aging, and how responses of genes are altered under various types of stress. Responses of older people to inflammation, toxic metals, and hormones often differ from those of younger people. Bowman, Yang and Adrian (1990) have studied the response of human genes to stress by inserting them into the mouse genome. In these studies, human genes were introduced into a fertilized mouse egg, and the expression and modulation of the human genes were followed during aging of the mouse. The responses of human genes in the mouse to stresses such as hormones, heavy metals, and inflammation were also analyzed in old mice. The expression of several human plasma proteins such as α1-antitrypsin, transthyretin, transferrin, amyloid-P component, C-reactive protein, and apolipoprotein A-1 was studied. These are useful approaches that are likely to throw light on the expression and modulation of human genes under various experimental conditions, which otherwise are not possible to study in humans. The above in vivo studies on the expression of different genes using different experimental models have been useful for the understanding of the aging process at the genetic level.

Gene expression in vitro

Hayflick and Moorhead's (1961) discovery of the limited replicative ability of human diploid fibroblasts (HDF) in in vitro culture
has stimulated much research on the in vitro senescence of cells. HDFs
in vitro are not subjected to the same environmental influence as HDFs
in vivo. Therefore, the data from these studies cannot fully explain the
mechanism of in vivo aging. Nevertheless these studies have provided
certain important insights into the mechanism of aging in vitro at the
genetic level. Two recent reviews are useful in this context (McCormick
& Campisi, 1991; Goldstein, 1990).

Animal cells appear to have the ability to replicate only for a limited
number of times, after which they stop dividing, senesce, and die. Therefore, they have a finite life-span phenotype; that is, eukaryotic cells have
an intrinsic limit to their proliferative capacity. This cellular senescence
may be a manifestation at the cellular level of organismal aging, because
for a given population of cells, the maximum number of times the cells
can divide (population doubling) is generally inversely proportional to
the age of the organism, and directly proportional to the maximum life
span of the species (Rohme, 1981). Also, cells derived from donors with
heritable premature aging syndromes, such as Werner's syndrome and
progeria, divide for a lower number of times and senesce sooner in
comparison to age-matched normal donors. Cellular senescence thus
appears to be a mechanism for restricting the growth of the organism
and production of tumors, because it is seen that genetic events that
confer an infinite life-span (immortal) phenotype to cells also increase
their susceptibility to tumorigenic transformation. The malignant tumor
cells generally have an immortal phenotype. These observations suggest
that the finite life-span phenotype is under genetic control.

Dominant nature of cellular senescence

Cell-fusion studies involving normal and immortal human cells
have shown that the hybrids have limited life spans (Harris et al., 1969;
Bunn & Tarrant, 1980; Pereira-Smith & Smith, 1983; Sager, 1989). This
suggested that the phenotype of cellular senescence is dominant, and
immortality results from recessive changes in growth-regulatory genes
of the normal cells.

Another important finding was that when a heterokaryon cell hybrid
was formed by having one old and one young HDF nuclei in a single

cytoplasm, initiation of DNA synthesis in the young nucleus derived from actively proliferating young HDF was inhibited, but its ongoing DNA synthesis was not inhibited (Norwood et al., 1974; Burmer, Ziegler & Norwood, 1982; Yanishevsky & Stein, 1980). If the senescent cells were treated with inhibitors of protein synthesis before fusion, inhibition of DNA synthesis in young nuclei was prevented. It appeared that this inhibitory effect was due to proteins. This conclusion was supported by the finding that when poly-A$^+$ RNA from senescent HDF was microinjected into proliferating HDF, DNA synthesis was inhibited (Lumpkin et al., 1986). These studies strongly suggest that cellular senescence is dominant, and that immortality results from recessive changes in growth inhibitory genes.

Studies on fibroblast cells in in vitro culture suggest that the decline in their proliferative capacity followed by senescence may be controlled by genes. Two phenomena are observed in fibroblast cells in culture: (1) Cells stop dividing; this is followed by (2) functional changes that lead to their senescence. Occasionally, however, from among the senescing cells arise immortal variants spontaneously. The frequency of appearance of such variants depends on the species. For rodents, this frequency is ~1/10^{-5} to 1/10^{-6} cells. For human and chicken cells, senescence is almost complete. Spontaneous immortalization of human fibroblasts has not been reported, and if immortal cells are formed, the rate may be as low as 1/10^{-12}. However, immortal human fibroblasts can be produced by exposure of the culture to carcinogens or viral oncogenes. The frequency of occurrence of immortal cells under such conditions is 1/10^{-6} to 1/10^{-7}. Wright, Pereira-Smith and Shay (1989) and Radna et al. (1989) have shown that human fibroblasts can be immortalized by the large T antigen of simian virus 40 (SV40). However, this immortalization requires the continued presence of the T antigen or induction of T antigen by dexamethasone. In its absence, the cells cease to proliferate, are arrested at the G1/S boundary, and lose the immortal phenotype. Thus the arrest at the G1/S boundary is bypassed by the property of T antigen to stimulate DNA synthesis. This suggests that T antigen is responsible for only one of at least two steps required for human fibroblasts to express the immortal phenotype.

The most consistent phenomenon observed in cellular senescence is that the finite life span is dominant and the immortal phenotype is recessive. This may be because immortalization requires the loss of both alleles of a dominant gene. Though this may be an explanation for the very low frequency and recessive nature of immortalization, it is unlikely that the senescent phenotype is under the control of a single locus.

Chromosome implicated in cellular senescence

The identification of the chromosome that is involved in senescence was studied by microcell fusion by which single human chromosomes are introduced into immortal human cell lines. Whole-cell fusion studies have shown that the tumorigenic phenotype of HT-1080, a fibrosarcoma cell line, can be suppressed by fusion with normal human fibroblasts. A comparison of nontumorigenic cells reveals that there is a correlation between the loss of chromosomes 1 and 4, and reexpression of tumorigenic potential (Benedict et al., 1984). Human chromosome 1 induces senescence in an immortal hamster cell line (Sugawara et al., 1990), whereas human chromosome 4 limits the proliferative life span of three immortal human tumor cell lines (Ning et al., 1991). In cell hybrids formed between HDF and an immortal Syrian hamster cell line, most cells exhibit a limited life span comparable to that of HDF, indicating that cellular senescence is dominant in these hybrids. The hybrid clones that do not senesce do not have both copies of human chromosome 1, whereas all other chromosomes are present in at least some of the immortal hybrids. Further, the introduction of a single copy of human chromosome 1 to the immortal hamster cell by microcell fusion causes typical signs of cellular senescence, whereas transfer of chromosome 11 has no effect on the growth of the cells. These data suggest that chromosome 1 may have a role in cellular senescence. Furthermore, Pereira-Smith and Smith (1988) found from cell-fusion studies that there are at least four complementation groups for immortalization. Thus, cellular senescence has a genetic basis. It would be of interest to determine whether the introduction of chromosome 1 into immortal cells causes methylation of certain genes in the cells that revert to the senescent phenotype.

Proto-oncogenes and cellular senescence

The ability of cells to proliferate depends on both positive (stimulatory) and negative (inhibitory) signals received from outside the cell. These signals exert their effects on cell proliferation ultimately through regulation of expression of specific genes. Proto-oncogenes encode positive growth regulators, whereas tumor suppressor genes are believed to encode growth inhibitors. A dynamic balance between stimulatory and inhibitory mechanisms is required for growth and proliferation of cells. On the other hand, for terminally differentiated and senescent cells that have irreversibly lost the ability to divide, stimulatory signals may be suppressed, or inhibitory controls may be constitutively active. Both these mechanisms may operate in senescent cells. Thus, the gradual decline in the proliferative ability of cells after several rounds of active proliferation may involve a shift in the balance between positive and negative regulatory signals, or a derangement in the homeostatic balance between these two types of regulatory signals. This may directly or indirectly have an effect on the critical genes whose expression may initiate the aging process (Kanungo, 1980).

The expression of several proto-oncogenes increases when fibroblasts are stimulated to proliferate by external factors (Hofbauer & Denhardt, in press). When quiescent fibroblasts are stimulated to proliferate by mitogens, three proto-oncogenes – c-*fos*, c-*myc*, and c-*ras* – are induced before DNA synthesis begins (Greenberg & Ziff, 1984; Armelin et al., 1984; Muller, Bravo & Burckhardt, 1984; Holt et al., 1986; Riabowol et al., 1988; Lu et al., 1989). If anti-sense RNAs of the proto-oncogenes are transfected or antibodies against their protein products are microinjected, proliferation of the cells is inhibited. Especially, their DNA synthesis is inhibited. The DNA content of senescent fibroblasts that have stopped dividing is the same as that of G1 phase cells (2n). External stimulatory signals fail to initiate DNA synthesis in these cells. What is the role of c-*myc*, c-*fos*, and c-*ras* in the senescent fibroblasts? Seshadri and Campisi (1990) found that the expression of c-*fos* is repressed at the transcriptional step in senescent fibroblasts, though the expression of several other genes is the same as those of early passage cells. However, for c-*myc* and c-*ras* proto-oncogenes, the basal and growth-factor inducible expressions do not change as the fibroblasts senesce in culture (Rittling et al., 1986; Seshadri & Campisi, 1990). Therefore, it appears that senescence is due to selective repression of a few genes and not due to the general breakdown of the entire transcription machinery.

The proto-oncogene, c-*fos*, encodes a nuclear protein. It is an essential component of the activator protein-1 (AP-1) family of transcriptional regulatory complexes. When the fibroblasts are stimulated by a mitogen, the expression of c-*fos* increases within minutes, but quickly returns to the low basal level that is characteristic of quiescent and proliferating cells. It is likely, therefore, that it may control the expression of other genes that are required for proliferation and whose expression follows that of c-*fos*. However, when the c-*fos* expression vector was microinjected into senescent fibroblasts, there was high level of Fos protein in the nuclei, but DNA synthesis was not stimulated (McCormick & Campisi, 1991). On the contrary, microinjection of SV40 T antigen vector stimulated DNA synthesis in all fibroblasts and immortalized the cells (Wright et al., 1989; Radna et al., 1989). Therefore, repression of c-*fos* expression in senescent cells may be one of the many regulatory changes that are responsible for irreversibly blocking proliferation. Besides the repression of c-*fos* proto-oncogene in senescent fibroblasts, thymidine kinase gene is also repressed (Chang & Chen, 1988). Whether it is the cause or consequence of growth arrest is not known.

Another important protein that has received attention in relation to fibroblast proliferation and senescence is *cdc2* (p34) kinase, which is implicated in the regulation of cell cycle. It was shown in yeast that it is essential for both initiation of DNA synthesis and completion of mitosis. It is a highly conserved protein both structurally and functionally. Its expression increases several hours after mitogenic stimulation of early passage human fibroblasts and several hours after the decline of c-*fos* expression (Lee et al., 1988). Expression of the *cdc2* gene is not seen after the fibroblasts senesce (Fig. 5.15; Stein et al., 1991). Young quiescent HDF have low levels of *cdc2* mRNA and protein. After serum stimulation, these cells produce *cdc2* mRNA and protein, and synthesize DNA and undergo mitosis. However, if senescent cells are serum-stimulated, no *cdc2* mRNA accumulates, and they do not undergo mitosis. If *cdc2* DNA is microinjected into senescent HDF, the cells still do not synthesize DNA. These cells are deficient in cyclins A and B, which are cofactors of the protein kinase activity of p34 protein. It is seen that the senescent HDF have little or no cyclin A and cyclin B (Fig. 5.16). These deficiencies may be responsible for the lack of DNA synthesis in senescent HDF. Whether or not the repression of c-*fos* and *cdc2* proteins occurs by the same or different mechanisms is not known. It is clear, however, that failure of senescent fibroblasts to respond to growth stimulatory signals involves a selective block of gene expression and not a global block to expression of genes.

The possibility of the presence of growth inhibitors in senescent fibroblasts has been explored. Maier et al. (1990) reported that the expression of the cytokine, interleukin IL-1, in human endothelial cells markedly increases as the cells undergo proliferative arrest and senesce in culture. If these cells are continuously exposed to IL-1 anti-sense

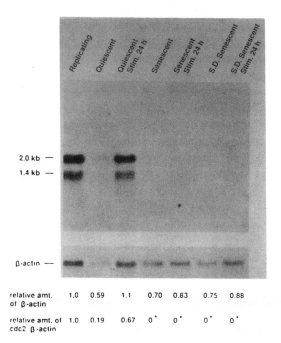

Figure 5.15. Steady-state levels of *cdc*2 transcripts in senescent, quiescent, and replicating IMR-90 cells. Poly-A⁺ RNA was prepared from IMR-90 cells in different growth states and analyzed for the presence of *cdc*2 transcripts by northern blot analysis. Amount of *cdc*2 mRNA in each sample was normalized to the amount of cytoplasmic β-actin in that sample. Relative amount of *cdc*2 β-actin is given at the bottom of the figure. (Stein et al., 1991)

oligomer DNA, the level of IL-1 protein declines and the proliferative life span increases. IL-1 has no signal sequence and is not secreted. If IL-1 is added exogenously, the growth of endothelial cells is inhibited. Therefore, the increase in IL-1 level may contribute to the arrest of proliferation and senescence of human endothelial cells. But the growth inhibitory effect of IL-1 is cell-specific. It acts, on the contrary, as a mitogen for fibroblast cells. Also, addition of the IL-1 anti-sense oligomer does not extend the proliferative capacity of the endothelial cells indefinitely, and removal of the oligomer results in the generation of a senescent phenotype. Hence, the genes responsible for inhibition of endothelial cells may be different from those required for fibroblasts, and senescence of human endothelial cells in vitro may be regulated by intracellular IL-1.

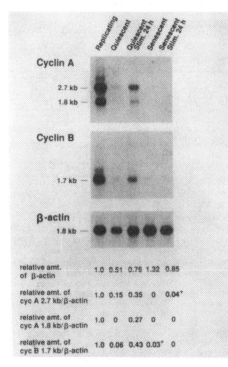

Figure 5.16. Steady-state levels of cyclin A and cyclin B transcripts in senescent, quiescent, and replicated IMR-90 cells. A northern blot of poly A⁺ RNA from IMR-90 cells in different growth states was prepared and analyzed. The same blot was probed sequentially for cyclin A, cyclin B, and β-actin transcripts. (Stein et al., 1991)

Response to heat stress

Response of cells to heat stress has been studied in cell cultures (Liu et al., 1989). Late-passage fibroblast cells showed lower heat-shock response. The cause for the lower response was analyzed by using IMR-90 fibroblast cells. The cells were heat shocked at 42°C for different periods, the cell extract was prepared, and centrifuged at 100,000 g. The supernatant was used for gel retardation assay to identify heat-shock proteins that bind to a 24-mer consensus cis-acting heat-shock DNA element (HSE). The consensus HSE element has alternating GAA or TTC blocks arranged at two nucleotide intervals, GAANNTTC. It was found that there was a distinct decrease in the level of HSE-binding protein in old cells. The molecular weight of the protein was 83 kDa as determined by the South–Western technique. It was shown that the HSP is involved in transcriptional activation of *hsp* genes in IMR-90 cells. A lower level of HSE binding protein, therefore, may be responsible for the decrease in the expression of the heat-shock gene (Choi et al., 1990).

Fargnoli et al. (1990) also studied the response of lung and skin fibroblasts to heat shock in cell cultures. The induction of HSP 70 mRNA and protein in fibroblasts of 24-month-old rats was significantly lower than that of 5-month-old rats when the cells were exposed to heat stress at 42.5°C for 90 minutes. Freshly excised lung tissue from old rats also showed a lower response. Thus senescence is controlled at the genetic level and is not a random event.

Inhibitors of cellular proliferation

Another interesting protein factor that inhibits cell proliferation is the retinoblastoma susceptibility gene (*Rb*) product. The loss of *Rb* function and absence of its protein predisposes the susceptible cells to tumorigenic transformation. The *Rb* gene codes for a protein, which is nuclear and is phosphorylated. It is relatively unphosphorylated in quiescent cells. When stimulated by a mitogen, it is phosphorylated. This is immediately followed by DNA synthesis. T antigen of SV40 and other oncogenes bind to unphosphorylated Rb protein. Therefore, underphosphorylated Rb protein presumably inhibits cell proliferation. This inhibition is reversed when Rb protein is phosphorylated and binds to T antigen and other oncogene products (Buchkovich, Duffy & Harlow, 1989). The finding that Rb protein fails to become phosphorylated in senescent fibroblast cells (Stein, Beeson & Gordon, 1990) suggests that senescent cells may contain a constitutively active growth inhibitory Rb protein, which prevents DNA synthesis. This is supported by the fact that oncogenes that immortalize cells encode proteins that bind and possibly functionally inactivate at least two cellular proteins that are tumor-suppressor gene products, p53 protein and Rb susceptibility gene product. It is of interest that *cdc2* kinase phosphorylates Rb protein, in addition to many other proteins (Pines & Hunter, 1990). Also, *cdc2* is induced when phosphorylation of Rb protein occurs just prior to DNA synthesis. The discovery that the 5' promoter region of c-*fos* gene has a negative-acting Rb response element (Robbins, Horowitz & Mulligan, 1990) suggests that there may be a connection between the repression of c-*fos* and expression of *cdc2*, and underphosphorylation of Rb protein in senescent cells.

The irreversible block to proliferation of cells in culture, especially HDF, resembles terminal differentiation. The cells are arrested at G1/ S boundary. Senescent HDF are generally longer and less motile. An important characteristic of senescent HDF is that a very small fraction of cells cycle and those that do spend a longer time in G1 phase. DNA synthesis is blocked in these cells. However, if these cells are infected with SV40, DNA synthesis is reinitiated (Gorman & Cristofalo, 1985). Both viral and host DNA replication require factors encoded by the host cell. Therefore, the DNA synthesis machinery of the host cell is not impaired but only turned off. Even though growth-stimulating factors such as epidermal growth factor (EGF) and platelet-derived growth factor (PDGF) fail to stimulate the division of senescent HDF, the membrane receptors for these growth factors remain intact with respect to their affinity. Also, autophosphorylation of tyrosine kinase remains intact (Goldstein & Shmookler-Reis, 1985).

Cause of senescence of fibroblasts

Two possibilities for the cause of senescence of in vitro cells have been considered by Holliday (1990). One is that cellular aging is a stochastic multistep or multiple-hit process in which defects in macromolecules or other cellular components accumulate to a point where cell division is no longer possible and death eventually follows. Although many defects may be repaired or removed, such mechanisms cannot be efficient enough. Therefore, there is an overall increase in defects at each generation. In the second possibility, it is assumed that age is determined by a program or a counting mechanism, which controls the number of cell divisions before proliferation ceases. So there is a positive regulation of cell division, but this is not necessarily associated with cell death. In both stochastic and programmed aging, Holliday argues that genes control the life span. In the former, the accuracy of macromolecular synthesis and the efficiency of repair or removal of defects is dependent on proteins or enzymes that are coded by genes. In the latter, there is an in-built genetic mechanism for specifying a proliferative life span.

The concept of programmed aging based on cumulative cell divisions was initially supported by the finding that cells cease replication because they produce an inhibitor of the initiation of DNA synthesis (Norwood et al, 1974; Stein & Yanishevsky, 1979; Rabinovitch & Norwood, 1980). The main conclusions that can be drawn from these studies are that (1) senescent cells are blocked at the G1 stage of the cell cycle; (2) senescent cells, when fused to young cells to produce heterokaryons, inhibit DNA synthesis over a 2- to 3-day period; and (3) in similar heterokaryon experiments, senescent cells also inhibit DNA synthesis in the nuclei of some immortal cell lines, showing that senescence is a dominant characteristic of cells with respect to proliferative ability. This suggests the presence of an inhibitor in senescent cells.

However, senescent cells, when fused to immortal cell lines such as HeLa cells or SV40 transformed fibroblasts, synthesize DNA (Burmer et al., 1982; Pereira-Smith et al., 1985). Therefore, it was proposed that the immortal cells produce an initiator of DNA synthesis, or repress the inhibitor so that the senescent cells can proceed from the G1 to S phase. Experiments with enucleated cytoplasts from senescent cells and with inhibitors of protein synthesis support this interpretation and suggest that the inhibitor is a protein.

What then causes the cessation of proliferation of HDF, which on reaching confluency, become quiescent and become blocked at G1? How do cells blocked at G1 differ from senescent cells? It is suggested that in quiescent cells the same inhibitor is present, but its effect is overridden by growth factors or mitogens in young cells. As the cells proceed through their life span, their ability to respond to growth factors declines. Terminal-phase III cells fail to respond to growth factors (Phillips, Kaji & Cristofalo, 1984).

The presence of an inhibitor of DNA synthesis has been shown by microinjection of poly-A^+ mRNA from senescent, quiescent, or young cells into young cell recipients, and measurement of the extent of inhibition of DNA synthesis by ^{3}H-thymidine incorporation into DNA. Significant inhibition was found with poly-A^+ mRNA from senescent cells, whereas that from quiescent cells caused considerably lower inhibition, and that from young cells did not (Lumpkin et al., 1986). Later studies have shown that a nuclear protein, statin, which is present only in senescent and quiescent cells may be the factor that inhibits DNA synthesis (Wang & Lin, 1986; Wang, 1987; Ching & Wang, 1988). It was suggested that the 57 kD statin is associated with the switch from a growing state to a nongrowing one, and blocks the transition from G1 to S, possibly by blocking DNA synthesis.

Data from several experiments do not support the DNA inhibition theory. In the heterokaryons between young and senescent cells, ^{3}H-thymidine labeling experiments were done only for short periods (2–3 days). Whether the senescent nuclei are dominant over longer periods is not known. In heterokaryons between senescent HDF and HeLa cells or SV40-transformed cells, DNA synthesis is reactivated in the latter, but in the former, DNA synthesis is inhibited and the heterokaryons have limited life span. Skin fibroblasts of patients suffering from Werner's syndrome have a significantly lower population doubling (PD) in comparison to the age-matched normal controls. Werner's syndrome is a premature aging disease, which is inherited as an autosomal recessive mutation. This type of mutation implies a lack of function. However, when Werner's fibroblasts are fused with normal cells, DNA synthesis is inhibited in the normal nuclei of the heterokaryon (Tanaka et al., 1980). How does a recessive defect cause the premature appearance of a dominant phenotype? Furthermore, statin has a close homology with the protein elongation factor, EF-1, which is unlikely to have a direct effect on transition from cycling to noncycling state (see Warner & Wang, 1989, for review). The finding that the activity of EF-1 declines in phase III cells makes statin an unlikely candidate for inhibiting passage from G1 to S (Cavallius, Rattan, & Clark, 1986).

With regard to the phenotype of senescent HDF, the frequency of aneuploids, polyploids, and chromosome aberrations increases significantly (Thompson & Holliday, 1978). The rate of replicon elongation is significantly reduced (Petes et al., 1974). HDF are unable to maintain a constant level of 5mC in their DNA which may contribute toward their finite life span (Holliday, 1986; Fairweather, Fox & Margison, 1987; Catania & Fairweather, 1991).

Is the limited life span of cultured cells due to aging? There are two opposing views that are to be considered in this context. According to the inhibitor hypothesis, aging of cells is regulated by a specific mechanism; that is, the appearance of an inhibitor confers a dominant function on the senescent phenotype, and the finite life span is positively regulated. Why and after how many divisions of the cell this function is acquired need to be addressed. If cessation of cell division means aging, then are the differentiated cells such as neurons and skeletal muscle cells, which stop dividing at a very early stage of development, senescent cells? These differentiated cells stay on and perform specific functions efficiently for a long period, which in higher mammals is several years. That these cells are aged cannot be accepted because the criteria to distinguish between an aged and young cell is not just the ability to divide. It should include the ability to carry out various functions. Another possibility is that the limit to growth has been imposed as a barrier against uncontrolled proliferation leading to tumors (Dykhuizen, 1974). This also implies that a limited life span has a positive function, which does not necessarily have a relationship to aging or senescence.

The opposing view is the one originally proposed by Hayflick (1965), who believes that the limit to fibroblast growth in culture is a manifestation of aging at the cellular level, and understanding the mechanism of aging in cultured cells may give an insight into the mechanism of aging of cells in vivo. The maximum life span of several animals correlates well with the number of PD of their fibroblasts in vitro (Rohme, 1981). There has been, during mammalian evolution, a continued increase in the maximum life span. Whereas the maximum life span of the earliest mammals is believed to have been about 1.5 years (Cutler, 1975) that for humans is about 100 years. The prolongation of the life span could have been achieved by the development of mechanisms whereby the normal cell functions are maintained for longer and longer durations. This could be attained by the evolution of an increasingly efficient homeostatic machinery for maintaining various functions, especially transcription, translation, and turnover of the macromolecules that carry out various functions in cells after they have ceased dividing. Ultimately, as with any dynamic system, the corrective/repair mechanisms not being 100% perfect, entropy increases and a gradual breakdown of the homeostatic machinery occurs, leading to senescence as proposed by Kanungo (1975, 1980).

Despite the fact that there seems to be a positive correlation between the PD number of fibroblasts of some species and their maximum life span, the number of divisions may not have any relationship with the senescence phenomenon. In higher vertebrates the cells cease to divide early in development and differentiate to perform specialized functions. It is not known how many PDs the neurons and skeletal muscle cells have undergone before they cease division. Apparently, it is these cells that contribute to senescence of the organism more than the cells that continue to divide throughout the life span, such as bone marrow and epithelial cells. It is necessary to find out the number of times neurons and skeletal muscle cells divide in vivo before they differentiate, and to see if the cessation of their division is also controlled by chromosome 1 or chromosome 4. If it is found to be so, then the finding that one of these two human chromosomes has a dominant role in the inhibition of cell division would be useful in understanding differentiation and cessation of cell division. We may then have to look for genes that cause senescence of these cells.

Studies on similar lines have been carried out on fibroblast cells derived from patients suffering from Werner's syndrome (WS) who show premature aging and death. These patients also exhibit impaired somatic growth and an early appearance of age-related pathologies such as atherosclerosis, malignancy, osteoporosis, soft-tissue calcification, insulin-resistant diabetes mellitus, lenticular cataracts, and skin atrophy and ulceration (Epstein et al., 1966). It is a rare autosomal recessive disease. The WS mutation has pleiotropic effects, and patients and their cells show many differences from those of normal persons. Goto et al. (1992) have shown a close linkage of the WS mutation to the tissue plasminogen activator gene on chromosome 8.

Human diploid fibroblasts (HDF) from WS patients grow more slowly than normal cells, develop senescent morphology earlier, and have highly reduced replicative ability in comparison to the age-matched controls (Goldstein, 1978). Murano et al. (1991) constructed a cDNA library from late-passage HDF to determine whether any genes are overexpressed in these cells in comparison to those of normal HDF. By cross hybridization and partial DNA sequencing, they identified 18 independent gene sequences, nine of them known and nine unknown. The known genes include $\alpha_1(I)$ and $\alpha_2(I)$ procollagens, fibronectin (FNT), ferritin heavy chain, insulinlike growth factor binding protein-3 (IGFBP-3), osteonectin, plasminogen activator, inhibitor type 1, thrombospondin, and crystallin. Among the nine unknown sequences, two were novel. Northern blot studies showed higher levels of these 18 mRNAs in WS HDF in comparison to those of the normal HDF. Of these five genes, $\alpha_1(I)$ procollagen, FNT, and IGFBP-3 showed higher mRNA levels both in WS HDF and late-passage normal HDF than in early passage cells. Thus, senescence of both WS HDF and normal HDF is accompanied by overexpression of a few genes, which are similar, and play a role in senescent arrest of cellular replication in late passage cells, and in the genesis of WS. It is likely that in normal biological aging and accompanying diseases, such genes may be expressed.

Changes in gene expression

Giordano and Foster (1989) made a cDNA library from mRNA of late-passage WI-38 cells, and differentially screened with that of early passage cells. A 2.2-kb transcript was found that increased in late-passage cells. It was homologous to EF-1. This decreased in serum-starved cells. The role of this transcript in aging is not known.

Decline in proliferative ability may not only be due to the decline of transcription of specific genes, but also to transcription factors. Hence, Riabowol, Schiff and Gilman (1992) studied the activity of transcription factor AP-1. AP-1 is one of a family of closely related proteins of about 40 kDa that bind to the consensus sequence 5′ TGANTCA 3′. The proto-oncogene, c-*jun*, belongs to the AP-1 gene family. AP-1 alone binds to DNA poorly, but in association with the Fos protein it binds to the site effectively as a heterodimer (Fos-Jun/Fos-Ap1). Fos protein has a highly conserved positively charged DNA-binding domain and leucine zipper motifs. It cannot form homodimer and cannot bind to DNA. Riabowol et al. (1992) found that AP-1 activity was required for human cells to proliferate in response to serum, and that its activity was reduced in the cells prior to the total cessation of cell division and attaining complete senescent phase (Fig. 5.17). Also, the induction of Fos protein was reduced prior to cessation of cell division. Furthermore, the composition of AP-1 changed, and more of jun–jun homodimer was formed instead of Fos-jun heterodimer. This may be due to posttranslational modifications of Fos protein such that it cannot form heterodimer with jun protein. It is suggested that changes in AP-1 activity may contribute to the inability of fibroblasts to proliferate.

The studies on cells in vitro, especially on human fibroblasts, have provided useful information on two specific events in the life span of cells: (1) Senescence is a dominant character, and (2) the proliferative ability of human fibroblasts is controlled by a gene(s) located in chromosome 1. However, cessation of cell division is not senescence. So the cause of senescence even in cultured cells remains unanswered. Moreover, the type of senescence the fibroblasts undergo in vitro, the biochemical and the molecular changes that occur after they cease to proliferate and enter phase III may or may not be similar to the changes that occur in them in vivo. Hence it is necessary to study, using appropriate markers, the aging of cells in vivo. Following the life span of fibroblasts, hepatocytes, neurons, and cardiac and skeletal muscle cells in their in vivo environment would be useful in understanding the process of senescence.

Reactivation of genes in old age

Genes in X chromosomes

A gene that is dormant in the adult can be activated later in life; this was first demonstrated by Cattanach (1974). During the early

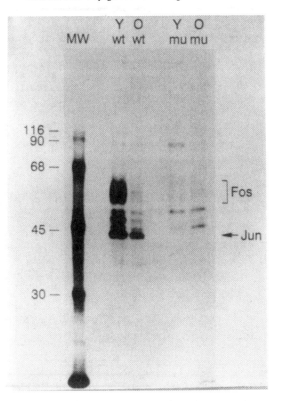

Figure 5.17. DNA affinity precipitation assay of AP-1-binding protein in young and old fibroblasts. Young (Y lanes, MPD = 42) and old (O lanes, MPD = 77). Hs68 fibroblasts were deprived of serum for 48 hours then placed in a medium containing 10% FBS and (^{35}S)-methionine and incubated with 0.5 μg of biotinylated wild-type (wt lanes) or mutant (mu lanes) AP-1 oligonucleotides. Bound proteins were recovered with streptavidin–agarose beads and analyzed by electrophoresis on a 12.5% polyacrylamide gel followed by fluorography. (Riabowol, Schiff & Gilman, 1992)

development of female mammals, one of the two X chromosomes is randomly inactivated in each cell. This inactivation is stably inherited by mitotic division, so that all descendants have the same active X chromosome (either paternal or maternal), and the same inactive one. Hence, each adult female is a mosaic with regard to the X chromosome. This shows that although both X chromosomes have the same genetic information and reside in the same environment, one of them remains inactive throughout the life span. An autosomal gene translocated into an X chromosome is also subject to inactivation if that X chromosome becomes inactive.

Tyrosinase gene in the mouse, which is responsible for hair pigmentation, is autosomal and is located in chromosome 7. In the mutant mouse, in which the gene is translocated to the inactive X chromosome and is inactivated, the mouse becomes an albino. Cattanach (1974) found that as the mouse grows older, the tyrosinase gene becomes reactivated and the animal's hair becomes pigmented. This is the reverse of what is found in normal aging when dark hair becomes white. How this gene is reactivated in an inactive X chromosome while other genes remain inactive is not known. Whether or not other genes of the inactive X chromosome are reactivated in old mice needs investigation.

Wareham et al. (1987) studied the X-linked *spf* (sparse fur) mutation in the gene coding for ornithine carbamoyl transferase (OCT). A mutation in the OCT gene located in the inactive X chromosome produces an abnormal OCT. The normal enzyme tests histochemically positive at pH 7, but the enzyme of mutant mice tests negative at pH 7. Using a histochemical method, Wareham et al. (1987) found that the gene is reactivated in mice with increasing age. Reactivation is not linear with time, but accelerates with aging. The reactivation of the gene is not gradual but an "all-or-none" process because no intermediately staining cells are found. Thus, the OCT gene becomes fully active or remains inactive. It is necessary to know if this reactivation is due to demethylation. The X-linked HPRT gene in humans is, however, not reactivated (Migeon, Axelman & Beggs, 1988); neither is the gene for glucose-6P-dehydrogenase located on this chromosome demethylated. If changes in DNA methylation are related to age, it does not appear to be applicable to germ-line cells.

It is likely that the methylation status of −CCGG− sequences at critical sites may turn the gene "on" or "off". The demethylating agent, 5-azacytidine (AZT), can reactivate genes in the inactive X chromosome in somatic cell hybrids (Jones, 1986), and there is convincing evidence that X-chromosome inactivation is related to differential methylation of cytosine in the DNA of the two X chromosomes (Monk, 1986). It would be interesting to know which −CCGG− sequence in the inactive OCT gene undergoes demethylation, and whether this is due to an increase in the level of demethylase. Whatever the mechanism might be, it is clear that the reactivation of the gene is due to its destabilization by a random event. If such a reactivation is random, it is difficult to comprehend how such events could cause "normal" aging because it is more or less a gradual process. The normal mouse that does not have this mutation also ages. Hence random destabilisation or dysdifferentiation cannot lead to "normal" aging.

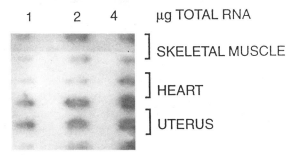

1   2   4   μg TOTAL RNA

] SKELETAL MUSCLE

] HEART

] UTERUS

Figure 5.18. Levels of α-skeletal actin mRNA in the skeletal muscle, heart, and uterus of young (25 weeks) and old (135 weeks) female rats as determined by slot–blot hybridization. Level is higher in the heart of old rats, but lower in skeletal muscle and uterus of same-age animals. (Jaiswal & Kanungo, 1990)

Such reactivation does not appear to occur in the genes located in the inactive X chromosome in the human female. The X-linked hypoxanthine phosphoribosyltransferase (HPRT) gene was tested in women who are heterozygotes for mutations in this locus. Heterozygotes who are more than 10 years old have excess HPRT in their skin fibroblasts, but this excess does not increase with age. The silent locus does not reactivate spontaneously in culture, but only in response to AZT, a potent inhibitor of methylation. This age-related reactivation is not a feature of all X-linked loci, and may have species-, tissue- and locus-specific determinants. Also, inactivation may be more stable in genes that are constitutively expressed in all tissues than in genes like those for OCT, which have limited expression.

Increased activation of other genes

Certain genes that have low activity in an adult tissue become increasingly active in the old organism, and certain genes are expressed in nontarget tissues in the old. Jaiswal and Kanungo (1990) measured the level of α-skeletal actin (α-SKA) mRNA by RNA–DNA slot–blot hybridization using a cDNA probe for α-SKA. They found its level to be higher in the heart of old rats (Fig. 5.18) in comparison to that of young rats. However, nuclear run-on transcription was not found to be different. α-SKA is the major isoform of the skeletal actin of fetal cardiac ventricle, whereas a different isoform is predominant in the adult heart (Izumo et al., 1988). Under experimental aortic coarctation in the adult, α-SKA mRNA is rapidly re-expressed followed by hypertrophy of the heart. Hypertrophy occurs due to increase in cell size without cell division. It is an adaptive process employed by postmitotic myocardial cells against increased hemodynamic pressure overload. The continuous pressure overload of the heart throughout life span may stimulate increased expression of α-SKA gene, which in turn may lead to hypertrophy of the heart. This may lead to lower contractibility of the heart and affect its function. This is consistent with the finding that c-*myc*, c-*fos* and *hsp* 70 genes are activated first as an early response to pressure overload, and this is followed by overexpression of the α-SKA gene (Izumo, Nadal-Ginard & Mahdavi, 1988).

Cutler (1982) reported that certain genes are unexpectedly expressed in nontarget tissues in old age. Such deranged differentiation may upset the functions of tissues and cause aging. An increase in the mRNAs of α- and β-globin genes and also of murine leukemia virus (MuLV) in the brain and liver of long-lived mice was found by Ono and Cutler (1978), and Florine, Ono, and Cutler (1980). However, in their later studies Ono et al. (1985b) did not find any significant difference in the expression of globin genes between short-lived and long-lived mice in old age. The expression of MuLV was, however, higher in the short-lived mice in old age. So apparently age-related relaxation of MuLV gene occurs in old age. What significance the expression of MuLV gene has for aging is not known.

In later studies, Cutler and co-workers showed that the *myc*-family of oncogenes are expressed in several tissues of old mice (Semsei et al., 1989a). They also studied satellite DNA, which is known to be highly methylated and is believed not to be transcribed, in an effort to determine whether it is transcribed in old age. It was found that these sequences begin to be transcribed after adulthood only in the heart and not in other tissues of mice (Gaubatz & Cutler, 1990). This may be due to demethylation of these sequences that is reported to occur in old age (Howlett, Dalrymple & Mays-Hoopes, 1989). Whether there is tissue- and age-specific demethylation of satellite DNA sequences needs to be examined.

Amyloid precursor protein gene and Alzheimer's disease

The number of people suffering from Alzheimer's disease (AD) is increasing rapidly with the increase in the average life span. It is a disease of old age that generally occurs in people who are over 65 years. In the familial form of the disease (FAD), however, it occurs at younger ages, between 40–50 years. FAD is characterized by an autosomal dominant inheritance, and the disease appears to be triggered by a genetic defect (Tanzi George-Hyslop, & Gusella, 1991). The main symptoms of the disease are loss of short-term memory and cognitive functions. The neuropathological signs of AD are the presence of neurofibrillary tangles (NFTs) in the cytoplasm of neurons, extracellular senile plaques, and neuronal cell death.

NFTs consist of dense bundles of fibers called paired helical filaments (PHF). They have a fuzzy outer coat which contains the microtubule-associated protein, tau. Senile plaques contain a protein core surrounded by degenerating nerve terminals. The protein core is made up of β-amyloid, a peptide containing 40 amino acids, which is called βA4 peptide (Glenner & Wong, 1984). It is derived from an amyloid precursor protein (APP), which is an integral transmembrane protein (Selkoe, 1991a, 1991b; Nordstedt et al., 1991).

The single-copy APP gene is located in human chromosome 21. Its expression is regulated by GC-rich elements located in its promoter (Pollwein, Master & Beyreuther, 1992). There are at least three different isoforms of APP having 695, 751, and 770 amino acid residues of Mr~120 kDa. They are generated by alternative splicing of the pre-mRNA transcribed from the single-copy gene (Goldgaber et al., 1987). APP is posttranslationally modified by glycosylation, tyrosyl sulfation, and phosphorylation. The N-terminal region of APP is extracellular and the C-terminal is intracellular. The βA4 amyloid peptide (of 40 amino acid residues) spans from residue 597 to 634 of the APP protein. Of these 40 residues, 28 are in the extracellular domain of APP and 12 are in the membrane-spanning region.

NFTs and senile plaques appear in the brain of normal persons after 65 years of age, especially in the hippocampus and other areas associated with memory, as is seen in persons suffering from AD. For the most part, the distinction between the normal aging brain and AD is quantitative rather than qualitative. Therefore, production of NFTs and senile plaques may be due to alterations in the amounts of individual isoforms of APP or their proteolytic fragments. However, neither in normal aging nor in AD are there age-associated changes in the levels of total APP mRNA or total APP protein in the brain (Holtzman & Mobley, 1991).

APP and βA4 peptide play a central role in the pathogenesis of AD. Familial AD is associated with a point mutation in the coding sequence of the APP gene near the βA4 domain (Goate et al., 1991). Individuals with trisomy 21 (Down syndrome) invariably develop AD pathology after 30 years of age. Thus, chromosome 21 and APP gene are associated with AD.

APP is normally cleaved within the βA4 region and the extracellular domain is released (Sisodia et al., 1990). This cleavage precludes amyloidogenesis and leaves the transmembrane and intracellular domains of the protein for degradation by the cell. The proteolytic events that generate the βA4 peptide from the APP are not known. It is speculated that mutations close to the normal APP cleavage site may alter its binding to the processing proteolytic enzyme. Mutations in the coding region of the APP gene may alter processing of APP, leading to the production of amyloidogenic fragments (Goate et al., 1991).

Nordstedt et al. (1991) have measured the levels of different APP holoprotein and C-terminal fragments in the brains of young, of nondemented aged, and of aged individuals with AD. Five species of APP were detected. Three with Mr of 106, 113, and 133 kDa, which are apparently mature and immature isoforms of APP holoprotein and two smaller proteins of Mr 15 and 19 kDa represent apparently the proteolytic C-terminal fragments of APP. The 133, 113, 106, and 15 kDa fragments were found in both gray and white matter, whereas the 19 kDa fragment was found in gray matter only. The levels of 113, 106, and 15 kDa species were not significantly different in the brains of young, normal aged, and AD individuals. However, the levels of 133 and 19 kDa species increased by two- to threefold with age. The 19 kDa fragment may be a product of the 133 kDa species and may have the βA4 domain and hence be amyloidogenic. The age-dependent increase either in a mature APP isoform and/or in a putative amyloidogenic fragment may explain why AD is associated with advanced age.

Even though the total level of APP in the brain does not appear to change during aging, the relative distribution of various species of APP and its C-terminal fragments apparently changes. An increase in the level of amyloidogenic fragment during aging or in AD may cause amyloid formation. Thus, the formation of APP 19 or a similar fragment is believed to be necessary but not sufficient for amyloidogenesis in AD. Since the levels of the putative amyloidogenic fragment in aged controls and AD patients are not different, a greater efficiency in the conversion of APP to βA4 amyloid peptide may be responsible for the accelerated formation of amyloid in AD and FAD.

Persons suffering from Down's syndrome have an extra chromosome 21. They also develop βA4 deposits at a relatively early age. It was also found that one form of familial AD (FAD) appears to be caused by a genetic defect that is also located in chromosome 21. However, a clear linkage of AD to markers on chromosome 21 in several different families, including some with late-onset (older than 65 years) forms of the disease has not been demonstrated. It appears, therefore, that AD is genetically heterogeneous and can arise in different forms. All these alterations somehow act by a common mechanism that involves increased deposition of amyloid β protein (Selkoe, 1991b).

Goate et al. (1991) found a mutation in exon 17 of the APP gene in persons suffering from FAD. This mutation causes the substitution of an isoleucine for valine at residue 717 within the transmembrane region immediately C-terminal to the βA4 domain. However, this mutation is not seen in all cases of FAD and in late-onset AD (see Tanzi, George-Hyslop, & Gusella, et al., 1991, for review). It has been suggested that the slightly more hydrophobic isoleucine replacing valine at residue 717 may alter anchoring of APP in the membrane. This may prevent normal cleavage within the βA4 domain. Another mutation that has been detected is a substitution of glutamine for glutamic acid at position 22 of the βA4 region.

An interesting discovery is that APP has a stretch of amino acids in the N-terminal extracellular domain that inhibits the proteases that cleave proteins. If proteolytic cleavage occurs at amino acid 15 or 16 within the amyloid segment, then there is no deposition of amyloid and no senile plaque is formed. However, when proteases cleave on either side of the amyloid segment, then the amyloid region (40 amino acids) is released and is deposited as amyloid (Fig. 5.19). It is this proteolytic cleavage that is enhanced in old age and in AD, causing deposition of amyloid in the brain. Whether this is a different protease (secretase) that cleaves the APP on either side of the amyloid region or the same protease is not known. It is likely that the activity of this protease is increased both in old age and in AD.

It has been reported that the amyloidogenic fragment is formed in lysosomes rich in hydrolases (Estus et al., 1992). Thus, there seems to be more than one way of producing amyloidogenic fragments in neurons. Therefore, when the membrane APP is not cleaved by secretase, it is taken up by lysosomes where it is cleaved in another way, that is, on either side of the amyloidogenic fragment that leads to amyloid formation. Thus, there appears to be a shift in the balance from cleavage in the membrane to cleavage in lysosomes in old age and in AD. The exact role of the two mutations in the APP gene is not known. It is likely that the mutations are responsible for the transfer of APP to

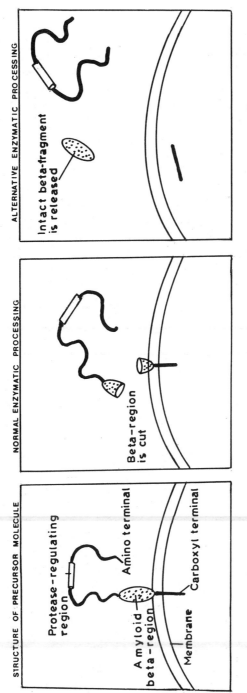

STRUCTURE OF PRECURSOR MOLECULE

Protease-regulating region

Amino terminal

Amyloid beta-region

Membrane

Carboxyl terminal

NORMAL ENZYMATIC PROCESSING

Beta-region is cut

ALTERNATIVE ENZYMATIC PROCESSING

Intact beta-fragment is released

Figure 5.19. β-Amyloid precursor protein (β-APP) is a membrane–spanning molecule that can be processed in more than one way (left). Normally, enzymes cut β-APP in middle of the β-region (center). Sometimes enzymes liberate the intact β-fragment by cutting only at its end (right). These fragments can then be secreted and accumulated in tissues. Some forms of β-APP contain a protease regulator, which may be involved in the protein's normal function. (Selkoe, 1991a; © 1991, Scientific American, Inc.)

lysosome and its cleavage by hydrolases to generate βA4. βA4 accumulates in lysosomes initially and is then secreted outside the cell for deposition as amyloid. This is corroborated by the finding that the extracellular senile plaques that contain amyloid also contain lysosomal enzymes. Another possibility is that the secretase level may decrease in old age, shifting the balance toward the lysosomal cleavage. Or the activity of secretase may change in such a manner as to alter the site at which it cleaves APP preferentially.

Amyloid formation leading to deposition of senile plaque is linked to the way its precursor, APP, is cleaved. The two possibilities for its abnormal cleavage are that (1) the mutation in the protein is responsible for the change in its cleavage site, and (2) a lysosomal protease other than the secretase cleaves it at abnormal sites. The common feature in both the familial AD and normal old persons beyond 65 years is that there is activation of an enzyme. What the signal is for its activation, whether the secretase level decreases because the transcription of its gene decreases, and whether other proteases are involved in this abnormal cleavage are questions that need to be answered before a breakthrough can be made in controlling the onset of AD.

The identification of the APP gene has been a major discovery. Studies on alterations in its expression and the processing of its mRNA, especially its alternative splicing, need to be carried out. How and why the site of cleavage of the APP protein changes in old age and in FAD, leading to the release of the βA4 amyloid fragment, have remained an enigma so far. The reason for the accumulation of tau to form the PHF also remains to be elucidated. This topic has assumed immense importance because of the rapidly increasing number of old people who suffer from Alzheimer's disease.

Methylation

No other change in genes during aging has received as much attention as the methylation of their cytosine moieties. This is not surprising because this appears to be the only notable change in the primary sequence of the DNA. Hence it has been used to explain the mechanism of age-related changes in genes. Despite the fact that a direct correlation has been seen between methylation and inactivation of several genes, there are certain genes in which no changes in the methylation status occur and yet their expression decreases with age. Also, there are examples of genes in which methylation decreases, and their expression also decreases. Moreover, it is not known whether changes in cytosine methylation are the cause or effect of aging.

Methylation of the −CCGG− sequences and CpG doublets in the genome and in specific genes has been implicated in overall transcription, and transcription of specific genes. Slagboom, de Leeuw and Vijg (1990) studied the effect of methylation of the tyrosine aminotransferase (TAT) gene on its transcription in female rats. The TAT mRNA level of the liver of 24- and 36-month-old rats was nearly 65% lower than that of 6-month-old rats. It was also seen by Msp I/Hpa II digestion followed by Southern hybridization that six CpG sites in the TAT gene were hypomethylated in the rat. Thus hypomethylation lowers transcription of the TAT gene in the liver. TAT mRNA is not present in the brain and spleen, nor is the gene hypomethylated in old age in these organs. The ornithine transcarbamylase mRNA level, however, does not change with age. On the other hand, the albumin mRNA level of 24-month-old rats is 80% higher than that of 6-month-old rats.

The role of DNA methylation in lowering transcription has been studied by several workers (Jones & Taylor, 1980; Razin & Riggs, 1980; Doerfler, 1983; Riggs et al., 1985; Bird, 1987; Cedar, 1988; Boyce & Bird, 1991; Hergersberg, 1991). The cytosine residues of mammalian genomes, especially in the CpG doublets, is heavily methylated. It was shown for many genes that if this sequence is methylated, transcription is repressed (Razin & Riggs, 1980). Analysis of the human X-linked housekeeping gene, phosphoglycerate kinase I (PGK), by genomic sequencing shows that every cytosine of all CpG doublets is methylated in the inactive X chromosome, while no 5mC is found in the same sequence in the active X chromosome (Pfeifer et al., 1990). Transcription of methylated genes introduced into cells and demethylation of genes using 5-azacytidine have substantiated this conclusion. Two models have been proposed to explain the mechanism of action of methylated and demethylated CpG doublets. From the "direct" model it is postulated that transcription factors are unable to bind if the site is methylated, and hence the gene is repressed. Certain transcription factors have been shown to be sensitive to methylation, as for example, the cAMP responsive element (CRE) binding protein (CREB) (Iguchi-Ariga & Schaffner, 1989). However, not all transcription factors are methyl sensitive. Sp1 binds to its site and activates the gene regardless of methylation of the site (Höller et al. 1988). Moreover, very few recognition sites for transcription factors have CpG.

There are several experimental evidences to support the second (indirect) model. In this model, it is postulated that methylated DNA is bound to a nuclear protein(s), which secondarily prevents transcription factors from interacting with the gene. Murray and Grosveld (1987) found that no specific methylation site in the promoter was involved in the repression of the globin gene. Methyl-CpGs in different regions of the promoter were sufficient to repress the gene. These researchers concluded that a minimal methylation-free zone was required for expression. It was postulated that proteins with an affinity for methylated CpGs are responsible for repression. Methylated DNA has indeed been shown to interact differentially with nuclear components. Upon transfection of different methylated and non-methylated constructs into mouse L cells, the methylated constructs preferentially assemble into relatively nuclease-resistant chromatin (Keshet et al., 1986). Also, methylated CpGs are themselves protected against nonspecific nucleases and against restriction enzymes that recognize CpG (Solage & Cedar, 1978; Antequera et al., 1989). These data indicate that nuclear factors bind to methylated CpGs, leading to the formation of an "inactive" chromatin structure. Such a protein factor (CpG-binding protein, MeCP-1) has been identified (Meehan et al., 1989). Promoters of four genes – *jun,* thymidine kinase (tk), globin, and phosphoglycerate kinase (PGK) – have a good distribution of CpGs and also adopt an altered nuclease-resistant conformation. Methylation of the promoter causes repression of the genes both in the in vitro system containing nuclear extract, and in in vivo systems (Figs. 5.20 & 5.21). This is supported by the fact that MeCP-1-deficient cells show greatly reduced repression of methylated gene.

These studies have shown that inhibition of transcription by DNA methylation is an indirect effect. The key experiments that support this conclusion are that the transcriptional inhibition that is seen at low template concentrations can be overcome either by increasing the concentration of methylated template or by addition of methylated competitor DNA. The methylated competitor DNA may mop up the mediator of inhibition, leaving the template free to interact with the transcriptional machinery. This also indicates that the "direct" effect of CpG methylation, if any, on the repression of transcription is relatively small.

Methylated DNA in the nucleus has lower accessibility to nucleases. Both PGK and α-globin gene promoters adopt an altered chromatin structure when methylated. Both genes have CpG islands and are normally nonmethylated in animal cells. However, CpG islands of the PGK gene are methylated on the inactive X chromosome (Hansen, Ellis &

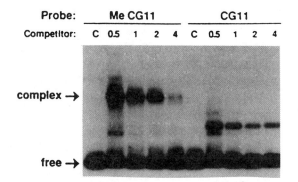

Figure 5.20. Band-shift assay to examine methyl CpG-binding protein (MeCP-1). It is present in HeLa cell nuclear extracts and can bind to the constructs that show indirect inhibition of transcription in vitro. About 0.1 ng of the methylated (MeCG11) or nonmethylated (CG11) radiolabeled oligonucleotide probe was incubated with 5 μg of HeLa cell extract in the presence of increasing amounts of nonspecific competitor DNA. Lane C shows a control band shift in the absence of extract. The prominent band present only in lane CG11 is due to an unknown protein that binds nonmethylated CG11. (Boyes & Bird, 1991; © 1991 Cell Press)

Gartler, 1988), and α-globin gene gets methylated in permanent cell lines (Antequera et al, 1990). These methylated genes are resistant to Msp I, which shows that methylated sites in the chromosome are made inaccessible by association with nuclear proteins. When methylated, transcription from these promoters is inhibited by binding to the nuclear protein, MeCP-1. Hence MeCP-1 or a similar protein is involved in maintaining both the inaccessibility of methylated CpG islands for transcription factors and gene activity.

How then are promoters, which are not rich in CpG, repressed? There are examples of genes that are repressed by the presence of only one or a small number of CpGs (Ben-Hattar & Jiricny, 1988). In this context it is of interest that a gene with a few methyl CpGs can be reactivated in the presence of strong trans-activators (Weisshaar et al., 1988; Bednarik, Cook & Pitha, 1990), while genes with a high density of methyl CpGs, such as methylated CpG islands, are stably locked into an inactive state. This difference reflects differing affinities of the promoters for MeCP-1 and differences in the degree of expression of a gene. Strong promoters may be able to overcome the weak binding of MeCP-1 to a poorly methylated gene, allowing flexibility in the repression/expression of genes. If a gene is heavily methylated, strong promoters may not prevail against the tight complex formed with MeCP-1.

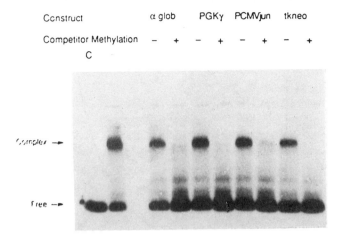

Construct   α glob   PGKγ   PCMVjun   tkneo

Competitor Methylation   −   +   −   +   −   +   −   +

Figure 5.21. Band shift of MeCG11 was performed as in Figure 20 in the presence of 2 μg of nonspecific DNA to give the band shift shown (−, lane 2). Constructs of α-globin, phosphoglycerate kinase (PGK), *jun*, and thymidine kinase (tk), 100 ng each, were tested for their ability to compete for the band shift in either their methylated (+) or nonmethylated (−) form. (Boyes & Bird, 1991; © 1991 Cell Press)

Alterations in mitochondrial DNA

The importance of mitochondria in energy production in the cell has caused several workers to examine its DNA and the expression of its genes during aging. Mitochondrial DNA (mtDNA), which has the added advantage of being small in comparison to that of nuclear DNA, has no introns, is not complexed with proteins, and replicates autonomously. A mitochondrion may have two or three copies of mtDNA, and each cell may have many thousands of mitochondria (Robin & Wong, 1988). Mammalian mtDNA is a circular 16.5 kbp dsDNA. Except for a short noncoding region, the remaining mtDNA codes for two rRNAs (12S and 16S), 22 tRNAs, and 13 mRNAs for 13 enzymes of the electron transport system. The other proteins required for oxidative phosphorylation are coded by nuclear DNA. The mutation ratio of mtDNA is higher than that of nuclear DNA (Linnane et al., 1989). It is reported that it has no proofreading or DNA repair system (Grivell, 1989). Any damage to mtDNA is expected to impair energy production. Hence damage, modifications, mutations, and deletions in mtDNA during the life span may affect energy production in the organism and contribute to aging.

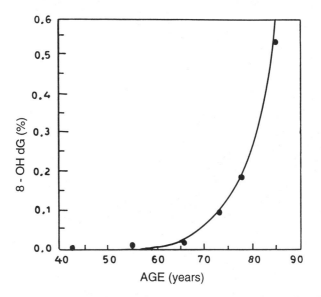

Figure 5.22. Levels of 8-OH-dG in mtDNA in humans of various ages. Percentages of the amount of 8-OH-dG in relation to that of total dG are plotted against the ages of subjects. 8-OH-dG accumulated almost linearly with age after age 65, whereas under 55 it remained at less than 0.02%. (Hayakawa et al., 1991b).

In the light of the importance of mitochondria in the metabolism of cells and organisms, several workers have recently begun studies on mtDNA. Trounce, Byrne, and Marzuki (1989) and Yen et al. (1989) have shown that oxidative phosphorylation in human mitochondria decreases with age. Hayakawa et al. (1991a) found that if mice are administered azidothymidine, the guanine residue is modified to 8-OH deoxyguanosine (8-OH-dG). Using ultramicro-HPLC combined with mass spectrometry, they also showed that the 8-OH-dG in mtDNA of the human diaphragm muscle increases with age (Fig. 5.22; Hayakawa et al., 1991b). In subjects below 55 years, the 8-OH-dG level in mtDNA was below 0.02% of the total dG level, but in subjects over 65, it continued to increase with age at a rate of 0.25%/10 years reaching 0.51% at 85. Concomitantly, it was found that multiple deletions of mtDNA occurred, which could be detected by electrophoresis, after amplification by PCR, as was seen for mtDNA of human heart (Sugiyama et al., 1991). It is speculated that the lower level of 8-OH-dG seen in the young may be because of higher replication rate of mtDNA that dilutes out the 8-OH-dG content. In older persons, its replication rate may be slower, which may lead to accumulation of 8-OH-dG. The presence of this modified guanine may cause deletions in mtDNA. Such modified bases may also alter the binding of the trans-acting factors needed for expression/inhibition of genes and replication as has been suggested by Kanungo and Saran (1991) (see also Chapter 4).

Sugiyama et al. (1991) made the interesting discovery that mtDNA of the myocardium of all subjects over 70 years have a 7,436-bp deletion. Using the PCR technique, they have shown that the percentage of mtDNA with this deletion increases in old age. It is 3% at age 80 and 9% at age 90. This segment of DNA codes for five subunits of complex I, and one subunit each of complexes III, IV, and V. It is significant that the mtDNA of the striatum of the brain of patients suffering from Parkinson's disease and of elderly control subjects have a 4,977-bp deletion (Ikebe et al., 1990). An increasing level of mtDNA with this deletion would decrease the ATP production capacity of this important organelle, as well as overall metabolism. It would be of interest to examine such deletions and the accumulation of 8-OH-dG in the mtDNA of the brain, which is a highly aerobic tissue. A comparison with the mtDNA of anaerobic tissues like the skeletal muscle and liver may give important information about the reasons for the decline in the overall decrease in metabolism seen during aging.

Free radicals are constantly being produced in mitochondria due to the high rate of oxidation in this organelle. It needs to be determined whether or not free radicals, especially the superoxide radical ($\cdot O_2^-$) and the hydroxyl radical ($\dot{O}H^-$) are the cause of the modification of guanine to 8-OH-dG which causes deletions in the mtDNA. The presence of 8-OH-dG may affect the incorporation of dA, dT, dC, and dG during replication. Hence its presence in mtDNA may cause point mutations, increase deletions, and promote supercoiling, thereby affecting transcription. However, it is intriguing why specific segments of DNA are deleted, as for example 7,436 bp in the heart and 4,977 bp in the brain of patients suffering from Parkinson's disease.

The deletion of 4,977 bp of mtDNA in the liver of old men is either between 8,469 and 13,447 bp or between 8,482 and 13,460 bp (Yen et al., 1991). The frequency of deletion of this segment increases with age, being five out of eight between 31 and 40 years, nine out of eleven between 41 and 50 years, and in all men above 50 years (Fig. 5.23). This deletion is not found in the mtDNA of the liver of stillborn children or the blood cells in normal subjects of all ages. There is a 13-bp repeat (ACCTCCCTCACCA) at the junction region. Since no introns are

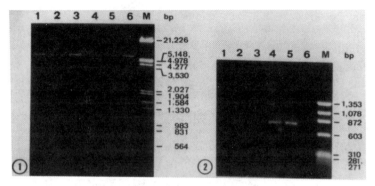

Figure 5.23. Electrophoretogram of the specific 4,977-bp deletion in the liver mtDNA in Chinese subjects of various ages. Lane 1 – blood cell mtDNA of a 79-year-old subject; lane 2 – liver mtDNA from stillborn baby; lane 3 – liver mtDNA of 35-year old; lane 4 – liver mtDNA of 57-year old; lane 5 – liver mtDNA of 79-year old; lane 6 – liver mtDNA of a 28-year old. M – φX174 DNA digested by Hae III. (Yen et al., 1991)

present in the mitochondrial genes, the 4,977-bp fragment that is deleted may code for several proteins. The loss of such a large DNA fragment, which is likely to code for several enzymes (proteins), may decrease mitochondrial function. It would be of interest to find out which proteins this fragment codes for, and why only this fragment is deleted. Cortopassi and Arnheim (1990) have found that some mtDNA of the brain and heart of old men have a 520-bp fragment deleted. Thus deletions of large fragments in mtDNA is of general occurrence during aging.

Other interesting age-related changes in mitochondria have been observed. Gadaleta et al. (1990, 1991) found that the steady-state levels of 12S rRNA and mRNA for subunit 1 of cytochrome oxidase in the brain and heart of senescent rats are significantly lower than those of adult rats. No such change is seen in the liver. If senescent rats are administered acetyl-L-carnitine, their RNA levels rise to those of the adult (Fernandez-Silva et al., 1991). Mitochondria purified from the cerebral hemisphere of old rats have 50% lower amounts of mitochondrial RNA than those of young rats. This is due to a lower rate of transcription.

The mammalian heart is a highly aerobic tissue and is dependent on ATP production by its mitochondria. Biggs et al. (1991) have found that the number of mitochondria in the heart declines in rats as they grow old. The level of ubiquitin, which is involved in the degradation of proteins and is a component of the inner mitochondrial membrane, declines with age (Biggs et al., 1991). Since cytochrome c is also a major component of the inner membrane, Biggs et al. (1991) measured the cytochrome c mRNA level of the heart of young (12 months) and old (24 months) rats. The mRNA level in the old animals was 22% lower than that of the young, though no change in the cardiac β-actin mRNA, Ca^{2+}/calmodulin protein kinase II mRNA and 18s rRNA was found. A lower level of cytochrome c mRNA, therefore, may contribute to a lower production of ATP and efficiency of mitochondria.

Of relevance in the context of the damage that occurs to mtDNA as a function of age is the generation and role of free radicals in mitochondria. The major source of superoxide radical ($\cdot O_2^-$) is the mitochondrial electron transport system (Chance, Sics & Bovaris, 1979), because even though two electrons are transferred from the substrate to the respiratory chain, the reduction of ubiquinone occurs by a single electron transfer. The ubisemiquinone thus generated, though largely oxidized by cytochrome b, may also react with O_2 to generate $\cdot O_2^-$. The yield of $\cdot O_2^-$ is about 1% of the O_2 used in mitochondria, and approximately 10^7 molecules/mitochondria/day may be produced (Richter, 1988). Most of the $\cdot O_2^-$ is converted by mitochondrial Mn-SOD to H_2O_2. However, this detoxifying effect is not total. Moreover, H_2O_2 is not totally converted by mitochondrial catalase and glutathione peroxidase to $H_2O + \frac{1}{2}O_2$. The fraction of H_2O_2 that is not converted to $H_2O + \frac{1}{2}O_2$ reacts with O_2 to produce an extremely active $\dot{O}H^-$ radical. H_2O_2, $\cdot O_2^-$, and OH^- can react with several types of biomolecules including DNA.

Based on these deleterious effects of reacting oxygen species in mitochondria, Miquel (1991) has proposed the "extranuclear somatic mutation" hypothesis of aging. He argues that mtDNA synthesis occurs at the inner mitochondrial membrane near the sites of formation of the reactive oxygen species that are the cause of the production of deleterious molecules such as malonaldehyde and lipoperoxides. mtDNA is naked, unlike nuclear DNA which is protected by histones, and lacks excision repair. So it may not be able to counteract the damaging effects of malonaldehyde and lipoperoxides. Mitochondrial genome codes for hydrophobic proteins of the inner mitochondrial membrane. So, loss of fragments of mtDNA or mutated mtDNA that may result due to the above molecules may impair its replication and, thereby multiplication of the organelle. Thus, an irreversible decline in the bioenergetic ability may follow, with a concomitant loss of physiological performance of the cell.

In support of the above view, Miquel and co-workers (see Miquel,

1991, for review) found that a considerable amount of age pigment is derived from mitochondria of *Drosophila melanogaster*. It is of interest in this context that the mitochondria of a specific region of the brain – the substantia nigra, which is implicated in Parkinson's disease – is deficient in the respiratory chain. Despite the view that accumulation of highly reactive oxygen species in mitochondria is likely to damage mtDNA and other molecules in mitochondria, it is necessary to explain why the level of SOD, which is coded by a nuclear gene, decreases with age. Also, the reason for the increase in catalase and glutathione peroxidase is not easily explained by these arguments. Moreover, the nuclear genome is far in excess of the mtDNA and also codes for proteins that are far in excess in number and types. Therefore, it is unlikely that alterations in mtDNA cause aging of cells. If mtDNA were the sole cause of aging, one would expect that neurons and cardiac muscle cells, which have a large number of mitochondria and have a high rate of oxygen consumption, would have a short life, and skeletal muscle cells, which have a far lower number of mitochondria and have low oxygen consumption, would have a longer life. This does not appear to be so.

Editing and splicing of RNA

The expression of a gene into its final protein product requires several steps: (1) transcription of the pre-mRNA that includes initiation, elongation, and termination; (2) processing of pre-mRNA to mRNA that includes capping, poly-A$^+$ tailing, and splicing; (3) translocation of mRNA through the nuclear membrane; and (4) translation of the mRNA into the protein. Recently, it has been shown that certain mRNAs undergo editing before they are translated. Alterations in the activities/levels of factors required for any of the above steps may affect the rate of synthesis of a protein and its level. Information on all the steps for any particular protein is not available in relation to aging. Presently, attention is focused on the transcription of specific genes and the modulation of transcription, which are beginning to yield useful information. Two other steps that have the potential to influence the types and levels of proteins being synthesized are editing and splicing of mRNA. These two steps are described below.

Editing of mRNA

Surprises still keep coming up in biology. After it was discovered that RNA serves as an enzyme, the discovery that the message that the mRNA carries for synthesis of a specific protein undergoes editing came as a great surprise. Editing appears to play an active role in the transfer of the message. Benne et al. (1986), while studying the expression of *cox* II (cytochrome oxidase) gene of *Trypanosoma* mitochondria, surprisingly found that the mRNA of the gene had nucleotides that did not totally correspond to the gene. The sequence analysis of the cDNA derived from the mRNA showed that corrections were accomplished by inserting uridine nucleotides in some places and deleting them in others. These changes took place at very specific sites, not at random. This was corroborated by Feagin, Abraham and Stuart (1988), who found that extensive additions and deletions occur in the transcript of *cox* III gene of *Trypanosoma* mitochondria before it is translated. In one transcript, 550 uridines were added and 41 were removed to make the mature mRNA. Overall, 60% of the message was edited, indicating that the mRNA is edited before translation.

The discovery by Blum et al. (1990, 1991) of short RNA sequences, less than 40 nucleotides long, in *Trypanosoma,* and the presence of their genes in a minicircle DNA of the parasite gave the first insight into how the transcripts may be edited. These short RNAs are called "guide RNAs" (gRNA). Tails consisting of uridine nucleotides are added at the 3' end of the gRNA, just as adenosine nucleotides are added at the 3' ends of pre-mRNAs. This is the source of uridine nucleotides that are added to the *Trypanosoma* mRNA.

RNA editing has some similarity with RNA splicing in which introns are removed from pre-RNAs (Cech, 1991). According to this hypothesis, the gRNA aligns itself with the unedited RNA. The uridine tail of the gRNA then invades the unedited RNA, base-pairing with it by means of G:U. The RNA is then split into two. A new bond forms between one of the ends and the uridine at the tip of the tail. Then the unattached end of the mRNA undergoing editing attacks at the newly attached uridine and forms a bond with it. This transesterification leads to re-joining the RNA and releasing the tail with one less uridine. The gRNA is now ready to initiate another round of reactions. Blum et al. (1991) obtained experimental evidence for this suggestion when they found that the gRNA is joined by a stretch of uridine nucleotides to mRNA. Thus uridine tail of gRNA is the source of uridine.

The following mechanism of mRNA editing has been suggested (Fig. 5.24): (1)A few or hundreds of uridines are inserted or deleted probably by transesterification, instructed and carried out by short gRNAs (Blum et al., 1991). (2) In a mitochondrial transcript of a slime mold, single cytidine (C) residues are added at multiple sites by an unknown mech-

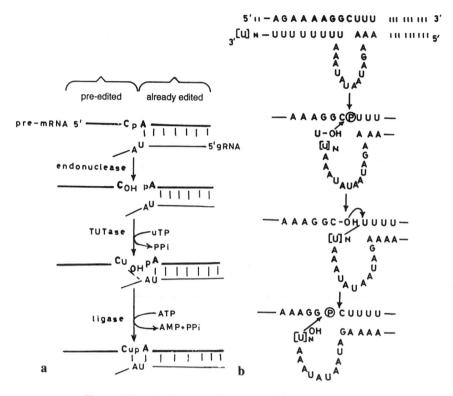

Figure 5.24. (a) Trans-esterification model of RNA editing. Guide RNA (gRNA) splits mRNA to insert uridine nucleotides. (Cech, 1991) (b) Mechanism of RNA editing. [Adapted from Cech (1991) by Hoffman (1991)]

anism. (3) Cotranscriptional polymerase shuttering results in the insertion of guanosine (G) residues in a transcript in a RNA virus. (4) In transcripts of vertebrate mitochondria, UAA and UGA stop codons are reconstituted by polyadenylation. A similar phenomenon is seen in nuclear transcript (see Cattaneo, 1992, for references). (5) Editing of mammalian apolipoprotein B mRNA, which is transcribed from a nuclear gene, involves conversion of residues. C is converted to U by deamination. This causes tissue-specific generation of a stop codon. (6) In many mitochondrial transcripts, conversion from C to U and U to C occurs during editing.

Conversion of C to U in apolipoprotein mRNA is determined by conserved sequences flanking the editing site (Hodges & Scott, 1992). In mitochondrial transcripts, the sequences flanking the various editing sites are completely divergent (Walbot, 1991). It has been found that the genomic DNA sequences encoding all six subunits of the glutamate receptor involved in synaptic transmission in the mammalian brain contain a glutamine codon (CAG), even though an arginine codon (CGG) is found in the cDNAs made from the mRNAs. The transcripts of one of the subunits are edited with 100% efficiency; three other subunits are not edited; and transcripts of the other two subunits are edited with 40% and 80% efficiency, respectively (Sommer et al., 1991). It is speculated that the A-to-G conversion may be a consequence of reverse transcription.

The details of the molecular mechanism of RNA editing are not yet known. What is known is that it is specific to certain mRNAs and to certain regions of these mRNAs. It is precise, since it ultimately produces a single, translatable transcript. RNA editing is posttranscriptional and proceeds in a general 3' to 5' direction (Abraham, Feagin & Stuart, 1988). The grNAs are complementary (allowing G:U base pairing) to portions of edited sequence and are believed to provide information that specifies the edited sequences (Blum et al., 1990).

Koslowsky et al. (1991) have analyzed the partially edited transcripts of NADH dehydrogenase 7 (ND7) and ATPase 6 (A6) cDNAs to assess the role of grNAs in editing in *Trypanosoma*. These transcripts contain blocks of partially edited sequences called junctions at the transition between the fully edited and unedited regions. The partially edited blocks overlap one another. Each junction is a mosaic of unedited, partially edited, and fully edited sites. Thus, editing does not progress in a strictly consecutive 3'-to-5' direction and sites are re-edited. The grNA appears to be progressively realigned with the mRNA being edited.

The occurrence of mRNA editing suggests that apparent coding errors such as "wrong" residues or frameshifts are corrected by this process. The editing may involve insertion and deletion of hundreds of residues, sometimes even creating new reading frames (Benne et al., 1986; Stuart, 1991). It appears to be a universal phenomenon as its occurrence has been shown in chloroplasts and mitochondria of higher plants, mitochondria of protozoa and vertebrates, and nuclear mRNA of mammalian brain (Cattaneo, 1991, 1992).

The occurrence of RNA editing whereby mRNAs may undergo changes in their sequences, and hence in their codons means that the synthesis of a protein does not correspond strictly to the sequence of a gene. Despite the prediction of Orgel (1963) nearly three decades ago that, due to errors in information transfer, altered proteins are synthesized with increasing age and are the cause of aging, no such proteins have been detected so far. In the light of the existence of mRNA editing one may look into the possibility of alterations in the nucleotide sequences of mRNAs as a function of age. If mRNA editing leads to the formation of termination codons at premature sites of an mRNA, no protein or a protein with a shorter length would be synthesized. There are several codons that may be converted to termination codons (UAA, UAG, and UGA) by the substitution of only one U. In this case, the search for altered proteins carrying a few substituted amino acids during aging may not bear fruit. A more meaningful approach may be to purify mRNAs, prepare cDNAs from them, and compare their sequences with those of the original transcripts to determine the extent of editing. This will throw light on the types of changes that occur at this crucial step of information transfer during aging.

Thus, RNA editing is an important mode of control of gene expression. Moreover, the existence of many types of mRNA editing indicates that the information content of mRNA is flexible. Alterations occurring in the gRNA or in any one or more types of the editing processes during aging may alter the production of mRNAs that are finally translated.

Alternative splicing of pre-RNA

Alternative splicing of a pre-mRNA is known to produce different mRNAs and isoforms of a protein. Most pre-mRNAs transcribed from split genes are spliced in such a way that the original sequences of the exons are maintained in the mature mRNAs; this is termed constitutive splicing. Some pre-mRNAs are, however, spliced in more than one way, thereby yielding a family of structurally related mRNAs that are translated into a family of protein isoforms. This form of pre-mRNA processing is called alternative splicing. Such splicing is seen in organisms ranging from *Drosophila* to humans, and occurs in several types of transcripts that encode varieties of proteins. This is a means of diversifying the output from a single gene without altering its genomic organization. In some cases, alternatively spliced mRNAs are produced concurrently in the same tissue, and the several protein isoforms may perform the same or different functions. For example, four myelin basic protein isoforms derived from a single gene are all components of the myelin sheath of the central nervous system (Singer & Berg, 1991). Some gene transcripts are spliced differently in various tissues. For example, the single mammalian calcitonin gene expresses calcitonin in the thyroid, but in the brain calcitonin-gene-related protein isoform is expressed each from its own distinctively spliced mRNA.

Three types of alternative splicing have been discovered. In type I, different promoters of the same gene are used to produce different pre-mRNAs with distinctive 5′ proximal regions of varying lengths. The vertebrate myosin light-chain (MLC) gene, which has nine exons, is transcribed from two promoters located ~10 kbp apart. The longer pre-mRNA contains all exons, but the shorter one lacks the first exon. Alternative splicing generates two nearly equal-sized mRNAs that differ in the 5 exons. During splicing of the longer pre-mRNA, exon 1 is joined to exon 4, ignoring the 3′-splice sites of exons 2 and 3. For the shorter pre-mRNA, exons 2, 3, and 5 are joined during splicing, and exon 4 is ignored (Fig. 5.25).

Type II alternative splicing involves pre-mRNAs that have different and variable length 3′ proximal sequences, usually, because the transcript is polyadenylated at different sites. The calcitonin gene produces two mRNAs. One is generated from a shorter pre-mRNA, and the other encodes the calcitonin-gene-related peptide. All four exons of the shorter pre-mRNA are retained. The longer pre-mRNA has two additional exons, 5 and 6, which are retained in the calcitonin-gene-related peptide mRNA, but exon 4 is lost. The 3′ splice site in intron 4 is the preferred pattern for the 5′ splice site of intron 3. Thus there is tissue specificity for the use of different polyadenylation signals for determining the splice sites.

In type III, identical pre-mRNAs generate different functional mRNAs. All the exons are present in the pre-mRNA, and splicing selections are made between existing introns. The fast skeletal muscle protein troponin T gene has 18 exons as well as two splicing patterns that generate α and β groups of troponin T mRNAs that contain either exon 16 or 17, respectively. This results from mutually exclusive splicing of exons 15, 16, and 18 or 15, 17, and 18. The choice of which exon is used is developmentally regulated. Exon 16 is specific to adult troponin T mRNA, and exon 17 is used in both embryonic and adult mRNAs. The other alternative splicing sites are exons 4, 5, 6, 7, and 8, which give rise to 32 possible combinatorial alternative splices. The pattern ranges from complete exclusion of 4–8 exons to retention of all 5 exons in α and β groups of troponin T mRNAs. Although an authentic 5′ splice site usually splices to the normal 3′ splice site at the end of the same intron, it may use an alternate 3′ splice site within the intron,

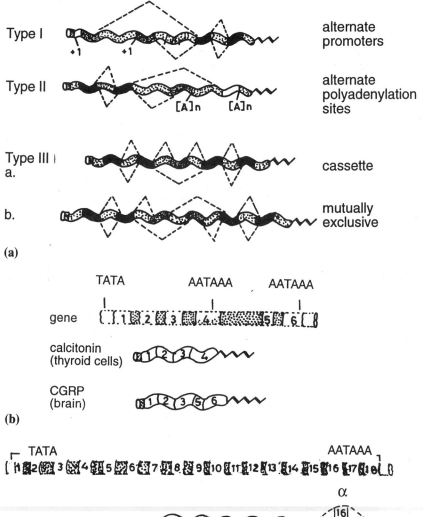

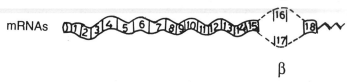

Figure 5.25. (a) Alternate patterns of splicing of pre-mRNAs. Dashed lines connecting exons above and below the RNAs indicate various patterns of splicing, depending on whether the pre-mRNAs have alternate 5′ ends, alternate 3′ ends, or identical 5′ and 3′ ends. (b) Alternately spliced mRNAs from pre-mRNAs with alternative polyadenylated 3′ ends. Calcitonin gene exons (numbered) and introns (stippled), its promoter's TATA, and its polyadenylated signals (AATAAA) are shown. Two different spliced mRNAs, one encoding calcitonin and the other encoding a calcitonin related protein (CGRP), are

resulting in the elongation of the next exon. Alternatively, if an alternate 3' splice site is used in the next exon, it leads to truncation of that exon. Similar results occur if alternate 5' splice sites within introns or exons are used. The existence and use of such "cryptic" splice junctions necessitates that the alternate splices generate consistent translational reading frames.

The intriguing problem is how the correct pairs of exon junctions are chosen for joining. One explanation is that cis-elements in the pre-mRNA itself are involved. The other possibility is that trans-acting factors, proteins, or other RNAs influence the choice of introns to be spliced. There is apparently a hierarchy of splice-site preferences that determines the relative degree of choice of 5' and 3' splice sites. This may occur during spliceosome formation. There are examples in which mutations at one authentic splice junction allow a normally unused site to be spliced. This may occur where identical pre-mRNAs need to be spliced differentially depending on the tissue, physiological condition, and developmental stage.

In this connection, the data on the alternative splicing of fibronectin (FNT) pre-mRNA during aging assume much relevance. FNT is composed of three types of repeating homology units (type I, II, and III), though it is coded by a single gene located on human chromosome 2 (see Hynes, 1985, for review). Alternative splicing of the rat (Schwarzbauer et al., 1983) and human (Kornblihtt, Vibe-Pedersen & Baralle, 1983) FNT mRNA is responsible for the production of 12 and 16 isoforms of FNTs in the rat and human, respectively. Alternative splicing occurs at each of the three exons – extra domain A (ED-A), ED-B, and type III connecting sequence (III CS). The FNT isoforms produced from the alternatively spliced mRNAs differ in properties (Schwarzbauer, 1991). Ratios of FNT mRNAs with or without a given exon were determined in several rat tissues and IMR-9 human fibroblasts during aging in vivo and cellular senescence in vitro (Magnuson et al., 1991). Statistically significant changes were found in both in vivo and in vitro aging. All

formed from transcripts that are polyadenylated at the first or second polyadenylated signal, respectively. (c) Alternatively spliced troponin T mRNAs from pre-mRNAs with identical 5' and 3' ends. Exons (numbered) of gene for fast skeletal troponin T and introns (stippled), as well as transcription signal TATA, and polyadenylation signal, AATAAA, are shown. Two groups of spliced mRNAs result from alternative inclusions of exons 16 (α) or 17 (β). Each of these has 32 different mRNAs, depending on which of the exons 4–8 are retained in the splicing. (Singer & Berg, 1991) three alternatively spliced exons are spliced out at a higher frequency as the animals and cells age. However, major alterations leading to the absolute predominance of one form over another are not seen during aging. Thus, the machinery for alternative splicing appears to be preserved during aging. Whereas ED-A and ED-B mRNAs in adult tissues ranged from 0 to 25% and 0 to 10%, respectively, those of fibroblast cells in culture ranged from 50% to 60% and 15% to 25%. Several factors such as serum deprivation, growth factors, and retinoic acid affect alternative splicing. Thus in vivo alternative splicing is not the same as that of in vitro splicing.

FNT synthesis is increased in human skin fibroblasts after extensive passage in culture (Chandrasekhar, Sorrentino & Mills, 1983). The FNT synthesized by late-passage cells is slightly longer, and its ability to bind to collagens I and II is lower. Burke and Danner (1991), using human foreskin fibroblasts, JAS 3, found an 8-fold increase in the expression of ED-A splicing variant in the late passage cells relative to the sum of all other variants. ED-A variant expression is also enhanced during wound healing (Hynes, 1990) and in tumor cells (Borsi et al., 1987). This study provides an example of the posttranscriptional changes that occur to produce different forms of mRNA and isoforms of the FNT protein. Whether the larger isoform of FNT is due to posttranslational modification such as glycosylation is not known.

It is likely that alternatively spliced transcripts of other genes as well may undergo change with age, and result in the production of different isoforms of proteins that have different properties. An important gene that has been implicated in Alzheimer's disease and that is prevalent in old age (> 65 years) is the amyloid precursor protein gene. Its transcript is alternatively spliced to give three mRNAs (Goldgaber et al., 1987). More genes need to be studied to determine whether this is a general phenomenon that occurs for genes whose transcripts undergo alternative splicing for the production of different isoforms of the same protein, because quantitative and qualitative changes in the isoforms of proteins during aging may contribute to the aging process significantly.

Thus, both RNA editing and splicing show the tremendous flexibility that the information-transfer system has, and how gene expression can change under various conditions. Not only is splicing not absolutely fixed by the sequences at the splice junction, but there is a possibility that the sequences at these sites and elsewhere in the mRNA may change through RNA editing to provide alternative splice sites. RNA editing may not be restricted only to mRNAs. There is thus an enormous possibility for changes in the amount of proteins and their isoforms being synthesized. This new area needs to be explored to find out if any age-related changes occurring in these processes contribute to functional decline in old age.

Telomere DNA

A novel finding is the relationship between telomeres present at the tips of chromosomes and senescence (Greider & Blackburn, 1987; Allshire et al, 1989; Morin, 1989; Zakian, 1989; Yu et al., 1990; Boeke, 1990; Blackburn, 1990, 1991; Greider, 1990; Moyzis, 1991; Harrington & Greider, 1991), particularly in the unicellular ciliates. Telomeres are specialized DNA–protein structures that comprise the chromosomal termini. They are required for stabilization of chromosomes, and allow complete replication of the 5 ends of chromosomal DNA. The chromosomes end with simple repeats, d(TTGGGG) in *Tetrahymena* and d(TTAGGG) in humans. This G-rich strand is oriented 5 to 3 toward the chromosome terminus. Repeats are added de novo by telomerase, which is an unusual DNA polymerase. It is a ribonucleoprotein having the enzyme and a RNA with the sequence AACCCC, which serves as the template for the synthesis of the telomere DNA repeat. It synthesizes the G-rich strand of telomeric DNA. Telomerase actually acts as a reverse transcriptase because an internal RNA directs the synthesis of a DNA strand. This RNA sequence is highly conserved (Romero & Blackburn, 1991). Telomerase is capable of de novo addition of telomere sequences onto nontelomeric DNA in chromosomes, and thus can carry out chromosome healing (Yu & Blackburn, 1991; Morin, 1991; Greider, 1991). Telomere addition in ciliates appears to be sequence independent. No consensus sequence for telomere addition sites has been found. Under in vitro conditions, telomerase synthesizes its species-specific repeats onto the 3' end of a G-rich telomeric sequence primer and not onto primers lacking telomeric sequences.

Olovnikov (1973) proposed (1) that cells lose a small amount of DNA following each round of replication due to the inability of DNA polymerase to fully replicate chromosome ends (telomeres), and (2) that eventually a critical deletion causes cell death. This hypothesis has been borne out by the recent finding that telomeres of somatic cells act as a mitotic clock, shortening with age both in vivo and in vitro in each round of replication. Furthermore, since telomeres stabilize chromosome ends against recombination, their loss may explain the higher frequency of dicentric chromosomes seen in late-passage (senescent) fibroblasts. Sperm telomeres are longer than somatic telomeres and are not shortened with age. Telomeres are also longer in immortal, transformed human cells and tumor cell lines, but not in somatic cells.

Harley, Futcher, and Greider (1990) and Harley (1991) examined the telomere length in different strains of human diploid fibroblasts (HDF) in cell cultures. Five cell strains from donors, aged 0 (fetal), 24, 70, 71, and 91 years, were cultured up to senescence. The length of the terminal restriction fragments (TRFs) from highly digested genomic DNA was measured as a function of mean population doubling in Southern blots by hybridization to a telomeric oligonucleotide probe $(TTAGGG)_3$. The results show that the mean length of the TRF decreases in all five strains by 48 ± 21 bp per population doubling (PD). Other internal repetitive sequences do not decrease in length with replicative age in vitro. Hence the terminal TTAGGG repeats are lost in a replication-dependent manner.

It is of interest that when fibroblasts from aged donors are examined in early passage, the loss of telomere length is 14 ± 6 bp per year of the donor. The mean length decreases by about 50 bp per PD. The mean telomere length decreases by about 2 kbp before the cells die. It calculates to about 40 generations (PD), based on 50 bp per generation. Significantly, the initial lengths of the terminal restriction fragment of 70- and 90-year-old donors are shorter than those of a 24-year old, newborn, or fetus. In addition, the hybridization signal to the probe decreases with increasing cell doubling, indicating true shortening of the repeat. The EST-1 mutant of yeast has a defect in telomere elongation that leads to a senescence phenotype. The mutant shows no immediate loss of viability, but rather a slow progressive death of cells (Lundblad & Szostak, 1989). Thus, telomere length appears to have a relationship with senescence of HDF and yeast. This provocative but speculative idea needs further study; especially whether or not loss of the complete telomere from a chromosome leads to unequal recombination and defective DNA synthesis needs to be investigated.

Since human sperm telomeres are longer than somatic telomeres (Cross et al., 1989), it was proposed that long telomeres are synthesized in germ cells as telomerase is active in these cells. Its activity is turned off in somatic cells, and sequences are lost from the chromosome ends with each round of replication. Transformed cells have a rather stable, though heterogeneous, length of telomeres. Whether the shortening of telomeres is the cause or effect of aging is not known, but certainly it is a good biomarker to use for aging studies, provided all types of cells exhibit such a phenomenon.

Cultured human fibroblast cells derived from embryos are known to divide about 50 times. The number of times other cells divide in vivo is not known. Whether telomere length can be used as a timer/biomarker for longevity will depend on the universality of this phenomenon. It has been speculated that the presence of telomerase in somatic, nondividing, and differentiated cells may be deleterious because it may cause cells to divide (Boeke, 1990). HeLa cells contain telomerase, and possibly all transformed cells also have the enzyme, which may preclude their senescence. This implies that one of the several types of transformations and perhaps also carcinogenesis may be due to activation of telomerase, or a factor that is required to maintain the length of the telomere in the germ line. If the somatic cells maintain long telomeres, the organism may be prone to spontaneous neoplasia. The mechanism by which telomeres are shortened and its relationship to aging in humans is an interesting problem for study. One possibility may be that the complete loss of one or more telomeres from the chromosomes may give a negative signal that may prevent DNA synthesis and lead to the differentiated state, the arrest of cell division, and senescence. Chromosomal aberrations are indeed observed in senescent HDF (Sherwood et al., 1988). Human telomeric DNA has been cloned and is yielding useful results (Riethman et al., 1989). Identification of the gene for telomerase, its cloning, and the production of transgenic mammals with active telomerase may provide further insight into the aging problem.

Yu et al. (1990) have demonstrated that the RNA component of telomerase is essential for its function. They introduced mutations in the template region of the cloned telomerase RNA gene of *Tetrahymena*. When these mutations were expressed in vivo, the telomere sequence and length were correspondingly altered.

Several lines of evidence point toward a relationship among telomere, aging, and cell immortalization. Sperm telomeres are longer than somatic cells (Harley, 1991), which could be possible if somatic cells lack telomerase. In that case the cells would gradually lose telomeric DNA due to incomplete replication or degradation of ends. This suggestion is supported by the finding that broken chromosomes of plants are healed in zygotic tissues but not in terminally differentiated tissues. Mutations in the yeast gene EST1 (ever shorter telomeres) (Lundblad & Szostak, 1989) and in the *Tetrahymena* telomerase RNA (Yu et al., 1990) led to altered telomere maintenance, and most of the mutants grew poorly and eventually died. EST1 is homologous to reverse transcriptase, and thus may be a component of yeast telomerase.

On the basis of the above findings, Harley (1991) has proposed a hypothesis in which telomere loss acts like a mitotic clock. This hypothesis postulates that: (1) Telomerase is active at some stage of gametogenesis and maintains telomere length in germ cells between generations of an organism. (2) During differentiation of most, if not all somatic tissues, telomerase is repressed. When somatic cells divide, loss of telomeric DNA occurs due to incomplete replication. (3) When a certain length of telomere length is reached on one or more telomeres in a dividing cell, a "checkpoint," analogous to those regulating cell cycle events, is signaled, which evokes the Hayflick limit. The cells stop dividing, which may be due to the fact that the shortened telomere may fail to bind to nuclear envelope or certain telomere-binding proteins. (4) Immortalization of a cell involves activation of telomerase.

Conclusions

Studies on the types of changes that occur in genes during aging and the identification of the genes that may be crucial for aging have been going on only for a decade. The number of genes in an adult eukaryote is enormous, and to identify the few crucial genes that change in expression and initiate aging, and the reason why they begin to change, is indeed difficult. Several approaches, model systems, and techniques are being used in attempts to solve these problems. Cell death occurs during the embryonic stage. A defined number of cells are programmed to die at specific locations of the embryo. These cells undergo rapid structural and functional changes, senesce, and die. One gene that seems to have a distinct role in the death of cells during insect development is the gene for polyubiquitin. Its expression rapidly increases before cells die. It is an obvious candidate for initiating the dissolution of cellular structures since ubiquitin is required for degradation of proteins. What triggers its expression is, however, not known.

How do the senescence and death of cells in adulthood compare with that in an embryo? Two types of cells are present in an adult: bone marrow and epithelial cells, which are premitotic and continue to divide throughout the life span; and neurons and skeletal and cardiac muscle cells, which are postmitotic and stop dividing at a very early stage of development. There is an apparent resemblance between the senescence and death of premitotic cells and embryonic cell death because in both cases cells at specific locations are earmarked for death, and their life span is short. Is the polyubiquitin gene also involved in their death, and do they undergo apoptosis? The postmitotic cells, however, have a long life span, and the timing of their senescence and death does not appear to be fixed. Certain cells live till the end of the life span of the organism, and certain cells die earlier. Is the trigger for their senescence the same as that of premitotic cells? Moreover, each type of postmitotic cell differs from another because it expresses certain unique genes that give the cell its own differentiated status. Do these unique genes have any role in its longevity, or do all cells, both premitotic and postmitotic, have a common set of genes whose expression/repression leads to aging?

Messenger RNAs from young and old rats have been used to construct cDNA libraries, and by subtractive hybridization, certain genes including T-kininogen, have been found to increase in expression in old age. Though the importance of these genes in aging is not obvious, such studies are useful in identifying genes that are either down-or up-regulated in differentiated tissues. Several questions remain to be answered. Are the genes responsible for senescence of premitotic and postmitotic cells the same or different? Is the trigger for aging of various differentiated cells the same or different? Are the same genes that are responsible for the cessation of cell division in the developing organism and the proliferation of fibroblast cells in vitro also responsible for their senescence and death, or another set of genes is involved? Studies on genes that encode enzymes/proteins and genes that are expressed under stresses of different kinds such as inflammation and high temperature are beginning to show how the expression of certain genes may be altered as a function of age. Such changes may destabilize the homeostatic balance of various functions, contribute to the decline of activity of the organism, and decrease its ability to adapt to changes in intrinsic and extrinsic factors.

It is not enough, however, just to know if a gene is expressed more or less after adulthood. What is of importance is to find out why the changes occur. This may reveal if there is a common molecular mechanism at the genetic level for the senescence of all kinds of cells, premitotic and postmitotic. This will require studies of the promoter region of genes which change in expression, an identification of their cis-acting elements and the trans-acting factors that regulate their expression, and alterations in these factors. Such studies have been undertaken for the fibronectin gene. Slight changes in the level or modification of trans-acting factors may produce profound effects on the transcription of a gene, and thereby change the level of a protein/enzyme. Alterations in trans-acting factors may also reactivate certain genes, as is seen for genes in the X chromosome, and enhance expression of certain "undesirable" oncogenes in old age. Such reactivation or up-or-down regulation may be of crucial importance for triggering senescence and/or its progression. One gene of crucial importance is the amyloid precursor gene, which has been implicated in Alzheimer's disease which afflicts people over age 65. The next few years should see advances in the understanding of the regulation of a few specific genes, and hopefully the ones that are involved in aging. Transgenic animal model studies are a useful approach to this problem.

Cell-fusion studies have provided evidence that chromosome 1 of human fibroblasts is likely to have a locus that controls cell prolif-eration, and that senescence is a dominant phenotypic character. How-ever, we still do not know which chromosome(s) is responsible for senescence. Subtractive hybridization is a useful approach to identify genes whose expression is up or down regulated in cells in culture. However, genes involved in senescence in vitro may not be the ones that are involved in senescence in vivo as the environment of the cells in vivo is different. Nevertheless, useful insights have been derived from such studies.

It is still not known whether methylation of cytosine residues in the DNA is the cause or effect of aging. While identification of factors that bind to methylated sequences is yielding useful information in this area, other types of modifications also may occur in the bases of both nuclear and mitochondrial DNA. Their role in the regulation of genes remains to be explored. Whether such modifications also affect deletions in the DNA needs to be studied. On the other hand, how the telomere DNA, besides playing a role in the stabilization of a chromosome, may also control the functions of its genes or act as a clock for aging needs to be explored.

Of the several steps in the information-transfer process from the gene to the protein, splicing of transcripts and editing of mRNAs are two steps that may profoundly affect the levels of gene products. Changes in splicing during aging will affect the levels of different isoforms of a protein, or contribute to scrambling of exons of different genes, thereby affecting a large number of gene products. Changes in mRNA editing, on the other hand, may contribute to alterations within a protein molecule and affect its function. Vast gaps exist in our understanding of the types of changes that occur in genes during aging. With the help of new genetic engineering technologies, the next decade should see some of these gaps filled.

6
Theories of aging

All multicellular organisms have two striking characteristics: (1) they show a gradual decline in their adaptability to the normal environment after attaining reproductive maturity, and (2) all members of a species have a more or less fixed life span. These two characteristics are inherited, and hence genetically controlled. Rats, mice, and *Drosophila,* kept under controlled environmental conditions, live for periods characteristic of the species and strain.

Besides these characteristics, it is also known that the time required for all members of a species to reach reproductive maturity is the same, as for example, 10 weeks for rats and 12 years for humans. Likewise, the reproductive period of all individuals of a species is more or less the same. The various stages of development leading to reproductive maturity are precisely timed. Thus, it is likely that developmental changes up to the attainment of reproductive maturity are genetically controlled and occur according to a genetic program.

Medvedev (1990) has brought together all theories on aging, classified them, and made an analysis of each group of theories. We shall consider here only those theories that attempt to explain aging at the genetic level. These theories are based on genes as the primary sites at which changes occur to initiate the process of aging. Factors such as food, temperature, humidity, radiation, pollution, and various stresses, however, influence the rate of aging.

Before discussing these theories, certain characteristics of aging that point to genetic involvement need to be considered. Though the manifestations and the sequence of deteriorative changes may be different in organisms of different phyla, the end result is an increase in the susceptibility of all organisms to one or more diseases, or a derangement of one or more organs which is followed by death. Within a class, the deteriorative changes in all individuals appear to be similar, though their rates may differ. This is well illustrated in mammals (Table 1.1). Though

246

the maximum life span of humans is nearly 50-fold longer than that of the insectivore, *Sorex fumens* (life span ~2 years), their deteriorative changes are more or less similar except that they occur faster in the latter.

Another point of interest is that all individuals within a species live approximately for the same length of time under given environmental conditions. For example, rats live for 3.5 years in a colony, and *Drosophila* live for 30 days. The maximum life span is the survival potential of the species. The maximum life span of humans is ~100 years. However, all individuals of a species do not attain the maximum life span in the wild or under uncontrolled conditions or varying environments. Another factor that points to a genetic basis for aging and life span is that offspring of long-lived parents have longer-than-average life spans, and the life spans of identical monozygotic twins is generally the same (Table 1.2).

The duration of the developmental period, reproductive period, and maximum life span is also an important factor. For mammals at least, there seems to be a positive correlation between the time required for reaching reproductive maturity and the maximum life span (Table 1.1). The mammal with the longest life span, *Homo sapiens,* takes the longest time to reach sexual maturity. The next in order are Indian elephants, horses, and domestic cattle. There are exceptions, particularly among the small mammals, where adaptive changes have evolved in response to special habitats. In all mammals, as also in most animals, the reproductive phase occupies a significant period of the life span and is followed by a postreproductive phase. The Australian marsupial mouse is an exception, as is also the Pacific salmon (*Onchorhynchus*) and the octopus (*Octopus hummelincki*). In these species the individuals die after a single act of reproduction. It appears as if sexual activity causes drastic depletion of certain essential factors that are not replenished, and that may be necessary for sustaining life.

In this context it is significant that there is an inverse relationship between the rate of reproduction or fecundity and duration of the reproductive period and longevity. Smaller mammals like rodents have a higher fecundity, a shorter reproductive period, and a shorter postreproductive period in their natural habitat than larger mammals like elephants, horses, cattle, and humans, whose reproductive rate is far lower. It would appear that a rapid reproductive rate is accompanied by a rapid depletion of certain essential factors and consequent deteriorative changes in the organism. However, this is not always true since turtles lay large numbers of eggs and yet have long life spans (Spector, 1956). Also, the queens of social insects, the honeybee and termite, which lay large numbers of eggs, have long life spans.

A survey of the life spans of mammals during their evolution shows that there has been a gradual increase in the time required for them to reach reproductive maturity. This is especially evident in hominids. During their evolution spreading over 3 million years, the maximum life span of hominids has increased about twofold, the maximum rate of increase of about 14 years per 100,000 years having been attained about 100,000 years ago (Cutler, 1975). This could occur only by mutations in the genome and gene rearrangement. Whether genetic changes resulting in longer life span caused the slower generation time exhibited by these long-lived mammals, or whether the slowing down of the generation time resulted in extended longevity is not known.

Of primary concern to us here is the time of onset, rate, and duration of aging. Even though the time at which reproduction begins is more or less well defined for each species, it is not known when senescence actually begins. If the time at which reproduction ceases is taken as the time of onset of senescence (though the functional capabilities of practically all organs begin to decline much earlier, and the rate of reproduction slows down long before it completely ceases), then it is observed that in higher mammals the period of senescence has also been extended concomitant with the extended reproductive period. The senescence phase is of no particular advantage to animals so far as the perpetuation and evolution of their species are concerned, and yet this phase has been extended. Is it because a slow reproductive rate has a low deleterious effect on other vital functions which, therefore, continue to function for longer periods? Unfortunately, sufficient data on the postreproductive period and, in particular, the effect of reproduction on the activities of other organs, are not available for correlating the reproductive period with senescence. It is important to determine how each reproductive cycle influences the functioning of various organs. The sharp hormonal changes that occur during reproduction are likely to affect the functioning of various organs, as it is known that the reproductive hormones affect the expression of several genes in various organs besides those of reproductive organs.

Even though the primary site of aging is the genome, where changes may be initiated after the attainment of reproductive maturity, alterations in the cellular environment brought about by hormones, temperature, and nutrition may have significant effects on the degree and rate of expression of genes that may lead to deterioration in function of one or more organs and cause aging of the organism. This is why the time of onset of aging and its rate and duration in the postreproductive period may vary among different members of a population and species.

The above observations strongly suggest that senescence has a genetic basis. Hence, several workers have developed theories that attempt to explain senescence at the genetic level. These theories are discussed below.

Somatic mutation theory

This was the first theory that attempted to explain the cause of aging at the DNA level. Over 50 years ago, Ross and Scott (1939) reported that rats exposed to whole-body irradiation that was too low to produce any acute syndrome died earlier than the control rats. Then it was reported that the symptoms of aging and death of irradiated rodents (Sacher, 1956; Henshaw, 1957) and humans (Warren, 1956) were similar to those of normal individuals, except that there was higher incidence of neoplasia in the former. Thus it was concluded that irradiation accelerates the process of aging. Based on these data, Szilard (1959a, 1959b) proposed the "somatic mutation" theory of aging. According to this theory, mutations that occur randomly and spontaneously destroy genes and chromosomes in postmitotic cells during the life span of an organism and gradually increase its mutation load. The increase in mutations and loss of functional genes decrease the production of functional proteins. Cell death occurs when the mutation load in a cell increases beyond a critical level. So the number of postmitotic cells decreases, and the overall functional ability of the organism declines.

To test this theory, Stevenson and Curtis (1961) and Curtis (1963, 1964) irradiated mice by X-ray (400 rad) or treated them with chemical mutagens such as nitrogen mustard (0.125 mg), and determined their survival. The frequency of chromosomal aberrations such as failure of sister chromatids to separate during mitotis and chromosome breaks was determined in the regenerating liver of young and old mice. They also carried out partial hepatectomy by carbon tetrachloride in mice of different ages and the number of chromosome aberrations were determined in metaphase chromosomes arrested by colchicine.

The main findings from the above experiments were that (1) the life-shortening effect of X-rays was dose dependent (Fig. 6.1); (2) the frequency of chromosomal aberrations was higher in the regenerating liver cells of irradiated mice; (3) nitrogen mustard did not produce any dose-

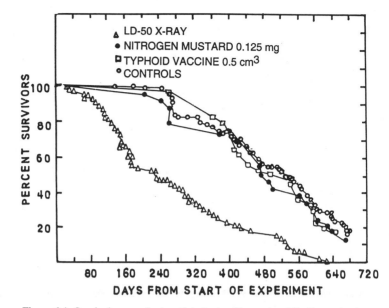

Figure 6.1. Survival curves for irradiated mice (2-months old). Curves begin at 30 days after treatment to eliminate the effects of acute mortality. Data indicate that single massive but nonlethal doses of noxious chemicals do not decrease life expectancy, whereas single massive but nonlethal doses of X-rays do so markedly. (Curtis, 1963; by permission of *Science*, © 1963 AAAS)

dependent response in shortening the life span, nor in the frequency of chromosomal aberrations; and (4) short-lived A/HEJ strain mice (average life span of 395 days) had more chromosomal aberrations than long-lived C57BL/6J mice (average life span of 600 days) (Crowley & Curtis 1963). Curtis (1963) suggested that random ionizing radiations damage postmitotic cells more than the premitotic cells. In the former, the mutagenic effects accumulate, and the cell has no means of eliminating them since it does not divide. In the premitotic cells, the damaged cells are eliminated and are replaced by undamaged cells. Also, germ line cells are more resistant to chromosomal damage, and hence the normal perpetuation of the species is not affected. *Drosophila melanogaster* were similarly exposed to a single dose of X-ray early in life. No immediate effects were seen, but X-rays had a life-shortening effect. The survival curve shifted to the left with increasing dose, showing a dose–effect relationship (Fig. 6.2; Lamb, 1977).

The effects of X-rays on the life span of the moth (*Habrobracon*), in

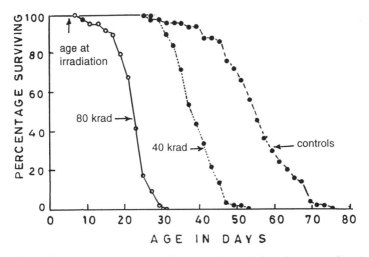

Figure 6.2. Survival curves for irradiated male *Drosophila melanogaster.* (Lamb, 1977)

which the females are diploid (2N) and the males are either diploid (2N) or haploid (1N), were studied by Clark and Rubin (1961) and Clark, Bertrand and Smith (1963). Both diploid and haploid males have the same life span. The haploid insects are, however, more sensitive to ionizing radiation than diploid insects. This shows that somatic mutation cannot be the cause of aging, because if this were true, the diploid insect should have a longer life span than the haploid insect, since the former has twice the number of genes and hence is more resistant to ionizing radiation. Also, the diploid males and females should have a similar life span. Data indicating that haploid insects are more sensitive to ionizing radiation than diploid insects show that the damage caused by radiation is repaired more effectively if the chromosome number is higher, and the repair is not dependent on sex. Thus, the decrease in the life span caused by radiation is different from that caused by normal aging. These data, therefore, are not consistent with the prediction of the somatic mutation theory.

Thompson and Holliday (1978) tested the theory using fibroblast cell cultures. When human fetal lung fibroblasts (strain MRC-5) were treated with colchicine for 3–6 hours, the surviving population contained nearly 60% tetraploid cells, which continued to divide. Their life span was not different from that of diploid cells. Also, their growth rate was similar. If accumulation of mutations or genetic damage were the cause of their death, then the tetraploid cells should have greater resistance and longer life as has been argued above for *Habrobracon,* but this is not true. Hoehn et al. (1975) also found that human skin diploid fibroblasts have the same life span as those of tetraploid cells. If mutations are involved in the senescence of cultured cells, they are not expressed during vigorous growth phase II, and hence must be recessive. It is likely that there is a gradual increase in the genetic load of recessive defects that may eventually lead to inactivation of one or more indispensable genes in both homologous chromosomes. Even so, the tetraploid cells with 4N chromosomes should be able to withstand more damage, and hence should have longer life spans. If mutations are deleterious, cells should die before mutations accumulate. The diploid and tetraploid cells should have similar life spans only if deleterious mutations occur toward the end of their life spans.

The data on the effects of ionizing radiation on spermatogonia and on various species of animals with different life spans also contradict this theory. The approximate LD_{50} for spermatogonia, humans, mice, and *Drosophila* are 50, 450, 500, and 6,400 roentgens, respectively. This is not consistent with the predictions of the theory in which germ cells are resistant to ionizing radiation. Also, according to this theory long-lived species should be relatively more resistant to radiation, but it has been shown that humans having approximately 50-fold and 1,200-fold longer life spans than mice and *Drosophila,* respectively, are more sensitive to radiation.

Though ionizing radiation has been shown to shorten the life span of mice (Lindop & Rotblat, 1961; Casarett, 1964) and *Drosophila* (Atlan, Miquel & Binnard, 1969; Miquel et al., 1972; Gartner, 1973), Strehler (1964) found that *Drosophila* exposed to ~4,500 rad actually lived longer than controls. A similar response was seen in germ-free mice (Walburg, 1966). The symptoms that develop in irradiated individuals whose longevity is shortened are different from those seen in normal aged individuals and unirradiated controls. Hence, ionizing radiation does not accelerate the normal aging process, but may cause early death by increasing the frequency of cancer that may arise from random damage to DNA. It should be emphasized that it is difficult to test the validity of the somatic mutation theory as there are no objective criteria for measuring the rate of somatic mutations in postmitotic cells. The only way such mutations have been measured is by evaluating the mortality rate. The latter may be due to several factors. Also, the reason the germ cells are supposed to be more resistant to ionizing radiation is not explained. The higher longevity of *Drosophila* (Strehler, 1964) and mice (Walburg et al., 1966) after exposure to low doses of ionizing radiation may be due to secondary effects; however, at higher doses the effects on mortality are evident. Thus, the cause of aging does not appear to be due to the accumulation of mutations. Life shortening due to ionizing radiation may be a nonspecific effect and may be due to "radiation syndrome," which is unrelated to natural aging.

Several workers have analyzed the possible changes in the DNA/ genome during aging to determine whether it has any relationship with aging. Since the mitochondrial genome (mtDNA) is relatively very small in comparison to the nuclear genome, several workers have studied its structural changes (see Chapter 5). One of the most important discoveries was that large segments of the human mitochondrial genome are deleted as the body ages. A 7,436-bp DNA segment is deleted from the myocardial mtDNA during the aging process. At age 80, 3% of mtDNA has this deletion, whereas at age 90, 9% has this deletion (Sugiyama et al., 1991). This segment of mtDNA codes for certain subunits of complexes I, III, IV, and V of the respiratory chain. It is conceivable that the loss of this segment would impair energy production by mitochondria.

Yen et al. (1991) have found deletion of a 4977-bp segment of DNA from the mtDNA of the liver in old persons. The deletion occurs at two specific sites, and the percentage of mitochondria with this deletion increases with increasing age. Such deletions also could impair mitochondrial function. Another significant change seen in mtDNA of the human diaphragm is the formation of 8-OH guanosine with increasing age. Its percentage increases at the rate of 0.25% per 10-year increase in age beyond 65 years. The presence of this modified base may be responsible for deletions in the mtDNA which may impair mitochondrial function (see Chapter 5). The nuclear genome (DNA) is very large as compared to mtDNA. It would be of value to find out if specific and consistent deletions occur in nuclear DNA, and whether they occur in repetitive or unique DNA sequences, or in introns or exons.

DNA damage and repair

Despite the fact that nuclear DNA is complexed with chromosomal proteins to form the chromatin complex, and is apparently less susceptible to damage, the DNA needs to dissociate itself from these proteins to carry out replication and transcription, and modulation of these processes by binding of other proteins to different regions of the DNA. At such times, either one or both DNA strands may break or undergo modifications. Though the DNA repair enzyme may repair the nicks, such repairs may not be 100% efficient.

The possibility of the occurrence of nicks in one or both DNA strands has been examined by several workers. The rationale for these studies is that if such breaks are located in a gene, they would affect their expression. Price, Modak and Makinodan (1971) measured, in vitro, the incorporation of ^{3}H-thymidine into DNA of the brain, liver, and heart of mice of various ages by autoradiography using calf thymus DNA polymerase. Increased thymidine incorporation was found in senescent mice, which showed that a higher number of single-strand breaks occur in old mice and expose more sites for DNA synthesis or repair. It is also likely that the level of DNA repair enzyme may decrease with age, which would leave the nicks unrepaired. This is consistent with an earlier finding of Samis, Falzone, and Wulff (1966), who showed higher incorporation of ^{3}H-thymidine into DNA of senescent mice.

Another method of assaying the occurrence of single-strand breaks in the DNA is to study the sensitivity of DNA to nuclease S1, which cleaves single-strand regions. Chetsanga et al. (1975) found that DNA in the liver of senescent mice (20 months) is more sensitive to nuclease S1 than that of 1- to 15-month-old mice. Also, alkaline sucrose gradient sedimentation of DNA shows that the DNA of the brain of old mice sediments in a polydisperse manner but that of young mice sediments in a monodisperse manner. The DNA of old mice undergoes degradation to small fractions of low M.W. by nuclease S1. These findings have provided good evidence that the number of nicks in the nuclear DNA increases with age.

DNA from the liver of old mice has a lower M.W. than that of young mice. However, no such difference is seen in the brain, spleen, and thymus (Ono, Okada & Sugahara, 1976; Polson & Webster, 1982). Also, no such difference is seen in young and old *Drosophila*. So it appears that the degree of DNA damage is tissue specific and species specific. Alternatively, in some tissues the repair mechanism is more efficient than others, and some species repair the damage in their DNA more efficiently than others. However, fibroblasts of patients suffering from Werner's syndrome or progeria do not show defective DNA repair even though the individuals show accelerated aging. Hence the degree of damage that occurs in different tissues and species may not be related to the aging process per se.

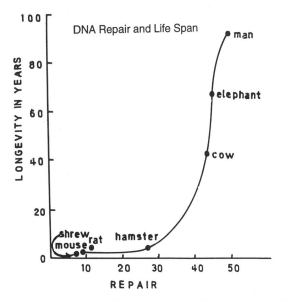

Figure 6.3. Relationship between longevity and incorporation of isotopic nucleotide into the DNA of fibroblasts under standardized conditions during repair following ultraviolet radiation. (Hart & Setlow, 1974)

The extent of unscheduled repair of damaged DNA in fibroblasts after their exposure to ultraviolet light was measured by Hart and Setlow (1974). Significantly, the fibroblasts of long-lived mammals repair the damage to a greater extent than short-lived mammals. There is a linear relationship between the log of the life span of the species and ^{3}H-thymidine incorporation into DNA of their fibroblasts (Fig. 6.3). Thus, it appears that the longer life span of a species may be due to more efficient repair capacity of its DNA. However, when ^{3}H-thymidine incorporation into DNA is measured in cultured fibroblast cells from the same species, lower incorporation occurs in late passage cells (Little, 1976). This shows that the level of the DNA repair enzyme decreases with age in all species.

Various types of damage occur in great number in the cellular DNA during the life span of an organism. Some of the more important ones are depurination, depyrimidination, deamination of cytosine, single-strand breaks (ssb), double-strand breaks, and O^6-methylguanine (Tice & Setlow, 1985). If these damages are not repaired, they will accumulate with advancing age and severely impair the expression of genes. Some of the damages are repaired rapidly. Single-strand breaks may arise from oxidative processes in the cell as well as enzymatic action near depurinated regions of DNA. Their repair is fast: ~200,000/hour in mammalian cells. Double-strand breaks are rare. They may arise due to ionizing radiations, by the action of chemicals like bleomycin, or by endonucleolytic attack on the strand opposite to an ssb site. They are repaired rather slowly. Ultraviolet irradiation may cause the formation of a dimer between adjacent pyrimidines in the same strand. Such lesions are recognized by specific endonucleases or glycosylases, which remove the damaged segment of DNA and replace it by the normal nucleotides using the opposite unaltered strand as a template. Modified nucleotides are removed after a nick in the backbone of the strand and the correct nucleotide is replaced. Also, repair can be effected by direct modification of the base (Tice & Setlow, 1985).

Popp et al. (1976) reported that persons accidentally exposed to radiation (in atomic bomb tests) as youths had isoleucine in their α- and β-globin chains of hemoglobin. Since isoleucine is not a normal amino acid in these polypeptide chains, its occurrence in these persons was due to a somatic mutation in the codon. However, its level did not increase with age.

Single-strand breaks are reported to accumulate in human fibroblast cells at passages close to senescence (Suzuki, Watanabe, & Horikawa 1980; Icard et al., 1979) and also in mouse cell strains (Beupain, Icard & Macieira-Coelho, 1980). Excision repair of DNA has been assessed after exposing fibroblast cells to UV light, and then estimating unscheduled DNA synthesis (UDS). This declined in late passage cells (Painter, Clarkson & Young, 1973). A positive and linear correlation between the number of apurinic sites in the *Paramecium* genome and the clonal age of the population has been observed by Holmes and Holmes (1986), which shows that in protozoa accumulation of DNA damage is also age-related.

Vijg et al. (1985) measured UV-induced DNA repair in early-passage fibroblasts obtained from skin biopsies of 4- to 44-month-old inbred rats. There was very little difference in the excision repair of individual rats regardless of age. Also, there was very little difference in the DNA repair in cells taken from the same rat at different ages. So the absence of inter- and intra-individual differences in the DNA excision repair system indicates that DNA repair system is not an important determinant of longevity. Goldstein (1971) and Painter et al. (1973) reported similar results from human fibroblasts, though late-passage human fibroblasts show a decrease in DNA repair.

Thus, even though DNA damage may contribute to the aging process in certain tissues, neither the damage nor the failure to repair the damage is a direct cause of aging.

Error theory

The error theory of aging was advanced by Orgel (1963), who postulated that errors occurring during information transfer steps like transcription and translation may cause accumulation of defective proteins and in turn cause aging. Such errors include the incorporation of wrong nucleotides into mRNAs during transcription. This may change the triplet codons and thereby cause incorporation of the wrong amino acids into proteins during translation, causing the proteins to be partially or totally inactive. Particularly, errors in proteins that are involved in protein synthesis, like the RNA polymerase II that is required for transcription and the aminoacyl-tRNA synthetases that are required for charging tRNAs with specific amino acids, would amplify such errors in cells. These errors may self-propagate and cause an exponential increase in defective enzymes and proteins, and this, in turn, may lead to "error catastrophe," resulting in senescence and the death of cells.

Incorporation of errors during the transcription or translation into enzymes that catalyze metabolic reactions may not have deleterious effects on the cell since the enzymes have short half-lives. The errors are erased as soon as the enzymes are degraded. If, however, an error is incorporated into RNA polymerase II or aminoacyl-tRNA synthetase, it would cause incorporation of errors into many other types of proteins as well. The level of error-containing proteins would, therefore, increase exponentially. For example, a defective RNA polymerase II may incorporate one or more incorrect nucleotides into various types of mRNAs that it transcribes, and cause changes in the codons for amino acids. Similarly, a defective aminoacyl-tRNA synthetase may charge a tRNA with a wrong amino acid which in turn may be incorporated into a protein. Thus, the error theory is based on the assumption that the information-transfer machinery is prone to error. In other words, the fidelity or accuracy of the machinery is not absolute, and once errors arise in enzymes for protein synthesis, they are amplified.

This theory was later modified by Orgel (1970), who proposed that even though the accuracy of the protein-synthesizing machinery is not absolute and allows introduction of errors, such errors may not always accumulate since the successive generations of protein-synthesis apparatus are discrete. According to him, if C_n is the error frequency in the nth generation of the protein-synthesis apparatus, R is the residual error frequency, and α is the proportionality constant between errors in the synthesis apparatus and errors in newly synthesized proteins, then

$$C_{n+1} = R + \alpha C_n$$

If $C_o = 0$, then

$$C_n = R (1 + \alpha + \alpha^2 + \cdots + \alpha^{n-1}).$$

If $\alpha > 1$, C_n increases indefinitely. If $\alpha >>> 1$, the error frequency increases exponentially and causes error catastrophe. However, α may not be >1 in all cases and, therefore, error catastrophe may not be inevitable. In such cases, a steady-state error frequency of $R/(1-\alpha)$ is reached (Orgel, 1973). These two types of situations may arise as follows: Suppose that the first generation of protein-synthesis apparatus that synthesizes aminoacyl-tRNA synthetases, ribosomal proteins, and other proteins has no errors. Since the fidelity of the apparatus is not absolute, the second generation would have some errors. When this latter generation is used for protein synthesis, the resulting third-generation proteins for the protein-synthesizing machinery would have more errors, and so on. If this process continues, one of the two following situations will arise: (1) error frequency may diverge resulting sooner or later in such large numbers of errors that the cell would no longer function due to error catastrophe; and (2) the error frequency would converge to a stable non-zero value or attain a steady-state level in which case no aging phenomenon would be seen. Orgel (1973) argues that the cell may also produce a protein-synthesis apparatus with less errors from an apparatus having more errors by means of enzymes that scavenge error-containing proteins (Goldberg, 1972), resulting in the stabilization of a small error frequency.

The validity of the error theory has been tested by several workers. Printz and Gross (1967) found that the *leu-5* mutant of the fungus, *Neurospora*, synthesizes a temperature-sensitive leucine-activating enzyme which at higher temperatures substitutes a leucine for other amino acids during translation. At low temperatures, protein synthesis proceeds normally, and the fungus has a normal life span. However, at a higher temperature (35°C) the mutant senesces early. Lewis and Holliday (1970) found that the fidelity of protein synthesis is decreased when the mutant is maintained at 37°C instead of at 25°C. However, the error frequency soon stabilizes and remains constant for a considerable time. The cells, however, begin to age after about 70 hours. These researchers further reported that the thermolability and specific activity of glutamate dehydrogenase of the fungus rapidly decrease at 37°C.

Holliday and Tarrant (1972) have argued that somatic mutations and errors in protein synthesis may not be distinguishable from each other. Errors in protein synthesis may arise due to somatic mutations. An error in the protein-synthesis apparatus may give rise to defective DNA polymerase which may, in turn, cause errors or mutations in DNA by introducing wrong nucleotides during replication.

Orgel (1973) pointed out that errors do not accumulate in germ cells, because if this occurred, the species would be wiped out. He postulated that "quality control" mechanisms may operate during oogenesis and early development, and lead to the rejection of ova and embryos having high levels of errors. He, however, does not say why and how such a quality-control mechanism, if there is one, fails to operate after the development period is over. Furthermore, if the occurrence of errors is the cause of cessation of cell division, senescence, and death, transformed or tumor cells should have no errors. It is unlikely that DNA replication and protein synthesis have absolute fidelity in transformed cells. If fidelity is increased on transformation, how is it brought about?

Using a different approach, Holliday and his co-workers studied the aging of cultured cells. Even though cultured human fibroblasts have a limited division potential, occasionally cells arise that divide indefinitely. Kirkwood and Holliday (1975), Holliday et al. (1977), and Kirkwood (1977) suggested that cultured cells are potentially immortal, but after a period of division, certain cells appear that become irreversibly committed to senescence and death. These cells multiply normally for a period, but after a certain number of divisions called the "incubation" period, M (number of cell divisions that elapse between commitment and death), they senesce and die. The uncommitted cells continue to divide.

What causes a cell to become committed and stop dividing? Is it a cytoplasmic factor or a gene? Occasionally, one finds fibroblast cells that continue to divide in culture, but they are abnormal cells, and often have an abnormal number of chromosomes. The commitment of cells to become differentiated and stop dividing occurs during development in vivo. Certain cells like neurons and skeletal and cardiac muscle cells stop dividing, become postmitotic, differentiate, and synthesize specific proteins to perform specific functions. Certain cells (for example, bone marrow and epithelial cells), on the other hand, continue to divide throughout the life span. Both types of cells are, therefore, committed, and such commitment is believed to occur by differential gene expression.

Holliday et al. (1977) and Kirkwood (1977) suggested that the commitment of cultured cells to senesce is due to errors, and that somatic cells have a diminished ability to regulate errors. The suggestion of Kirkwood (1977) that "to sustain the prospect of further evolutionary change and so improve its chance of ultimate survival, an organism must make occasional copying errors" appears to show that the organism knows that it must evolve and, therefore, makes errors. This is unacceptable. Certainly, had the fidelity of the information-transfer system been absolute, there would have been no evolution. Hoffman (1974) reported that the translation machinery is such that errors are not possible and hence error catastrophe cannot occur. Kirkwood and Holliday (1975) have pointed out the unlikelihood of such accuracy and have suggested that each protein-synthesizing machinery may have some activity even if it has errors, and this determines whether or not stability is attained. If the proportion of activity retained by erroneous adaptors, R, is low, stability will be achieved. If it is high, error catastrophe may result. This, according to them, is the basis of evolution.

Martin, Sprague and Epstein (1970) reported that cells in vivo retain their potential for division even in old age. Hence cells from even 90- to 100-year-old individuals are capable of 20–25 population doublings. Kirkwood (1977) suggested that as cultured cells become committed, the population size falls sharply. A similar situation may occur in vivo whereby organs lose cells and hence activity. This may be due to a shift in the protein-synthesizing machinery from a stable to an unstable state that allows errors to occur in proteins.

Hopfield (1974) suggested that errors can be avoided in replication machinery by expenditure of energy, either by proofreading or by the destruction of erroneous products. Scavenging enzymes may remove error-containing products. Accuracy in the germ line is vital for gene survival, but may not be so essential for somatic cells. If accuracy has to be maintained in somatic cells as well, then the energy cost would be too high. So Kirkwood (1977) has suggested that aging may be due to switching off of the mechanism responsible for high accuracy in the translation apparatus at or around the time of differentiation of somatic cells to save energy. Thus the somatic cells are left in an unstable state that causes a gradual increase in errors, which after a period, commit them to error catastrophe. However, we have observed instances of cell death (apoptosis) in the embryonic stage (Chapter 1). How do specific cells become committed to senescence at such an early age? These arguments do not explain why errors are made and what regulates the error level to account for the spectrum of changes seen in various types of cells, cells that completely stop dividing (neurons and skeletal and cardiac muscle cells), cells that continue to divide slowly (hepatocytes), and cells that proliferate throughout the life span (epithelial and bone marrow cells). Also, the aging of cells in vitro is not the same as the aging of cells in vivo, because the factors to which in vivo cells are exposed are enormous. Moreover, why fibroblasts from chickens and humans have a finite life span, and why those from rabbits, hamsters, rats, and mice generally have infinite life spans is not easily explained by the error theory. One would have expected the latter to have a lower population-doubling potential if it is related in any way to the life span.

The experimental findings of Kanungo and his co-workers contradict the error theory. Kanungo and Gandhi (1972) compared malate dehydrogenase (MDH) from the liver of young and old rats immunologically, and found no age-related differences between the two. Hence, there is apparently no substitution of amino acids in MDH as the rat ages. Kinetic, electrophoretic, and immunological studies on acetylcholinesterase of the brain (Moudgil & Kanungo, 1973b) and alanine aminotransferase (AAT) of the liver (Patnaik & Kanungo, 1976) of young and old rats also do not show any apparent differences in the enzymes with increasing age. Further support for this conclusion came from the finding that tryptic peptide maps of both actin and myosin heavy chain of the skeletal muscle of young, adult, and old rats are the same (Srivastava & Kanungo, 1979). Therefore, no changes occur in the primary structure or amino acid sequence of enzymes/proteins as a function of age of the rat. Hence errors do not appear in proteins as an animal ages.

Studies on aldolase of the liver of mice (Gershon & Gershon, 1973b), and superoxide dismutase (SOD) of the liver, brain, and heart of rats and mice (Reiss & Gershon, 1976) also showed that the antigenicity, K_m, K_i, and the electrophoretic mobility in polyacrylamide gel and SDS-polyacrylamide gels were the same for both young and old animals. Furthermore, isoelectrofocusing studies of these enzymes did not show any differences in their electrical charges (Goren et al., 1977). However, the enzyme from the old animals had lower specific activities and higher temperature sensitivities. These changes have been attributed to post-translational modifications of amino acid side chains. Modifications such as phosphorylation, adenylation, deamidation, acetylation, carbamylation, glycosylation, and oxidation of −SH groups have been shown to occur in the side chains of amino acids in various proteins. Enzymes with these modifications may have a lower specific activity, which accumulates with age.

The K_m, K_i, M.W., and electrophoretic mobility of aldolase of the free-living nematode, *Turbatrix aceti,* have also been reported to be similar in young and old animals (Zeelon, Gershon & Gershon, 1973). However, the aldolase had a lower antigenicity and specific activity. Rothstein and his co-workers studied the following enzymes from young and old *T. aceti:* isocitrate lyase (Reiss & Rothstein, 1975), enolase (Sharma, Gupta & Rothstein 1976; Sharma & Rothstein, 1978), triose-P-isomerase, and phosphoglycerate kinase (Gupta & Rothstein, 1976). In each case the M.W., K_m, thermal stability, and electrophoretic mobility were found to be similar. The catalytic activities were, however, found to be lower in old animals, which they suggested could be due to a certain percentage of completely inactive molecules, and not due to misincorporation of wrong amino acids into the proteins. Two-dimensional mapping of proteins of the superior cervical ganglia of young and old rats was carried out using isoelectric points in one dimension and M.W. in the other (Wilson, Hall & Stone, 1978). No differences were observed. Also, two-dimensional gel electrophoresis was carried out on proteins of *Caenorhabditis elegans* and *Drosophila* of different ages, and early- and late-passage human fibroblasts in culture. No difference in the proteins of young and old were seen (Rothstein, 1987). Isocitrate lyase, enolase, and 3-phosphoglycerate kinase were also analyzed in homogeneous populations of *T. aceti* of different ages. The three enzymes had reduced catalytic activity in old animals. In addition, the enolase in the old animals differed in antigenicity from that of the young. On the basis of these data, Rothstein (1979) suggested that the differences observed in old animals may be due to conformational changes and not due to amino acid substitutions.

Studies on glucose-6P-dehydrogenase (G-6P-DH) have, however, shown differences between the enzymes in young and old animals. Menecier and Dreyfus (1974) reported differences in the antigenicity of G-6P-DH of erythrocytes of young and old humans. The frequency of somatic mutations as judged by alterations in G-6P-DH was found to be higher in old animals (Fulder & Holliday, 1975). However, Yagil (1976) reported that electrophoretic mobility, electroimmunodiffusion, and temperature sensitivity of the enzyme of the liver are the same in young and old mice.

The possibility of error occurrence in cultured fibroblasts as they age was studied by several workers. Holliday and Tarrant (1972) reported an increase in the proportion of inactive and heat-labile G-6P-DH in late-passage MRC-5 cells. Lactate dehydrogenase (LDH) of late-passage MRC-5 fibroblasts was reported to be antigenically different from that of early-passage cells (Lewis & Tarrant, 1972). Also, immunologic and catalytic properties, and electrofocusing data of phosphoglycerate kinase, M2-type pyruvate kinase, and G-6P-DH of senescent fibroblasts in culture are not different from those of early-passage cells. Goldstein and Moerman (1975a, 1975b) reported similar changes in three enzymes of cultured human skin fibroblasts.

The fidelity of DNA polymerase of late-passage cells was reported to be considerably decreased when tested with a synthetic template. Also, the replicon elongation was found to decrease (Linn et al., 1976; Murray & Holliday, 1981; Krauss & Linn, 1986). However, Silber et al. (1985) found no significant age-related changes in the fidelity of DNA polymerase of partially hepatectomized liver. Thus, DNA polymerase of late-passage fibroblasts has low fidelity, and the enzyme of rapidly dividing liver cell has high fidelity. Srivastava et al. (1991) carried out similar studies on DNA polymerase of the liver of 6-, 16-, and 26-month-old rats. The fidelity of the enzyme was lower in old rats. Moreover, in diet-restricted rats, fidelity was higher at the three ages. It must be mentioned here that in all these experiments, synthetic DNA was used as a template to test enzyme activity. The degree of infidelity of DNA polymerase reported by these workers is so high that it is unlikely to occur in normal cells as it would be catastrophic for these cells. The infidelity observed may be due to the use of synthetic polynucleotides as templates. Of interest in this context is the finding that in neurons of mice, DNA synthesis (which is largely repair synthesis) is carried out by DNA polymerase β. This enzyme is error prone in copying ϕX 174 DNA. However, the error frequency of the enzyme from neurons of young and old mice is not different (Rao et al., 1985). Furthermore, the replication of virus in late-passage cells is not different from that of early-passage cells. Therefore, no infidelity in virus replication occurs as cells senesce in culture.

One of the crucial enzymes needed for information transfer – RNA polymerase – was studied by Evans (1976) in early- and late-passage cells. No differences in thermolability and specific activity were found. Misreading of poly-U in vitro by crude preparations of *Escherichia coli* obtained at various times of growth was studied in the presence of dehydrostreptomycin, which causes incorporation of errors during translation. A significant increase in poly-U misreading occurred, but this increase reached a stable value within a few generations of growth in the presence of the antibiotic (Garvin, Rosset & Gorini 1973). The in vivo studies of Edelmann and Gallant (1977) also gave similar results. Measurement of mistranslation (incorporation of cysteine into the flagellin of *E. coli* in the presence of streptomycin) showed that error frequency increased up to a level as high as 50-fold higher than normal and then stabilized or converged. Thus even if errors occur during translation, stable states of translational accuracy are attained as suggested by Goel and Ycas (1975). Therefore, the occurrence of errors during information transfer is unlikely to be the primary cause of aging.

There is thus an overwhelming amount of data to show that the substitution of incorrect amino acids in proteins do not occur in any significant amount with increasing age either in vivo or in vitro. Furthermore, errors are random processes, and if they occurred, one would expect to find proteins having a spectrum of amino acid substitutions. No such proteins have been identified. Moreover, the more-or-less fixed life spans of various species, the well-timed events during the life span, and the gradual decline in function with increasing age cannot be explained on the basis of errors that occur at random, unless one invokes the concept of regulation of errors by genes. This would contradict the basic tenet of the error theory. Hence, the increase in errors in functional molecules with increasing age is unlikely to be the cause of aging.

What then is the cause of the decrease in the specific activity and increase in thermolability of the enzymes found in old organisms? It has been suggested that such differences could arise by posttranslational modifications, such as the glycosylation and methylation of proteins, which do not change their net electric charge (Gershon, 1979). This may occur if the $T_{1/2}$ of enzymes increases with age due to a lower rate of degradation. The molecules may then remain for longer periods and, therefore, will have a greater probability of undergoing covalent modifications by transferases, transaminases, and other enzymes. However, the $T_{1/2}$ of proteins would increase if the levels of proteases decrease or the polyubiquitin level decreases. Then one needs to explain why the levels of degradative enzymes decrease in old age. Thus, such suggestions for aging like errors and posttranslational modifications are only circular and do not explain the basic cause of aging.

Disposable soma theory

Because of the lack of experimental support for the error theory, Holliday and his colleagues (Kirkwood & Holliday, 1979, 1986; Holliday, 1988) put forward the "disposable soma" theory to explain why and how aging takes place. They have argued that the adult organism allocates portions of the available energy for (1) maintenance and repair of the nonreproductive tissues, or soma, such as repair of DNA damage, accuracy of macromolecular synthesis, degradation of defective proteins, wound healing, and immune response, and (2) reproduction. If more energy is diverted for somatic maintenance, less is available for reproduction. In the former situation the organism would have a long life and reproduce for a longer period. If little energy is available for maintenance, it will live for a short period, and so it must reproduce at a rapid rate for its perpetuation. Though theoretically, if maintenance and repair of damages are perfect, the soma will stay on and the organism will not senesce; this is not so.

Hence, Holliday and his colleagues proposed that the optimum level of investment in somatic maintenance and repair is always less than the level required for indefinite somatic survival, and hence the soma senesces. Since all organisms are subjected to mortality in their natural environment, there is no advantage in investing all the energy for maintaining the soma. Thus natural selection favors the strategy of investing only sufficient energy in somatic maintenance and repair so that the soma remains viable throughout its normal expectation of life in the wild. For individuals protected from the usual mortality that occurs in the old, aging will occur as accumulated damage to the soma eventually takes its toll. So there is a trade-off between rapid growth and reproduction that favors the preservation of an organism's genes, and survival of the soma that may or may not benefit reproduction. If a species has a high mortality rate in the wild, then it is counterproductive to invest in maintenance of the soma, and it is better to invest in rapid reproduction. The alternative strategy is to develop resistance or better adaptation to environmental hazards, and thus reduce the likelihood of death from accidents and starvation. Thus, an extended survival of the soma will increase fecundity. Such a strategy has evolved in humans and elephants who have slow growth rates, slow reproduction, and increased longevity (Kirkwood & Holliday, 1979). In short, the overall aging process is the result of the eventual failure of maintenance mechanisms.

The disposable soma theory, however, does not explain how the genes are involved in this process, because ultimately it is the genes that specify not only the fixed life span of a species, but also the duration of its various phases.

The above arguments of Holliday and his colleagues were put forth earlier by Kanungo (1975) in the "gene regulation" theory of aging (see later in this chapter) in which he had propounded that destabilization of the homeostatic functioning of genes occurs after the attainment of reproductive maturity. Later Kanungo (1980) elaborated and extended this theory to explain that destabilization of gene function leads to accumulation/depletion of factors that may stimulate the expression of undesirable genes such as oncogenes or suppress desirable genes.

Dysdifferentiation hypothesis

The "dysdifferentiation hypothesis" was proposed more recently by Cutler (Cutler, 1985; Dean, Socher & Cutler 1985; Kator et al., 1985) to explain the mechanism of aging. According to this theory, aging occurs as a result of the gradual drifting away of cells from their proper state of differentiation as a function of time. The effect is cumulative and self-aggravating in nature. That is, cells become progressively more distant from their proper differentiated state as the animal gets older, and all functions of the organism decline more rapidly than the sum of its constituents as suggested by Rosen (1978). It is suggested that the same mechanisms and factors that are responsible for differentiation of a cell are also responsible for its dysdifferentiation. Cutler (1985, 1991) states that ordered differentiation leads to the creation of the organism; proper differentiation is maintained to the extent in time that is necessary to ensure the evolutionary success of the organism; and then aging begins to set in as a result of the slow random loss of this differentiation. As examples of dysdifferentiation they reported the expression of retroviruses (MuLV and MMTV) in the mouse brain (Dean et al., 1985), and *myc* oncogenes (*N*- and *c-myc*) in various tissues of old mice (Semsei et al., 1989).

Dysdifferentiation implies that the cell retraces its path. That is, a differentiated skeletal muscle cell should become a myoblast, and neurons should go back to the preneuronal state. No such visible changes in any type of differentiated cell have been reported after the attainment of adulthood. It is not necessary to invoke alterations in the differentiated state of a cell to account for alterations in its various functions.

Subtle changes in the levels of receptors, trans-acting factors, and effectors may bring about significant changes in the expression of genes, and thereby alter cell and organ function. This was suggested in the gene regulation theory (Kanungo, 1975, 1980; see later in this chapter). It was not suggested that aging is a programmed phenomenon.

Gene regulation theory

All multicellular organisms have two important characteristics: (1) Their life span includes development, a reproductive period, and senescence. They undergo a gradual decline in their adaptability to their normal environment after attaining reproductive maturity. During the life span, they show well-timed events, such as in humans, birth at 9.5 months, maturity at 12 years, cessation of growth at 20 years, and menopause in females at around 45 years. After attaining reproductive maturity, they reproduce once (as in *semelparous* species) and rapidly senesce and die, or reproduce several times (as in *iteroparous* species) at regular intervals, but at a gradually decreasing rate, then senesce and die. (2) Different species have different maximum life span potentials, as for example, *Drosophila* (30 days), mouse (3 years), cat (20 years), elephant (70 years), and humans (100 years) (Table 1.1). Furthermore, all individuals of a species have a more or less similar life span; long-lived individuals have generally long-lived progeny (Table 1.2); identical twins have generally similar life spans (Chapter 1). In addition, there is a correlation between the duration of development and longevity. Hence, senescence or aging should not be viewed as an isolated and independent phase in the life span of organisms, but should be considered together with development and adulthood phases. These earlier phases may not only influence the organism's longevity but also the rate, duration, and mode of its senescence.

There are abundant data to show that early development is the result of sequential activation and repression of specific genes. For example (1) various globin chains that are coded by different genes appear sequentially during the gestation period in humans (Fig. 2-1; Zuckerkandl, 1965). The hemoglobin of the fetus during 1–2 months of gestation is $\alpha_2\epsilon_2$. This is followed by $\alpha_2\gamma_2$ which is predominant during the rest of the fetal life, and is called fetal hemoglobin (HbF). The adult hemoglobin, HbA, comprising $\alpha_2\beta_2$ appears just before birth. While the α gene is activated early during fetal life and stays active throughout life, the ϵ, γ, and β genes are activated one after the other sequentially.

Also when the gene for γ is activated, that for ε is repressed, and the gene for γ is repressed when the β gene is activated. Furthermore, these genes remain active for different durations: ε for ~2 months, γ for ~8 months and β throughout the life span following birth. The factors responsible for switching on these genes in a programmed and sequential manner are not known, but it has been suggested that the varying oxygen tensions in the fetal blood may contribute to their sequential expression.

(2) The protein patterns of the larval tissue of *Drosophila melanogaster* change as development proceeds (Arking, 1978, see Arking, 1991 for other references). This shows sequential changes in the expression of corresponding genes. Each developmental stage differs from the earlier stage with respect to a few proteins.

(3) The isoenzymes of lactate dehydrogenase (LDH) are not only tissue specific, but also change during development (Markert & Ursprung, 1962). Later Kanungo and Singh (1965) and Singh and Kanungo (1968) showed that the proportion of M_4-LDH decreases with age in the heart, brain, and skeletal muscle. Since the two subunits of LDH, H and M, are coded by separate genes, these studies show that the gene for the M subunit is repressed with increasing age. Why and how this occurs is not known. It is likely that lowering of oxygen tension in the cells due to its decreasing permeability may have a role in gene repression/expression. Alanine aminotransferase (AAT) is a dimer with two subunits, A and B. Studies on cytoplasmic AAT (cAAT) of the liver of the rat show that the gene for A is active in the early period of the life span. As the rats age, the A gene is repressed and the B gene is activated. Only B subunits form the dimeric enzyme in old age (Fig. 2.2; Kanungo & Patnaik, 1975). There are several other isoenzymes that show sequential changes in their expression during the life span (Kanungo, 1980). Since the isoenzymes differ in their catalytic rates, the appearance of different isoenzymes as the animal develops, matures, and ages seems to indicate a functional significance.

(4) There are abundant data to show that some enzyme levels decrease with age, some increase, and some remain unchanged. But what is of significance in this context is that these levels can be reversed by various hormones that act at the genetic level. The levels of cholineacetyltransferase (CAT) and acetylcholinesterase (AChE), which decrease with age in the brain of rats, can be raised and brought back to adult levels by administration of 17 β-estradiol (James & Kanungo, 1978). Likewise, the level of pyruvate kinase in the heart of old rats can be restored to that of adult heart by administering steroid hormones (Chainy & Kan-

ungo, 1978a). These studies show that the levels of enzymes are reversible and can be modulated by different effectors.

(5) Based on immunological, tryptic digestion, and kinetic studies (Kanungo & Gandhi, 1972; Chainy & Kanungo, 1978a; Srivastava & Kanungo, 1979), it was proposed by Kanungo (1975, 1980) that the primary structures of proteins synthesized in the old organism are the same as those of the young organism. Hence, no changes occur in the nucleotide sequences of the corresponding genes. So the changes in the levels of enzymes observed at various ages are due either to transcription of the genes or translation of their mRNAs. Later, Ono et al. (1985), using Southern transfer and filter hybridization techniques, reported that no changes occur in the amplification and rearrangement of nine cloned sequences of DNA of mice.

Enzymes control various functions at the molecular and cellular levels which, in turn, influence the functions of organs and the organism. Enzymes are responsible for DNA synthesis (DNA polymerase) and hence control cell proliferation. Enzymes are responsible for their own synthesis, including transcription and translation. They also are responsible for the synthesis of trans-acting factors and hormones that act on genes and modulate their transcription. Therefore, alterations in the levels of these factors and hormones would also alter the expression of genes, and thereby alter the levels of enzymes, and in turn alter various activities of the organism. The differentiated state of a cell is derived by repression of specific genes and expression of others.

The question that needs to be answered is: Once cells have differentiated and the organism has reached the adult stage and has reproduced, why does its functional status begin to change, resulting in the decrease in reproductive ability and deterioration of all functions leading to senescence and death? Kanungo (1970) proposed the gene regulation theory of aging which was later presented as a model (Kanungo, 1975, 1980) to explain the following two main characteristics of aging: (1) deterioration of functional ability after reproductive competence is attained, and (2) fixed life span of all individuals of a species.

Deterioration of function

According to the gene regulation model for aging (Kanungo 1975, 1980), senescence occurs due to changes in the expression of genes as a result of reproduction and other adult activities. As shown in the model (Fig. 6.4), beginning from the zygote stage, activation and repres-

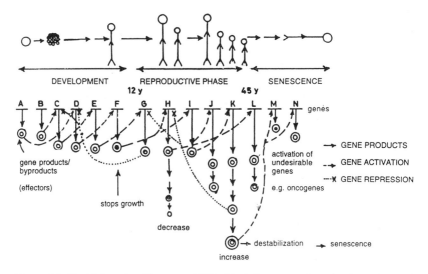

Figure 6.4. Model for aging (Kanungo, 1980). (Top) Representation of various phases of the life span, development, reproduction, and senescence. (Bottom) Number of active genes has been kept at a minimum, and genes that are permanently repressed are not shown for sake of clarity. Developmental and reproductive phases are dependent on unique genes, A-F and G-L, respectively. No specific genes for senescence are depicted in this model. Development occurs by the sequential activation of genes A-F, the product of gene A switching on gene B and so on. Some of the genes of late developmental phase, E and F, switch on certain unique genes G and H, belonging to the early reproductive phase. These genes, in turn, switch on sequentially other genes of the reproductive phase. The organism attains reproductive ability when required amounts of gene products are formed. Continued reproduction may cause depletion of certain factors which may be necessary for keeping certain essential genes active. Switching off of these genes may lead to deterioration of certain functions. Continued reproduction may also lead to accumulation of certain gene products (factors) beyond a certain level, resulting in activation of some undesirable genes, M and N, whose products may cause diseases such as autoimmune disorders or oncogenes causing cancer. Thus the decline in physiological functions that begins after a certain stage of reproductive phase may be due to destabilization of the functioning of the genes of reproductive phase or adulthood.

sion of specific genes cause development and growth of the embryo until a reproductively mature organism is formed. Products and by-products of the genes responsible for an earlier stage of development, on reaching critical levels, stimulate certain hitherto inactive genes and thereby advance the embryo to the next stage. It should be mentioned here that the steps are not linear as shown. As envisaged in the original model, there are several interacting and feedback processes and loops that operate, initiated by structural and regulatory genes, to give rise to a particular phenotype. Wilkins (1986) and Walford (1987) have essentially advanced the same views for explaining the mechanism of development and growth.

The above process goes on until certain genes are expressed which are responsible for the production of such sex hormones as 17 β-estradiol and testosterone, which confer reproductive ability to the organism. Reproductive function is of utmost importance for the perpetuation and evolution of the species. Some of the gene products of the reproductive phase, in turn, repress some of the genes that were active at an earlier stage and were responsible for growth. This gene repression is responsible for the cessation of further growth, a process that generally occurs in most animals after a short period of reproduction. The question that arises then is why the reproductive phase does not continue indefinitely.

The reproductive ability of an organism is the highest soon after reproductive maturity is attained. However, reproduction itself causes depletion of certain substances that are not replenished. Such substances may be of crucial importance for maintenance of gene expression required for reproduction and other adult functions. This process has been observed in semelparous animals such as the Pacific salmon, the Australian marsupial male mouse, and the female octopus, which reproduce only once, and then rapidly age and die. That reproduction depletes certain factors in these animals is substantiated by the finding that the life span of migrating salmon can be prolonged significantly by castration before the gonads develop (Robertson, 1961). Prevention of breeding in the male marsupial mouse increases its life span from 11 months to 2 years (Diamond, 1982). When certain plants are prevented from reproducing, they may continue to grow almost indefinitely (see Medvedev, 1990 for other examples). Indeed, it is known that in human females the level of 17β-estradiol gradually decreases after a short period of reproduction. On the other hand, certain substances may also accumulate as a result of reproduction since the animal may not be able to eliminate them.

The expression of genes, whether for growth or maintenance of adulthood and reproductive activity, depends on several factors, which are themselves the products or by-products of certain genes. The maintenance of optimum levels of these factors is required for the expression of the genes for which they act as modulators and / or trans-acting factors. Reproduction as well as other adult functions destabilize the homeostatic balance of these factors due to their depletion/accumulation. Also, nutrition and various types of stress to which the organism may be exposed to may destabilize the homeostatic balance of these factors. If a trans-acting factor falls below threshold level, it would fail to cause the expression of certain essential genes. If a factor accumulates and rises beyond a certain level, it may not only repress certain genes, but also may cause the expression of certain undesirable and harmful genes such as oncogenes. Thus, the homeostatic control of genes required for the reproductive phase or adulthood is destabilized or impaired. This leads to a gradual decline in the reproductive rate, which is one of the earliest and measurable functions to decline after the attainment of reproductive maturity.

Several examples can be cited in support of such destabilization or breakdown of the homeostatic balance of hormones and other products. 17β-Estradiol increases the level of oxytocin receptor in the rabbit uterus, and thereby increases its contractility in response to oxytocin. On the other hand, progesterone decreases the oxytocin receptor level, and also the contractility of the uterus (Nissenson, Flouret & Hechter, 1978). The two steroid hormones are known to bring about their effects by acting at the genetic level. Since their levels decrease with age, the contractility of the uterus that is mediated by oxytocin, is also expected to decrease.

The "sudden death" phenomenon described by Wodinsky (1977) in the octopus is a good example of how reproduction may destabilize certain factors in animals. The female octopus (*Octopus hummelincki*) lays eggs only once, broods them, reduces its food intake, and dies soon after the young hatch. If the two optic glands of the animal are removed after spawning, it does not brood, but continues to eat and grow. The longevity of the octopus is thus extended. It appears that certain factors present in the optic gland are responsible for brooding and cessation of feeding, followed by senescence and death. Removal of the optic gland or these factors delays senescence. The stress of formation and laying of eggs may deplete certain factors which may, in turn, cause the optic gland to release certain substances that are responsible for the physio-logical and behavioral changes.

The Japanese quail begins to lay eggs at about 10 weeks of age and continues to do so at an optimal rate until about 25 weeks when the number of eggs laid starts to decrease. In the 20-week-old quail, which lays eggs at optimal rate, the estradiol:progesterone ratio is 0.04. In the 42-week-old quail, which lays a considerably lower number of eggs, the ratio is 0.08 (Mahendra & Kanungo, unpublished data). Evidently, the shift in the ratio of the two hormones, which is due to decreases in the levels of both estradiol and progesterone, contributes to the decrease in the number of eggs laid. Estradiol is responsible for the production of the egg proteins. Decreasing the level of estradiol may have a role in the gradual decrease in expression of egg proteins, thereby lowering the ability of the bird to form and lay eggs (Gupta, Upadhyay & Kanungo, unpublished data).

The continuous hemodynamic pressure overload in the mammalian heart causes hypertrophy of the cardiac muscle, which is due to an increase in cell size without cell division. What is of significance is that the expression of the α-skeletal actin (SKA) gene is greatly increased in the heart of old rats (Jaiswal & Kanungo, 1990). α-SKA is a fetal protein and the re-expression of its gene in the old myocardium may cause hypertrophy of the heart muscle, thereby lowering its functional efficiency. The expression of c-*myc* and c-*fos* proto-oncogenes also increases in the heart as a function of age (Fujita & Maruyama, 1991; Jaiswal, 1988). Moreover, the proto-oncogenes, *ras* and *abl*, are also expressed in the old heart (Jaiswal, 1988). The proto-oncogenes c-*myc* and c-*fos* are expressed during the early developmental period and are implicated in cell proliferation. Their expression practically ceases after the cells stop dividing and differentiate. However, Semsei et al. (1989) found that c-*myc* gene is expressed in many tissues of old mice. Ono and Cutler (1978) had reported earlier that the c-type retrovirus (MuLV) is expressed in the brain of old mice. Rath and Kanungo (unpublished) have found that the expression of c-*myc* and c-*fos* in the brain of rats is far higher in old than in adult animals. The cells of these organs, therefore, appear to become more prone to tumorigenic transformation in old age. More work needs to be done to understand the implications of these findings.

The finding of Lumpkin et al. (1986) that senescent human diploid fibroblasts have far higher levels of poly-A$^+$ mRNAs that inhibit DNA synthesis in proliferation-competent cells supports the argument that undesirable genes are expressed in old age. When these poly-A$^+$ mRNAs are microinjected into young proliferating cells, DNA synthesis in the latter is inhibited. The level of inhibitor poly-A$^+$ mRNAs is 0.8% in senescent cells, and that of quiescent cells is 0.005%. The induction of antiproliferative mRNAs may be due to destabilization of the cellular machinery after several rounds of division.

According to the "gene regulation" theory, subtle changes occur in various factors after attainment of adulthood, not only due to reproduction, but also due to the nutrition of the animal and its exposure to

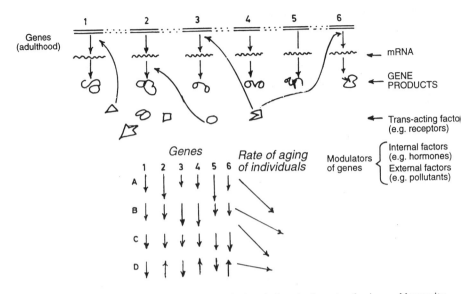

Figure 6.5. Depiction of individual variations in the rate of aging and longevity within a species. Individuals differ in their genotype, trans-acting factors that mediate gene expression, nutrition, stresses, and pollutants to which they are exposed. These affect the expression of genes in various ways in different organs. The sum total of the expression of an array of genes, say 1–6, affect the functioning of different organs. This accounts for the variations in the pattern and rate of aging of individuals, represented as slopes, within a species.

various types of stresses such as temperature, pollution, starvation, and radiation. For humans and domesticated and protected animals, in particular, psychological and social factors are also important. These factors vary not only for different populations of the same species living in different habitats, but also among individuals of the same population. They cause alterations in the levels of hormones and other trans-acting factors that interact with cis-acting elements of an array of genes that are essential for maintenance of adult functions including reproduction.

Alterations in the expressions of genes differ in different individuals of a species depending on their genotype and the stresses to which they are subjected. Hence, the rate and type of "normal" aging differ among individuals as shown in Figure 6.5. This figure explains why there are intraspecies variations, and why individuals of the same species show different aging patterns. Genotype, nutrition, and stress affect the expression of genes, mediated through receptors and effectors. Such changes alter the expression of genes and, therefore, the levels of enzymes and other proteins that they encode. This influences various activities of the organism and the rate of its aging. This is what accounts for the various rates of aging or slopes of aging, as shown in Figure 6.5, in different individuals within a species, namely, the intraspecies variation. As emphasized in the original model (Kanungo, 1975), aging is not programmed. It results from the destabilization of the homeostatic functioning of an array of genes that are essential for maintaining adult activities, including reproduction.

Determination of life span

The life span of an animal may be broadly divided into three phases: developmental, reproductive, and senescence. The initiation and duration of the developmental and reproductive phases depend on unique sets of genes, which are sequentially activated and repressed. The duration of these phases may vary within certain limits and may be influenced by intrinsic factors such as hormones and other effectors, and extrinsic factors such as nutrition and stresses. These may account for the variability in the durations of these phases in the life spans of individuals within a species.

The number of genes required for each phase of individual mammalian species may not differ significantly. The mammal with the longest life span, *Homo sapiens,* has about a 30-fold longer life span than that of the mouse, but the DNA content of the latter does not differ greatly from the former. It is also likely that the number and types of genes required for development and reproductive phases in these two species are similar. Then how and why do human beings live so much longer?

A new species that may evolve from a pre-existing one during evolution may have a longer or shorter life span. This may be due to genetic alterations such as mutations, deletions, translocations, rearrangements, and gene duplication in the genes affecting one or more phases or subphases of the life span, especially development and adulthood. This may alter the responsiveness of the genes to effectors, and result in shortening or lengthening of one or more phases/subphases, and thus alter the total life span. For example, if gene D (Fig. 6.4) is so altered that it takes longer time to be switched on by a previous gene product, and hence its by-product takes longer time to be produced, then there shall be an extension of the growth phase from that point on. On the other hand, gene D may so change that its product reaches its threshold level earlier and activates the next gene in the sequence earlier. Such a change in the gene may shorten the phase and thereby affect the total life span from that point on.

Likewise, a change in gene I of the reproductive phase may lengthen the phase and hence affect the life span. Such changes in one or more genes of the reproductive phase of the early ancestors of mammals that had short life spans may have played a role in the gradual prolongation of this phase, and caused the evolution of long-lived mammals. Shortening of the reproductive phase may occur if the gene I is so altered that it is activated faster. Such genetic changes may cause an organism to reproduce at an appropriate time in the wild, and may be selected during evolution on the basis of its selective advantage. Genetic changes that cause an organism to reproduce at an inappropriate time will be selected against as the offspring may not survive, and the force of natural selection shall be insufficient. Thus, those genetic changes that allow reproduction at an appropriate time will confer more survival and evolutionary potential on the species, and hence they will be selected, and help in the perpetuation of the species. Such alterations in gene function may have played a significant role in the selection process so essential for evolution.

It should be mentioned here that lengthening or shortening of the developmental period by genetic alterations may not necessarily influence the reproductive period. Alterations in the genes required for development may extend the period, but if no alterations in the genes required for reproductive phase have occurred then the duration of this phase shall not be changed. Once the early genes required for reproduction have been triggered by the gene products of the developmental period, the sequential operation of the genes for adulthood will continue, and its duration will not have any correlation with that of development. The findings of Johnson (1987) on the nematode, *Caenorhabditis elegans*, support this hypothesis. Johnson generated recombinant inbred lines of *C. elegans*, which had a 70% longer life span than the wild type. The lengths of developmental and reproductive periods were unrelated to increased life span, and development and reproduction were under independent genetic control.

The elucidation of the role of cis-acting elements in the promoter regions of genes in the regulation of gene expression provides an insight into how the expression of a gene may be increased/decreased or delayed/accelerated. The cis-acting elements may be viewed as modules that interact with trans-acting factors to become operational. There are many such modules – CCAAT, CACCC, GGGCGGG, CRE, GRE, ERE, PRE, for example – that bind to specific trans-acting factors.

Though the expression of a protein-coding gene depends on RNA polymerase II and various transcription factors, TFIID, TFIIA, TFIIB, TFIIE, TFIIF and TFIIH that bind to the start site, the regulation of its expression depends on one or more of the above modules. A transacting factor, on binding to the corresponding module, brings about conformational changes in the chromatin, and is believed to interact with the transcription factors and alter the rate of transcription. Thus the modular nature of gene expression offers an explanation for both deterioration of various functions leading to aging and extension or reduction of the life span of a species.

Strähle, Schmid, and Schutz (1988) used a plasmid having one or more 15-bp sequences containing GRE or ERE upstream of the TATA site of a thymidine kinase (tk) promoter and studied the expression of the reporter gene, chloramphenicol acetyltransferase (CAT), in MCF-7 cells with and without dexamethasone or estradiol. The presence of two GREs caused far greater stimulation of expression than one GRE. With four GREs, the expression increased still more, but the increase was not proportional to the number of GREs. When a CCAAT motif was placed 5' to the GRE, the induction was higher than with two GREs. If the CCAAT motif was subjected to mutation, then the induction was lower than with one GRE. If the GRE was placed at a longer distance from the start site, it was not sufficient for induction (Fig. 6.6). Hence, multiple GREs or a combination of GRE with other modules is necessary for induction. Thus clusters of repeating or heterogeneous modules may be a means by which evolution circumvents the problem of spontaneously arising motifs that may bind to trans-acting factors.

The gene for α_1-antitrypsin is a good example to illustrate interspecies differences in expression. In the mouse it is expressed in the adult liver and kidney (Kelsey et al., 1987). In humans, it is expressed in the gut, pancreas, and lung, in addition to the liver and kidney. When the human gene was placed in the mouse genome, it was expressed in a broad pattern characteristic of the human gene (Kelsey et al, 1987; Koopman, Povey & Lovell-Badge, 1989). So the broader tissue expression of α_1-antitrypsin gene in humans as compared to that of mouse is largely due to differences in cis-acting modules between the two species. It is argued that in addition to a set of common trans-acting factors to which both human and mouse α_1-antitrypsin genes respond, either the human gene must respond to a unique set of positive factors or the mouse gene must respond to a unique set of negative factors in order to elaborate the broader pattern of the human gene. It is also likely that regulatory differences between species could be achieved through changes in the

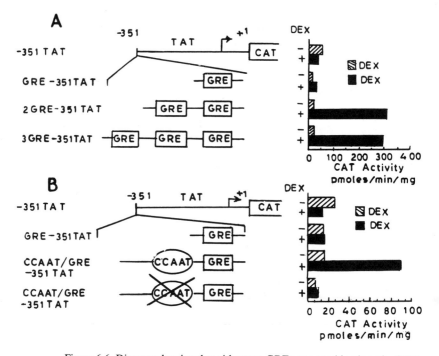

Figure 6.6. Diagram showing that either two GREs or a combination of a GRE with a CCAAT are required for induction from a promoter distant position. (A) Constructs containing either one, two, or three GREs at position −351 upstream of TAT promoter are transfected into Ltk⁻ cells and analyzed for CAT activity after treating the cells with 10⁻⁷M dexamethasone. The presence of two GREs greatly stimulates CAT activity, but the addition of another does not further increase activity, indicating saturation of the concerted effect. (B) Ltk⁻ cells are transfected with plasmids carrying the sequences as shown and then treated with 10⁻⁷M dexamethasone. In the second construct, CCAAT is placed 6 bp away from the 15-bp GRE. In the third construct, the CCAAT motif is shown destroyed by transversions. The expression of CAT greatly increases in the presence of a normal CCAAT. (Strähle, Schmid & Schutz, 1988)

concentrations of regulatory proteins and/or changes in the binding affinity of these proteins either through mutational or posttranslational modifications of the regulatory proteins, besides subtle alterations in the nucleotide substitutions in the promoter/enhancer elements (Cavener, 1992). A gene in one species may respond to a lower level of a regulatory protein, but to a higher level in a related species.

The number of cis-acting elements present in various genes studied so far is limited. Also the number of trans-acting factors is limited. But in different combinations they may bring about an enormous variety of regulatory changes in a gene present in different species. Moreover, the distances between these modules may add another dimension to this network. In addition, reversible modifications of bases in the cis-acting elements such as phosphorylation, acetylation, and methylation as proposed in Chapter 4 may influence the binding of trans-acting factors. Changes in the expression of even one regulatory protein may influence a large number of downstream genes. These may be the major mechanisms of evolution and longevity in organisms.

The validity of this hypothesis can be tested by studying the genes that are expressed during development and adulthood of two related species which vary significantly in the duration of these phases and in longevity, such as chimpanzees and humans. The maximum human life span is about three times greater than that of the chimpanzee. The human gene can be placed in the chimpanzee genome to examine when it is expressed, and which trans-acting factors are involved in its expression. The differences in the cis-acting elements and trans-acting factors of the two species can be examined. Involvement of a smaller number of modules and trans-acting factors in the chimpanzee would mean a less complex network and quicker activation of the gene. Several genes that are expressed during development or adulthood could be tested to provide an insight into the evolutionary significance of regulatory sequences in the promoter regions of genes and their role in longevity.

Thus, acceleration or deceleration of sequential activation/repression of genes may start at any point in the sequence of genes that regulate development and reproductive phases, depending on where one or more genes have been altered. A new species arises when one or more phases are reset. This results in a shorter or longer life span of the new species as compared to the ancestor from which it arose, and account for the differences in the life spans of related mammalian species (Table 1.1). The discrepancies seen in the relationships among the phases and maximum life span may be because resetting of one phase may not affect another phase.

Thus according to this model (Figs. 6.4, 6.5), the alteration of a gene that functions at a specific time of a specific phase not only changes the rate and duration of that phase from that point on, but also may alter the duration of the total life span. The work on human genetic diseases like progeria and Werner's syndrome supports this argument. Progeria is caused by a dominant mutation in an autosomal gene. Even though the newborn child appears normal and grows normally, by about age 6 growth ceases; the patient shows premature aging; and dies at about 12 years. Fibroblasts taken from these patients at age 9 undergo a far lower number of population doublings than age-matched controls. It appears that the gene(s) that is required for development has been altered to prevent complete development, and hence the child does not mature. Mutation in this gene possibly prevents the production of some essential factor(s) that may be necessary to switch on the next gene in the series or network. The development phase is thus greatly shortened and the reproductive phase is not triggered. The life span is shortened from that point in time when the mutant gene is expressed. Whether the mutant gene is expressed and, if so, whether it produces an undesirable product that causes premature aging need to be studied.

Werner's syndrome is also due to a mutation in an autosomal gene. In patients with this syndrome, the duration and process of development appear normal. Thus, it appears that the mutation is in a gene that acts after the mutant gene in progeria patients, and is required for the initiation and maintenance of reproduction. Here, it is the reproductive phase that is shortened and the total life span is also shortened. It appears that the mutant genes in progeria and Werner's syndrome have pleiotropic effects since they affect several organs and functions. Studies on these genes may yield valuable information on their role in longevity.

In summary, each species has a set of genes that controls development. Their sequential activation determines the duration of development, beginning from the zygote and ending at the reproductive stage. The reproductive stage is controlled by a set of genes whose expression controls reproduction and other activities in adults. The duration of the reproductive phase depends on the capacity of the organism to replenish the factors that become depleted due to reproductive activity. In other words, the better the maintenance of reproductive activity, the longer is its duration. The factors that become depleted may be essential for keeping certain genes active and for carrying out reproductive and other adult functions. Certain genes also remain repressed. Depletion of factors may switch on certain undesirable genes. Certain factors may also accumulate as a result of reproduction and activate/repress certain undesirable genes like oncogenes or overexpress or down-regulate certain essential genes.

Thus, the breakdown or loss of homeostasis of the levels of trans-acting factors and other by-products of genes in the cell, due to reproduction, nutrition, and stress of various kinds, destabilizes the expression of genes that are needed for maintaining adult functions. The destabilization or the noise in the system continues to increase with age as functioning of various organs deteriorates. The effect is cumulative and becomes exponential, because the effects of deterioration of two or more organs is far greater than the effect of deterioration of a single organ on the total performance of the organism. Indeed, because of the interdependence of organs, deterioration of one organ also affects the functioning of one or more organs. No unique genes causing aging are envisaged in this model, nor is it suggested that aging is programmed like development and reproductive phases. Whereas development and reproductive phases are under the control of unique genes that become activated at specific times, according to this model no unique gene(s) is required to cause aging. Hence aging is not programmed, as it would imply that specific genes evolved for the purpose and were selected during evolution of the species to become activated at a specific time after a certain period of reproduction.

It is suggested that normal aging is a passive process. It is merely a consequence of the organism attaining reproductive ability or adulthood even if it actually does not reproduce. It results from the destabilization of the homeostatic balance of gene products and other effectors that influence the expression of reproductive-phase genes. The loss of homeostasis is due to reproduction and other intrinsic as well as extrinsic factors. This can be simply put as follows: There are two water baths, A and B. A is kept in a room maintained at 25°C, and B is kept in a room maintained at 10°C. If the water in both the baths is required to be maintained at 37°C, there is a greater probability that B will break down sooner, as its heating coil will have to work longer, and hence shall be subjected to greater stress.

According to this theory, it is possible to prolong the reproductive phase or adulthood by preventing the destabilization of the homeostasis of the cellular environment or changes in the levels of the effectors that influence the expression of genes. Two experimental conditions that have been shown to extend the adult period are the restriction of caloric intake in rats, and the prevention of reproduction by castration in the Pacific salmon. Also, removal of the optic gland in the octopus rescues the animal from senescence. Other experimental methods may be adopted so that the factors that become depleted are replenished and those that accumulate are removed. This will allow maintenance of homeostasis of the cellular environment and prevent destabilization of gene function.

Indeed, there has occurred a gradual increase in the life spans and the reproductive phase, in particular, during the evolution of mammals.

The selection of species with longer reproductive phases and production of smaller number of offspring has been preferred over short reproductive periods and large numbers of offspring. A longer period of senescence is a consequence of a longer reproductive period, which has consequently extended the life span. Hence aging has not been selected during evolution, nor has it been programmed into the life span as development and reproduction have been. Besides being a consequence of extended reproductive period, it is likely that, especially in mammals, the enlargement of the brain that allows greater adaptability to environmental changes has also contributed to longer periods of senescence. This is more pronounced in animals, including humans, who are protected from the vagaries of nature. A secondary benefit of aging followed by death has been the elimination of unreproductive individuals from the population, which has made it possible for younger individuals to thrive and reproduce, and facilitates turnover of the species. This is the basis for the appearance of newer individuals with greater potential for selection and evolution.

7
Conclusions and future prospects

Biological research on aging is relatively a new field, at least in comparison to research on development. Nevertheless, its progress has been rapid, particularly during the last two decades, due to the utilization of sophisticated techniques of biochemistry, biophysics, molecular biology, and genetic engineering. However, aging is an enormously complex problem, as one has to examine a complex organization to find a few basic changes that cause aging. It is like looking all over a forest to locate a few worms that have started destroying the trees. In order to understand development, one proceeds from a relatively simple organization, step by step, to a complex organization. Development is like planting trees, both in space and time, to grow a forest. One can at least follow the steps. Though all the key steps of development are not yet deciphered, a few steps are beginning to be understood at the molecular level in certain organisms, such as how homeotic genes control segmentation in animals, how *myoD* gene regulates differentiation of skeletal muscle cell in vertebrates and how *ced*-3 and *ced*-4 genes cause the death of certain predetermined cells during the development of *Caenorhabditis elegans*.

During the past decade we have learned that the expression of several genes decreases after the attainment of reproductive maturity. The cause of such changes in expression of genes is being studied by analyzing specific sequences in their promoter regions, such as the DH-sites, methylation of cytosine and cis-acting elements, and the trans-acting factors that bind to these elements and modulate their expression. The idea is to control their expression by manipulating the levels of trans-acting factors. However, we do not know whether the changes in their expression are the cause or effect of aging. But we also know that aging has an effect on the expression of several genes when the organism is maintained on a calorie-restricted diet or is subjected to different types of stress such as temperature and inflammation. This is in conformity with the observation that the adaptability of organisms to various types of stress declines in old age. Again we do not know how these conditions alter the expression of the genes.

An important observation is that *semelparous* animals, such as the Pacific salmon, octopus, and Australian marsupial mouse, produce a large number of offspring only once and die soon after. Also, animals that have high fecundity have a short life span and short senescence period. They are eliminated in their natural habitat due to environmental hazards. In *iteroparous* animals, such as the highly evolved mammals, domesticated mammals, and humans, that reproduce several times, but produce only a few offspring at a time and take care of them during their early developmental stage, the senescence period is long. It appears that in semelparous animals certain factors become depleted so rapidly and drastically that they are unable to replenish them, and this causes rapid aging and death. In iteroparous animals, the depletion of such factors may be less drastic, and the animal is able to replenish them partially, at least, during the intervening period available between the cycles of reproduction. These factors may be required not only for the maintenance of reproductive activity but also for other adult functions.

Besides reproductive activity, extrinsic factors like nutrition, and stresses like starvation, temperature, radiation, infection, and pollution, which the animal encounters during the reproductive period, may contribute to the depletion of certain factors essential for adult functions. Psychological stress is an important factor in highly evolved animals, particularly mammals, that may alter the levels of neurotransmitters and hormones, and cause depletion/accumulation of certain factors needed for adult functions.

Indeed, it is known that the level of 17 β-estradiol decreases both in mammals and birds after a short period of reproduction. It is not only responsible for the expression of genes required for reproduction, but it also stimulates the expression of several other genes such as for AChE and choline acetyltransferase. Theoretically, therefore, if the level of 17 β-estradiol is prevented from alteration, it should be possible to prolong the reproductive period and some of the adulthood activities.

The senescence period is characterized by two other features. First, it is a gradual but cumulative process. That is, when an increasing number of functions and organs deteriorate, the effect on the total ability of the organism is cumulative. The rate of aging between 70 and 80 years is, generally, faster than the rate between 60 and 70. No specific identifiable functional or structural changes occur during this period unlike during development when differentiation, segmentation, organ-ogenesis, and cell death by apoptosis occur at specific times and become evident in both structural and functional changes. Also, the onset of the reproductive period is marked by the growth of reproductive organs and their accessory organs, initiation of reproductive activity, and cessation of growth of the organism. However, whatever structural changes occur during senescence, such as death of cells, loss of mitochondria, wrinkling of skin, loss of eyesight, and greying of hair, they vary greatly from organ to organ and from individual to individual within a species, both in the time of occurrence and the rate of their progress. So these changes cannot be used as biomarkers of aging. It is, therefore, unlikely that these events have been selected during evolution to cause aging.

Second, there are individual variations in functions and structures in the same population of a species, and also between populations of the same species living in different habitats. Hence, one finds intraspecies differences in the duration and rate of senescence among individuals and populations. This may be due to the differences in their genotype, and also the differences in the extrinsic and intrinsic factors to which they are subjected.

The above features suggest that aging is a passive process that occurs from the failure of the organism or the system to maintain various activities. All activities are controlled by enzymes and other proteins which, in turn, are coded by genes. The individual variations in the rate of aging and the variations in the causes of death suggest that aging is multigenic in character; that is, it is due to the failure of several genes to maintain their normal activity. The individual variations in the rate of aging seen in a species are due to differences in the degree of de-stabilization of the homeostatic levels of factors that are required, di-rectly or indirectly, for the expression of genes dedicated to adult functions. Variations in these factors depend on the genetic make-up of the individual, nutrition, and the intrinsic and extrinsic factors to which it is subjected during adulthood. Alterations in the levels of these factors affect the expression of specific genes. The degree of change in the expression of genes differs among individuals. Hence, one finds intraspecies variations in the rate of aging and duration of the senescence period.

It is suggested that no specific gene causes aging because no specific character initiates aging in all individuals of a species. Aging is also not programmed, because one does not find specifically timed changes or events after the reproductive period and during senescence. So, searching for a gene that causes aging may be futile. It may be worthwhile instead to determine which genes change in expression after growth is complete or after attainment of reproductive ability under different situations, and what factors contribute to such changes. One may then attempt to prevent these changes and expect that adulthood would be prolonged.

Another important feature is that all multicellular organisms, plants and animals, show a similar pattern of life span – development, adulthood and senescence – though individual variations within a species and between populations of the same species exist. Hence aging is a universal phenomenon. Studies on the developmental period of organisms belonging to different phyla have shown that certain genes that control specific events have been conserved during evolution. For example, the homeotic gene controls segmentation in animals as different as insects and mammals, and the *myoD* gene controls differentiation of skeletal muscle in vertebrates. It is probable that more such conserved genes exist that control specific events of development in the animal kingdom. On the basis of these observations, it is postulated that the basic mechanism of aging in animals is also the same; that is, it occurs due to the failure or breakdown or deterioration of the homeostatic functioning of genes that are required for the maintenance of reproductive and other adult functions as postulated in the "gene regulation" theory. No doubt, in different phyla and classes, different genes may begin to deteriorate in function and set the aging process in motion, but basically it is the breakdown of the regulation of gene expression that causes aging.

The presence of postmitotic and highly differentiated cells such as neurons and skeletal muscle cells in multicellular organisms has added a new dimension to the aging process. These cells stop dividing at an early age, and in highly evolved mammals a majority of them live as long as the animal lives. In several lower organisms, the cell number is fixed at birth. It has been suggested that it is the postmitotic cells that contribute to aging of the organism more than the premitotic cells. There is some basis in this suggestion because postmitotic cells, being thermodynamically unstable systems, are expected to undergo breakdown or deterioration in function sooner or later, depending on how efficient the system is in replenishing the ingredients it loses and the speed with which it repairs the structures and molecules that are damaged as a function of time.

No doubt, selection for extended longevity during evolution has necessarily proceeded in parallel with selection for effective functioning and maintenance of all organs because, if one organ breaks down too early in life, it would affect the functioning of the organism and its life span. Eventually, however, one of the organs begins to deteriorate earlier than others because of intrinsic and extrinsic factors that destabilize the homeostatic levels of gene products which, in turn, affect the regulation of specific genes in the organ. A striking example is the thymus which involutes when maturity is attained. Skeletal muscle cells deteriorate earlier than neurons. Deterioration of one organ will affect some other organs, which, in turn, may affect even more organs. Thus, the deteriorating effects become cumulative. Senescence is rapid or slow, depending on the rate of deterioration of one or more organs, and death occurs because one or more organs deteriorate to such an extent that vital functions cannot continue.

The powerful tool of subtractive hybridization should be useful for the identification of the genes whose expressions decrease or increase with age in various organs. Gel mobility shift assays may be helpful in identifying the cis-acting elements of these genes, and the trans-acting factors that bind to them and are involved in their expression. It may then be possible to administer appropriate hormones and effectors to raise the levels of these trans-acting factors to determine if the expression of these genes is restored.

Appropriate model systems need to be developed for studying the changes in the expression of genes during aging and under different stress conditions. Particularly, transgenic animal models containing genes from species with long and short life spans, as for example, a human gene that is expressed during development inserted into the mouse genome and vice versa, could throw light on the differences in the cis-acting elements and trans-acting factors involved in the determination of the duration of the development period in the two species. Such models may be developed for genes that are expressed only in adults, and those that cause progeria and Werner's syndrome. Generation of various types of mutants of various species, including mice, with short and long life spans and defects in various stages would be useful in identifying and understanding the role of specific genes in the aging process. Alternatives to fibroblasts are needed because these cells are premitotic. Though these cells have provided useful information, several activities of these cells cannot be extrapolated to those of postmitotic cells such as neurons and skeletal muscle cells. In the next decade these approaches should provide a clearer understanding of the mechanism of aging at the genetic level.

Several genes have been sequenced. This has aided in the identification of cis-acting elements in their promoter regions. Several trans-acting factors that bind to specific cis-acting elements have also been identified. It is increasingly being realized that the control of gene expression is modular in nature. One or more trans-acting factors, on binding to cis-acting elements in the promoter region of a gene, in different combinations influence the binding of RNA polymerase to the transcription start site and thereby transcription. Not only do several genes have common cis-acting elements or modules, but also studies so far have revealed that there are only a few types of cis-acting elements throughout the animal kingdom. The trans-acting factors are also limited. Particularly, the same helix-turn-helix, helix-loop-helix, zinc finger, and leucine zipper motifs are present in several trans-acting factors. They bind to specific cis-acting elements in the DNA and produce conformational changes. They occur throughout the animal kingdom. However, variations in the expression of different genes is brought about by different combinations of these modules and trans-acting factors. It should be possible to identify trans-acting factors that change in level after adulthood. Such molecules could be used as biomarkers of aging.

Aging as a subject of research has been one of the greatest challenges to biologists. Understanding the aging process will help us to develop methods for prolonging adulthood or postponing the onset of old age or providing a prolonged healthier old age. This may also defer the onset of old-age diseases to later ages, and thereby increase the work output of animals and humans, and improve the quality of life. While a prolonged debilitating senescence period is undesirable for all organisms including humans, aging followed by death has had a beneficial effect on life; that is, it has aided evolution of life as it has made the turnover of species possible.

The rapid increase in the number of old, retired, and nonworking people has made it imperative for molecular biologists to find the basic cause of aging at the genetic level. Identification of genes that fail to express or overexpress during normal aging and under various types of stress in old age and are the cause of the failure of the individual to adapt to stress may help to regulate those genes and prevent such derangement. Even though the genes involved may be different for different organs, it is likely that changes in only a few genes initiate the aging process and that the basic mechanism of aging is similar in all organisms. Identification of these genes and their modulators should provide a means for understanding why and how they change in expression.

Research is being done on the identification of the cis-acting elements of certain genes, which change in structure and conformation with age and the trans-acting factors that bind to them and modulate the expression of the genes. This may be followed by administration of appropriate effectors to maintain the levels of the trans-acting factors and prevent the changes in the expression of the genes that cause aging. The only authentic preventive measure that has thrown some light on this aspect is the restriction of caloric intake. Restricted diet not only prolongs longevity, but also has been shown to have an effect on the expression of certain genes when organisms are exposed to various types of stress such as temperature and imflammation. However, the mechanism by which this regulatory effect is brought about at the level of genes is not known.

The recent discovery that mRNAs are edited by small-guide mRNAs has added a new dimension, and has much potential for understanding the aging process. Is mRNA editing universal? Does mRNA editing change with age? Answers to such questions are needed to understand more about the information-transfer system in relation to aging.

The quality of human life in old age depends not only on the genetic make-up of the individual, but also how he/she is accepted/treated by the society and the government. Biologically speaking, no great changes occur in the individual soon after he retires from the job, but for him the social and psychological changes are drastic and immense. Even though we have argued here that there is no specifically timed event during aging and that it is not programmed, society has imposed an artificial well-timed and programmed event on its own people; that is, making people retire at 60 or 65. This is despite the fact that the persons have acquired in specific jobs the skills, expertise, and experience that are not found in younger individuals. In fact if early adulthood has vigor, vitality, and vibrance, then old age has expertise, experience, and wisdom. An old person is a rich resource center, and as such should be tapped as long as possible. This would improve the quality of life considerably, even as the molecular biologists continue their research on genes in an effort to defer the onset of old age.

References

Abbott, M. H., Murphy, E. A., Bolling, D. R. & Abbey, H. (1974). *Johns Hopkins Med. J.* 134: 1–16.

Abraham, J. M., Feagin, J. E. & Stuart, K. (1988). *Cell* 55: 267–72.

Adelman, R. C. (1970a). *J. Biol. Chem.* 245: 1032–5.

Adelman, R. C. (1970b). *Nature* 228: 1095–6.

Adelman, R. C. (1975). In *Enzyme Induction* (D. V. Parke, ed.), 303–11. Plenum Press, New York.

Agarwal, S. S., Tuffner, M. & Loeb, L. A. (1978). *J. Cell Physiol.* 96: 235–44.

Albright, S. C., Wiseman, J. M., Lange, R. A. & Garrard, W. T. (1980) *J. Biol. Chem.* 255: 3673–84.

Allan, G. F., Tsai, S. Y., O'Malley, B. W. & Tsai, M-J. (1991). *BioEssays* 13: 73–8.

Allan, J., Harborne, N., Rau, D. C. & Gould, H. (1982). *J. Cell Biol.* 93: 285–97.

Allan, J., Hartman, P. G., Crane-Robinson, C. & Aviles, F. (1980). *Nature* 288: 675–9.

Allan, J., Mitchell, T., Harborne, N., Bohm L. & Crane-Robinson, C. (1986). *J. Molec. Biol.* 187: 591–601.

Allfrey, V. G., (1970). *Fed. Proc.* 29: 1447–60.

Allfrey, V. G., Faulkner, R. & Mirsky, A. E. (1964). *Proc. Natl. Acad. Sci. (U.S.A.)* 51: 786–94.

Allshire, R. C., Dempster, M. & Hastie, N. D. (1989). *Nucl. Acids Res.* 17: 4611–27.

Ang, D, Liberek, K., Skowyra, D., Zylicz, M. & Gorgopoulos, C. (1991). *J. Biol. Chem.* 266: 24233–6.

Antequera, F., Boyes, J. & Bird, A. P. (1990). *Cell* 62: 503–14.

Antequera, F., Macleod, D. & Bird, A. P. (1989). *Cell* 58: 509–17.

Appels, R. & Ringertz, N. R. (1974). *Cell Diff.* 3: 1–8.

Arai, H., Mitsui, Y. & Yamada, M. A. (1983). *Mech. Age. Dev.* 23: 315.

Arceci, R. J. & Gross, P. R. (1977). *Proc. Natl. Acad. Sci. (U.S.A.)* 74: 5016–20.

Arendes, J., Zahn, R. K. & Muller, W. E. G. (1980). *Mech. Age. Dev.* 14: 49–57.

Arking, R. (1978). *Dev. Biol.*, 63: 118–27.

Arking, R. (1987). *Exptl. Gerontol.,* 22: 199–220.
Arking, R. (1991). *Biology of Aging: Observations and Principles.* Prentice Hall, Englewood Cliffs, N. J.
Armbrecht, H. J., Boltz, M., Strong, R., Richardson, A., Bruns, M. E. H. & Christakos, S. (1989). *Endocrinology* 125: 2950–6.
Armelin, J. M., Armelin, M. C. S., Kelly, K., Stewart, T., Leder, P., Cochram, B. H. & Stiles, C. D. (1984). *Nature,* 310: 655–60.
Atlan, H., Miquel, J. & Binnard, R. M. (1969). *J. Gerontol.* 24: 1–4.
Balhorn, R., Chalkley, R. & Granner, D. (1972). *Biochemistry* 11: 1094–8.
Balhorn, R., Reike, W. O. & Chalkley, R. (1971). *Biochemistry* 10: 3952–8.
Banerji, J., Olson, L. & Schaffner, W. (1983). *Cell* 33: 729–40.
Banerji, J., Ruscony, S. & Schaffner, W. (1981). *Cell* 27: 299–308.
Bartley, J., & Chalkley, R. (1970). *J. Biol. Chem.* 245: 4286–92.
Bednarik, D. P., Cook, J. A. & Pitha, P. M. (1990). *EMBO J.* 9: 1157–64.
Benedict, W. F., Weissmann, B. E., Mark, C. & Stanbridge, E. J. (1984). *Cancer Res.* 44: 3471–9.
Benezra, R., Davis, R. L., Lockshon, D., Turner, D. L. & Weintraub, H. (1990). *Cell* 61: 49–59.
Ben-Hattar, J. & Jiricny, J. (1988). *Gene* 65: 219–27.
Benjamin, R. C. & Gill, D. M. (1980). *J. Biol. Chem.* 255: 10502–8.
Benne, R. van den Burg, J. Brakenhoff, J. P., Sloof, P., van Boom, J. H. & Tromp, M. C. (1986) *Cell,* 46: 819–26.
Bentley, G. A., Lewit-Bentley, A., Finch, J. T., Podjarny, A. D. & Roth, M. (1984). *J. Molec. Biol.* 176: 55–75.
Benvenisty, N., Mencher, D., Meyuhas, O., Razin, A. & Reshef, L. (1985). *Proc. Natl. Acad. Sci.* (U.S.A.) 82: 267–71.
Berdyshev, G. D. (1976). *Interdisciplinary Topics in Gerontology* 10: 70–82.
Berg, J. M. (1990). *J. Biol. Chem.* 265: 6513–16.
Berkovic, S. F. & Mauritzen, C. M. (1977). *Biochim. Biophys. Acta* 475: 160–7.
Bernd, A., Batke, E., Zahn, R. K. & Muller, W. E. G. (1982). *Mech. Age. Dev.* 19: 361–77.
Bernues, J., Querol, E., Martinez, P., Barris, A., Espel, E. & Lloberas, J. (1983). *J. Biol. Chem.* 258: 11020–4.
Beupain, B., Icard, C. & Macieira-Coelho, A. (1980) *Biochim. Biophys. Acta* 606: 251–61.
Bieker, J. J., Martin, P. L. & Roeder, R. G. (1985). *Cell* 40: 119–27.
Biggs, R. B, Hanley, R. M., Morrison, P. R. & Booth, F. W. (1991). *Mech. Age. Dev.* 60: 285–93.
Birchenall-Sparks, M. C., Roberts, M. S., Rutherford, M. S. & Richardson, A. (1985). *Mech. Age. Dev.* 32: 99–111.
Bird, A. (1986). *Nature* 321: 209–13.
Bird, A. (1987). *Trends Genet.* 3: 342–7.
Bird, A. (1992). *Cell* 70: 5–8.
Blackburn, E. H. (1990). *J. Biol. Chem.* 265: 5919–21.

Blackburn, E. H. (1991). *Trends Biochem. Sci.* 16: 378–81.

Blake, M. J., Elad, S., Epharti, E., Fargnoli, J., Rott, R., Holbrook, N. & Gershon, D. (1991a) In *Liver and Aging* (K. Kitani, ed.), 213–23. Excerpta Medica, Amsterdam.

Blake, M. J., Udelsman, R., Feulner, G. J., Norton, D. D. & Holbrook, N. J. (1991b) *Proc. Natl. Acad. Sci. (U.S.A.)* 88: 9873–7.

Blum, B., Bakalara, N. & Simpson, L. (1990). *Cell* 60: 189–98.

Blum, B., Sturm, N. R., Simpson, A. M. & Simpson, L. (1991). *Cell* 65: 543–50.

Boeke, J. T. (1990). *Cell* 61: 193–5.

Boffa, L. C., Sterner, R., Vidali, G. & Allfrey, V. G. (1979). *Biochem. Biophys. Res. Commun.* 89: 1322–7.

Boffa, L. C., Vidali, G., Mann, R. S. & Allfrey, V. G. (1978). *J. Biol. Chem.* 253: 3364–6.

Bohm, L. Hayashi, H., Cary, P. D., Moss, T., Crane-Robinson, C. & Bradbury, E. M. (1977). *Eur. J. Biochem.* 77: 487–93.

Bonner, W. M. & Stedman, J. D. (1979). *Proc. Natl. Acad. Sci. (U.S.A.)* 76: 2190–4.

Boobis, A., Caldwell, J., Dematteis, F. & Davis, D (1985). *Microsomes and Drug Oxidation.* Taylor & Francis, London.

Borsi, L., Carnemolea, B., Castellani, P., Rosellini, C., Vecchio, D., Allemanni, G., Chang, S. E., Taylor, J. P., Pande, H. & Zardi, L. (1987). *J. Cell. Biol.* 104: 595–600.

Bose, R. & Kanungo, M. S. (1982). *Arch. Gerontol. Geriat.* 1: 339–48.

Bowman, B. H., Yang, F. & Adrian, G. S. (1990). *BioEssays* 12: 317–22.

Boyes, J. & Bird, A. (1991). *Cell* 64: 1123–34.

Bradbury, E. M. (1975). *Ciba Found. Symp.* 25: 131–42.

Bradbury, E. M. (1982). In *The HMG Chromosomal Proteins* (E. W. Johns, ed.), 89–110. Academic Press, New York.

Bradbury, E. M., Danby, S. E., Rattle, H. W. E. & Giancotti, V. (1975). *Eur. J. Biochem.* 57: 97–105.

Braddock, G. W., Baldwin, J. P. & Bradbury, E. M. (1981). *Biopolymers* 20: 327–43.

Breathnach, R., Mandel, L. & Chambon, P. (1977). *Nature* 270: 314–19.

Britten, R. J. & Davidson, E. H. (1969). *Science* 165: 349–57.

Brown, R. S., Sander, C. & Argos, P. (1985). *FEBS Lett.* 186: 271–4.

Bryan, P. N., Hofstetter, H. & Birnsteil, M. L. (1983). *Cell* 33: 843–8.

Buchkovich, K., Duffy, L. A. & Harlow, E. (1989). *Cell* 58: 1097–1105.

Bulos, B., Shukla, S. & Sacktor, B. (1972). *Mech. Age. Dev.* 1: 227–32.

Bunn, C. L. & Tarrant, G. M. (1980). *Exptl. Cell Res.* 127: 385–96.

Buratowski, S., Sopta, M., Greenblate, J. & Sharp, P. A. (1991). *Proc. Natl. Acad. Sci. (U.S.A.)* 88: 7509–13.

Buratowski, S., Zhahr, S., Guarente, L. & Sharp, P. A. (1989). *Cell* 56: 549–61.

Burdic, C. J. & Taylor, B. A. (1976). *Exptl. Cell Res.* 100: 428–33.

Burke, E. M. & Danner, D. B. (1991). *Biochem. Biophys. Res. Commun.* 178: 620–4.

Burmer, G. C., Ziegler, C. J. & Norwood, T. H. (1982). *J. Cell Biol.* 94: 187–92.

Buys, C. H. C. M., Osinga, J. & Anders, G. J. P. A. (1979). *Mech. Age. Dev.* 11: 55–75.

Byvoet, P., Shepherd, G. R., Hardin, J. M. & Noland, B. J. (1972). *Arch. Biochem. Biophys.* 148: 558–67.

Candido, E. P. M. & Dixon, G. H. (1972a). *J. Biochem.* 247: 3868–73.

Candido, E. P. M. & Dixon, G. H. (1972b). *Proc. Natl. Acad. Sci. (U.S.A.)* 69: 2015–19.

Candido, E. P. M., Reeves, R. & Davie, J. R. (1978). *Cell:* 14: 105–113.

Carey, M. (1991). *Curr. Op. Cell Biol.* 3: 452–60.

Carlin, C. A., Phillips, P. D., Knowles, B. B. & Cristafolo, V. J. (1983). *Nature* 303: 617–620.

Carmickle, L. J., Kalimi, M. & Terry, R. D. (1979). *Fed. Proc.* 38: 482 (Abs.).

Carrascosa, J. M., Ruiz, P., Martinez, C., Pulido, J. A., Stratrustegui, J. & Andres, A. (1989). *Biochem. Biophys. Res. Commun.* 160: 303–9.

Carter, D. B. & Chae, C. (1975). *J. Gerontol.* 30: 28–32.

Cartwright, I. L., Herzberg, R. P., Dervan, P. B. & Elgin, S. C. R. (1983). *Proc. Natl. Acad. Sci. (U.S.A.)* 80: 3213–17.

Casarett, G. W. (1961). *Adv. Gerontol. Res.* 1: 109–163.

Catania, J. & Fairweather, D. S. (1991). *Mut. Res.* 256: 283–93.

Cattanach, B. M. (1974). *Genet. Res.* 23: 291–306.

Cattaneo, R. (1991). *Ann. Rev. Genet.* 25: 71–88.

Cattaneo, R. (1992). *Trends Biochem. Sci.* 17: 4–5.

Cavallius, J., Rattan, S. I. S. & Clark, B. F. C. (1986). *Exptl. Gerontol,* 21: 149–57.

Cavener, D. R. (1992). *BioEssays,* 14: 237–44.

Cech, T. (1986). *Scient. Amer.* 255: 64–75.

Cech, T. R. (1983). *Cell,* 34: 713–716.

Cech, T. R. (1987). *Science* 236: 1532–9.

Cech, T. (1991). *Cell* 64: 667–9.

Cedar, H. (1988). *Cell* 53: 3–4.

Chainy, G. B. N. & Kanungo, M. S. (1976). *Biochem. Biophys. Res. Commun.* 72: 777–81.

Chainy, G. B. N. & Kanungo, M. S. (1978a). *J. Neurochem.* 30: 419–27.

Chainy, G. B. N. & Kanungo, M. S. (1978b). *Biochim. Biophys. Acta* 540: 65–72.

Chambers, S. A. & Shaw, B. R. (1984). *J. Biol. Chem.* 259: 13458–63.

Chambon, P. (1978). *Cold Spring Harbor Symp. Quant. Biol.* 42: 1209–34.

Chambon, P. (1981). *Scient. Amer.* 244: 60–71.

Chambon, P., Dierich, A., Gaub, M. P., Jakowler, S., Jongstra, J., Krust, A., LePennec, J. P., Oudet, P. & Reudelhuber, T. L. (1984). *Rec. Prog. Hormone. Res.* 40: 1–42.

Chance, B., Sics, H. & Bovaris, A. (1979). *Physiol. Rev.* 59: 527–605

Chandrasekhar, S., Sorrentino, J. A. & Mills, A. J. T. (1983). *Proc. Natl. Acad. Sci. (U.S.A.)* 80: 4747–51.

Chang, Z. F. & Chen, K. Y. (1988). *J. Biol. Chem.* 263: 11431–5.

Chapman, G. E., Aviles, F. J., Crane-Robinson, C. & Bradbury, E. M. (1978). *Eur. J. Biochem.* 90: 287–96.

Chatterjee, B., Majumdar, D., Ozbilen, O., Murty, C. V. R. & Roy, A. K. (1987). *J. Biol. Chem.* 262: 822–5.

Chatterjee, B., Surendranath, T. & Roy, A. K. (1981). *J. Biol. Chem.* 256: 5939–41.

Chaturvedi, M. M. & Kanungo, M. S. (1983). *Biochem. Internat.* 6: 357–63.

Chaturvedi, M. M. & Kanungo, M. S. (1985a). *Molec. Biol. Rep.* 10: 215–19.

Chaturvedi, M. M. & Kanungo, M. S. (1985b). *Biochem. Biophys. Res. Commun.* 127: 604–609.

Chen, C. C., Bruegger, B. B., Kern, C. W., Lin, Y. C., Halpera, R. M. & Smith, R. (1977). *Biochemistry* 16: 4852–5.

Chen, S. H. & Giblett, E. R. (1971). *Science* 173: 148–9.

Chetsanga, C. J., Boyd, V., Peterson, L. & Rushlow, K. (1975). *Nature* 253: 130–31.

Chetsanga, C. J., Tuttle, M., Jacoboni, A. & Johnson, C. (1977) *Biochim. Biophys. Acta* 474: 180–7.

Chicoine, L. G., Schulman, I. G., Richman, R., Cook, R. G. & Allis, C. D. (1986). *J. Biol. Chem.* 261: 1071–6.

Ching, G. & Wang, E. (1988). *Proc. Natl. Acad. Sci. (U.S.A.)* 85: 151–5.

Chodosh, L. A., Baldwin, A. S., Carthew, R. W. & Sharp, P. A. (1988). *Cell* 53: 11–24.

Choi, H.-S., Lin, Z., Li, B. & Liu, A. Y.-C. (1990). *J. Biol. Chem.* 265: 18005–11.

Christensen, M. E., Rattner, J. B. & Dixon, G. H. (1984). *Nucl. Acids Res.* 12: 4575–92.

Chuknyiska, R. S., Haji, M., Foote, R. H. & Roth, G. S. (1985a) *Endocrinology,* 115: 836–8.

Chuknyiska, R. S., Haji, M., Foote, R. H. & Roth, G. S. (1985b). *Endocrinology* 116: 547–51.

Chuknyiska, R. S. & Roth, G. S. (1985). *J. Biol. Chem.* 260: 8661–3.

Chung, S., Hill, W. E. & Doty, P. (1978). *Proc. Natl. Acad. Sci. (U.S.A.)* 75: 1680–4.

Ciechanover, A., Finley, D. & Varshavsky, A. (1984). *J. Cell Biochem.* 24: 27–53.

Clark, A. M., Bertrand, H. A. & Smith, R. E. (1963). *Amer. Naturalist* 97: 203–8.

Clark, A. & Rubin, M. A. (1961). *Radiation Res.* 15: 244–8.

Clarke, J. M. & Maynard Smith, J. (1955). *J. Genet.* 53: 172–80.

Cohen, J. J. & Duke, R. C. (1984). *J. Immunol.* 132: 38–42.

Colbert, D. A., Knoll, B. J., Woo, S. L. C., Mace, M. & O'Malley, B. W. (1980). *Biochemistry* 19: 5586–92.
Comings, D. E. (1968). *Amer. J. Hum. Genet.* 20: 440–60.
Conaway, R. C. & Conaway, J. W. (1991). *J. Biol. Chem.* 266: 17721–4.
Cook, J. R. & Buetow, D. E. (1982). *Mech. Age. Dev.* 20: 289–304.
Cortopassi, G. A. & Arnheim, N. (1990). *Nucl. Acids Res.* 18: 6927–33.
Courey, A., Holtzman, D. A., Jackson, S. P. & Tjian, R. (1989). *Cell* 59: 827–36.
Courey, A. & Tjian, R. (1988). *Cell* 55: 887–98.
Craine, B. L. & Kornberg, T. (1981). *Cell* 25: 671–81.
Crick, F. H. C. & Klug, A. (1975). *Nature* 255: 530–33.
Cross, S. H., Allshire, R. C., McKay, S. J., McGill, N. I. & Cooke, H. J. (1989). *Nature* 338: 771–4.
Crowley, C. & Curtis, H. J. (1963). *Proc. Natl. Acad. Sci. (U.S.A.)* 49: 626–8.
Csordas, A. (1990). *Biochem. J.* 265: 23–38.
Curtis, H. J. (1963). *Science* 141: 686–97.
Curtis, H. J. (1964). *Fed. Proc.* 23: 662–7.
Cutler, R. G. (1975). *Proc. Natl. Acad. Sci. (U.S.A.)* 72: 4664–8.
Cutler, R. G. (1982). In *The Aging Brain* (G. Giacobini et al., eds.), vol. 20: 1, Raven Press, New York.
Cutler, R. G. (1985). In *Molecular Biology of Aging* (R. S. Sohal et al., eds.) 307–41, Raven Press, New York.
Cutler, R. G. (1991). *Arch. Gerontol. Geriat.* 12: 75–98.
D'Anna, J. A., Tobey, R. A., Barnam, S. S. & Gurley, L. R. (1977). *Biochem. Biophys. Res. Commun.* 77: 187–94.
Darnell, J. E. (1983). *Scient. Amer.* 249: 89–99.
Darnold, J. R., Vorbeck, M. L. & Martin, A. L. (1990). *Mech. Age. Dev.* 53: 157–67.
Das, B. R. & Kanungo, M. S. (1985). *Biochem. Internat.* 12: 303–11.
Das, B. R. & Kanungo, M. S. (1986). *Molec. Biol. Rep.* 11: 195–8.
Das, B. R. & Kanungo, M. S. (1987). *Molec. Biol. Rep.* 12: 43–8.
Das, R. & Kanungo, M. S. (1979). *Biochem. Biophys. Res. Commun.* 90: 708–714.
Das, R. & Kanungo, M. S. (1980). *Ind. J. Biochem. Biophys.* 17: 217–21.
Das, R. & Kanungo, M. S. (1982). *Exptl. Gerontol.* 17: 94–103.
Davie, J. R. & Candido, E. P. M. (1978). *Proc. Natl. Acad. Sci. (U.S.A.)* 75: 3574–7.
Davis, R. L., Weintraub, H. & Lassar, A. B. (1987). *Cell* 51: 987–1000.
Davison, B. L., Egly, J. M., Mulvihill, E. R. & Chambon, P. (1983). *Nature* 301: 680–6.
Dean, R. G., Socher, S. H. & Cutler, R. G. (1985). *Arch. Gerontol. Geriat.* 4: 43–51.
DeLange, R. J., Farnbrough, D. M., Smith, E. L. & Bonner, J. (1969). *J. Biol. Chem.* 244: 319–44.

Diamond, J. M. (1982). *Nature* 298: 115–116.

Dickerson, R. E. (1983). *Scient. Amer.* 249: 94–111.

Dilella, D. G., Chiang, J. Y. L. & Steggles, A. W. (1982). *Mech. Age. Dev.* 19: 113–25.

Dingman, C. W. & Sporn, M. B. (1964). *J. Biol. Chem.* 239: 3483–92.

Dixon, G. H., Davies, P. L., Ferrier, L. N., Gedamu, L. & Iotrov, K. (1977) in *Molecular Biology of Mammalian Genetic Apparatus* (P. O. P. Ts'o, ed.) vol. 1: 335–79. North-Holland, Amsterdam.

Doerfler, W. (1983). *Ann. Rev. Biochem.* 52: 93–124.

Duerre, J. A. & Chakrabarty, S. (1977). *J. Biol. Chem.* 252: 8457–61.

Dunn, G. R., Wilson, T. G. & Jacobson, K. B. (1969). *J. Exptl. Zool.* 171: 185–90.

Dykhuizen, D. (1974). *Nature* 251: 616–18.

Dynan, W. S. & Tjian, R. (1985). *Nature* 316: 774–8.

Edelmann, P. & Gallant, J. (1977). *Proc. Natl. Acad. Sci. (U.S.A.)* 74: 3396–8.

Ehrlich, M. & Wang, R. Y.-H. (1981). *Science* 212: 1350–7.

Elgin, S. C. R. (1990). *Curr. Op. Cell Biol.* 2: 437–45.

Elgin, S. C. R. & Weintraub, H. (1975). *Ann. Rev. Biochem.* 44: 725–74.

Ellis, H. M. & Horvitz, H. R. (1986). *Cell* 44: 817–29.

Emerson, B. M. & Felsenfeld, G. (1984). *Proc. Natl. Acad. Sci. (U.S.A.)* 81: 95–9.

Epstein, C. J., Martin, G. M., Schultz, A. L. & Motulsky, A. G. (1966). *Medicine* 45: 177–221.

Estus, S., Golde, T. E., Kunishita, T., Blades, D., Lowery, D., Eisen, M., Usiac, M., Qu, X., Tabina, T., Greenberg, B. D. & YounKin, S. G. (1992). *Science* 255: 726–30.

Evans, C. H. (1976). *Differentiation* 5: 101–5.

Fairweather, S., Fox, M. & Margison, P. (1987). *Exptl. Cell Res.* 168: 153–9.

Fargnoli, J., Kunisada, T., Fornace, A. J., Schneider, E. L. & Holbrook, N. J. (1990). *Proc. Natl. Acad. Sci. (U.S.A.)* 87: 846–50.

Feagin, J. E., Abraham, J. M. & Stuart, K. (1988). *Cell* 53: 413–22.

Felsenfeld, G. (1992). *Nature* 355: 219–24.

Felsenfeld, G. & McGhee, J. (1983). *Nature* 296: 602–3.

Felsenfeld, G. & McGhee, J. D. (1986). *Cell* 44: 375–7.

Feng, J., Irving, J. & Villeponteau, B. (1991). *Biochemistry* 30: 4747–52.

Ferioli, M. E., Ceruti, G. & Comolli, R. (1976). *Exptl. Gerontol.* 11: 153–6.

Fernandez-Silva, P., Petruzzella, V., Fracasso, F., Gadaleta, M. N. & Cantatose, P. (1991). *Biochem. Biophys. Res. Commun.* 176: 645–53.

Finch, C. E. (1972). *Exptl. Gerontol.* 7: 53–67.

Finch, C. E., Foster, J. R. & Mirsky, A. E. (1969). *J. Gen. Physiol.* 54: 690–712.

Finch, J. T. & Klug, A. (1976). *Proc. Natl. Acad. Sci. (U.S.A.)* 73: 1897–1901.

Finch, J. T., Lewit-Bentley, A., Bentley, G. A., Roth, M. & Timmins, P. A. (1980). *Phil. Trans. Roy. Soc. (London)* 290B: 635–8.

Finch, J. T., Lutter, L. C., Rhodes, D., Brown, R. S., Ruston, B., Levitt, M. & Klug, A. (1977). *Nature* 269: 29–36.

Finley, D., Ciechanower, A. & Varshavsky, A. (1984). *Cell* 37: 43–55.

Fischer, J. A. & Maniatis, T. (1986). *EMBO J.* 5: 1275–89.

Fleming, J. E., Walton, J. K., Dubitsky, R. & Bensch, K. G. (1988). *Proc. Natl. Acad. Sci. (U.S.A.)* 85: 4099–4103.

Flores, O. H., Killen, M., Greenblatt, J., Burton, J. F. & Reinberg, D. (1991). *Proc. Natl. Acad. Sci. (U.S.A.)* 88: 9991–10003.

Florine, D. L., Ono, T. & Cutler, R. G. (1980). *Cancer Res.* 40: 517–23.

Florini, J. R. (1975). *Exptl. Aging Res.* 1: 137–44.

Freemont, P. S., Lane, A. N. & Sanderson, M. R. (1991). *Biochem. J.* 278: 1–23.

Friedman, V., Wagner, J. & Danner, D. B. (1990). *Mech. Age. Dev.* 52: 27–43.

Fry, M., Loeb, L. A. & Martin, G. M. (1981). *J. Cell. Physiol.* 106: 435–44.

Fry, M. & Weisman-Shomer, P. (1976). *Biochemistry* 15: 4319–29.

Fucci, L., Oliver, C. N., Coon, M. J. & Stadtman, E. R. (1983). *Proc. Natl. Acad. Sci. (U.S.A.)* 80: 1521–5.

Fujita, T. & Maruyama, W. (1991). *Biochem. Biophys. Res. Comm.* 178: 1485–91.

Fulder, S. J. & Holliday, R. (1975). *Cell* 6: 67–75.

Gabius, H. J., Graupner, G. & Cramer, F. (1983). *Eur. J. Biochem.* 131: 231–4.

Gadaleta, M. N., Petruzzella, V., Renis, M., Fracasso, F. & Cantatore, P. (1990). *Eur. J. Biochem.* 187: 501–6.

Gadaleta, M. N., Petruzzella, V., Renis, M., Fracasso, F., Garrell, J. & Campuzano, S. (1991). *BioEssays,* 13: 493–8.

Gafni, A. (1983). *Biochim. Biophys. Acta* 742: 91–9.

Gama-Sosa, M. A., Wang, R. Y.-H., Kuo, K. C., Genrke, C. W. & Ehrlich, M. (1983). *Nucl. Acids Res.* 11: 3087–95.

Garrell, J. & Campuzano, S. (1991). *BioEssays* 13: 493–8.

Gartner, L. P. (1973). *Gerontologia* 19: 295–302.

Garvin, R. T., Rosset, R. & Gorini, L. (1973). *Proc. Natl. Acad. Sci.* (U.S.A.) 70: 2762–6.

Gasser, S. M., Laroche, T., Falquet, J., Boy de la Tour, E. & Laemmli, U. K. (1986). *J. Molec. Biol.* 188: 613–29.

Gaubatz, J. W. & Cutler, R. G. (1978). *Gerontology* 24: 179–207.

Gaubatz, J. W. & Cutler, R. G. (1990). *J. Biol. Chem.* 265: 17753–8.

Germond, J. E., Hint, B., Qudet, P., Gross-Bellard, M. & Chambon, P. (1975). *Proc. Natl. Acad. Sci. (U.S.A.)* 1843–7.

Gershey, E. L., Vidali, G. & Allfrey, V. G. (1968). *J. Biol. Chem.* 243: 5018–22.

Gershon, D. (1979). *Mech. Age. Dev.* 9: 189–96.

Gershon, D. & Gershon, H. (1970). *Nature* 227: 1214–17.

Gershon, H. & Gershon, D. (1973a). *Mech. Age. Dev.* 2: 33–42.

Gershon, H. & Gershon, D. (1973b). *Proc. Natl. Acad. Sci. (U.S.A.)* 70: 909–13.

Gibson, G. E., Peterson, C. & Jenden, D. J. (1981). *Science* 213: 674–6.

Gilbert, W. (1978). *Nature* 271: 501.

Gilbert, W. (1985) *Science* 228: 823–4.

Gill, G. & Tjian, R. (1991) *Cell,* 65: 333–40.

Gillies, S. D., Morrison, S. L., Oi, V. T. & Tonegawa, S. (1983). *Cell* 33: 717–28.

Giordano, T. & Foster, D. N. (1989). *Exptl. Cell. Res.* 185: 399–406.

Glenner, G. G. & Wong, C. W. (1984). *Biochem. Biophys. Res. Commun.* 122: 1131–35.

Goate, A., Chartier-Hartin, M. C., Mullan, M., et al. (1991). *Nature* 349: 704–6.

Goel, N. S. & Ycas, M. (1975). *J. Theo. Biol.* 55: 245–82.

Goldberg, A. L. (1972). *Proc. Natl. Acad. Sci. (U.S.A.)* 69: 422–6.

Goldgaber, D., Lerman, M. I., McBride, O. W., Saffiotti, U. & Gajdusek, D. C. (1987). *Science* 235: 877–80.

Goldknopf, I. L., Wilson, G., Ballal, N. R. & Busch, H. (1980). *J. Biol. Chem.* 255: 10555–8.

Goldstein, S. (1971). *Proc. Soc. Exptl. Med.* 137: 730–4.

Goldstein, S. (1978). In *The Genetics of Aging* (E. L. Schneider, ed.), 171–224. Plenum Press, New York.

Goldstein, S. (1990). *Science* 249: 1129–33.

Goldstein, S. & Moerman, C. J. (1975a). *Nature* 255: 159.

Goldstein, S. & Moerman, C. J. (1975b). *New Engl. J. Med.* 292: 1306–9.

Goldstein, S. & Shmookler-Reis, R. J. (1985). *Nucl. Acids Res.* 13: 7055–65.

Goodfriend, T. L., & Kaplan, N. O. (1964) *J. Biol. Chem.* 239: 130–5.

Goodfriend, T. L., Sokal, D. M. & Kaplan, N. O. (1966). *J. Molec. Biol.* 15: 18–31.

Goren, R., Reznick, A. Z. Reiss, U. & Gershon, D. (1977). *FEBS Lett.* 84: 83–6.

Gorman, S. D. & Cristofalo, V. J. (1985). *J. Cell. Physiol.* 125: 122.

Gorovsky, M. A., Pleger, G. L., Keebat, J. B. & Johmann, C. A. (1973). *J. Cell. Biol.* 57: 773–81.

Gorski, J., Welshons, W. V., Sakai, D., Hansen, J., Walent, J., Kassis, J., Shull, J., Stack, G. & Campen, C. (1986). *Recent Prog. Hormone Res.* 42: 297–322.

Goss, J. R., Finch, C. E. & Morgan, D. G. (1991). *Neurobiol. Aging* 12: 165–70.

Goto, M., Rubenstein, M., Weber, J., Woods, K. & Drayna, D. (1992). *Nature* 355: 735–7.

Greenberg, M. E. & Ziff, E. B. (1984). *Nature* 311: 433–7.

Gregerman, R. I. (1959). *Amer. J. Physiol.* 137: 63–4.

Greider, C. W. (1990) *BioEssays* 12: 363–9.

Greider, C. W. (1991) *Cell* 67: 645–7.

Greider, C. W. & Blackburn, E. H. (1987) *Cell* 51: 887–98.
Gresik, E. W., Wenk-Salamore, K., Onetti-Muda, A., Gubits, R. M. & Shaw, P. A. (1986). *Mech. Age. Dev.* 34: 175–89.
Grimes, S. R. & Henderson, N. (1984). *Exptl. Cell Res.* 152: 91–7.
Grivell, L. A. (1989). *Nature* 341: 569–71.
Gross-Bellard, M. & Chambon, P. (1975). *Cell* 4: 281–300.
Grosschedl, R. & Birnsteil, M. L. (1980). *Proc. Natl. Acad. Sci. (U.S.A.)* 77: 7102–6.
Grosveld, F., van Blom A. G., Greaves, D. R. & Kollas, G. (1987). *Cell* 51: 975–85.
Groudine, M. & Weintraub, H. (1982). *Cell* 30: 131–9.
Gruenewald, D. A. & Matsumoto, A. M. (1991). *Neurobiol. Ageing* 12: 113–21.
Grunstein, M. (1990). *Trends Genetics* 6: 395–400.
Guigoz, Y. & Munro, H. N. (1985). In *Handbook of the Biology of Aging* (C. E. Finch & E. L. Schneider, eds.) 878–93. Van Nostrand Reinhold, New York.
Gupta, S. K. & Rothstein, M. (1976). *Biochim. Biophys. Acta* 445: 632–44.
Gurley, L. R., D'Anna, J. A., Barham, S. S., Deaven, L. L. & Tobey, R. A. (1978). *Eur. J. Biochem.* 84: 1–15.
Gustafsson, L. & Pärt, T. (1990). *Nature* 347: 279–281.
Ha, I., Lane, W. S. & Reinberg, D. (1991). *Nature* 352: 689–95.
Hahn, H. P. von (1963). *Gerontologia,* 8: 123–31.
Hahn, H. P. von (1970). *Exptl. Gerontologia* 5: 323–34.
Hahn, H. P. von & Fritz, E. C. (1966) *Gerontologia* 12: 237–50.
Hahn, S., Buratowski, S., Sharp, P. A. & Gurante, L. (1989). *Cell* 58: 1173–81.
Haji, M. & Roth, G. S. (1984). *Mech. Age. Dev.* 25: 141–8.
Haji, M., Rumiana, S., Chuknyiska, R. S. & Roth, G. S. (1984). *Proc. Natl. Acad. Sci. (U.S.A.)* 81: 7481–4.
Haldane, J. B. S. (1942). *New Paths in Genetics*. Harper, London.
Hall, J. C. (1969). *Exptl. Gerontol.* 4: 207–22.
Hanahan, D. (1985). *Nature* 315: 115–22.
Hanaoka, F., Sayato, J., Hardwick, J., Hsieh, W. H., Liu, S. H. & Richardson, A. (1983). *Biochim. Biophys. Acta* 652: 204–17.
Hancock, R. (1978) *Proc. Natl. Acad. Sci. (U.S.A.)* 75: 2130–4.
Hansen, R. S., Ellis, N. A. & Gartler, S. M. (1988). *Molec. Cell. Biol.* 8: 4692–9.
Harley, C. B. (1991). *Mut. Res.* 256: 271–82.
Harley, C. B., Futcher, A. B. & Greider, C. W. (1990). *Nature* 345: 458–60.
Harman, D. (1956). *J. Gerontol.* 11: 298–300.
Harman, D. (1981). *Proc. Natl. Acad. Sci. (U.S.A.)* 78: 7124–8.
Harrington, L. A. & Greider, C. W. (1991). *Nature* 353: 451–4.
Harris, H., Miller, O. J., Klein, G., Worst, P. & Tachibana, T. (1969). *Nature* 223: 363–8.

300   *References*

Harrison, S. C. (1991). *Nature* 353: 715–19.
Hart, R. W. & Setlow, R. B. (1974). *Proc. Natl. Acad. Sci. (U.S.A.)* 71: 2169–73.
Hayaishi, O. (1976). *Trends Biochem. Sci.* 1: 9–10.
Hayakawa, M., Ogawa, T., Sugiyama, S., Tarama, M. & Ozawa, T. (1991a). *Biochem. Biophys. Res. Commun.* 176: 87–93.
Hayakawa, M., Torii, K., Sugiyama, S., Tanaka, M. & Ozawa, T. (1991b). *Biochem. Biophys. Res. Commun.* 179: 1023–29.
Hayes, J. J., Clark, D. J. & Wolfee, A. P. (1991). *Proc. Natl. Acad. Sci. (U.S.A.)* 88: 6829–33.
Hayflick, L. (1965). *Exptl. Cell Res.* 37: 614–36.
Hayflick, L. & Moorhead, P. S. (1961). *Exptl. Cell Res.* 25: 585–621.
Hebbes, T. R., Thorne, A. W. & Crane-Robinson, C. (1988). *EMBO J.* 7: 1395–1402.
Heifetz, S. R. & Smith-Sonneborn, J. (1981). *Mech. Age. Dev.* 16: 255–63.
Heikkinen, E. & Kulonen, E. (1964). *Experientia* 20: 310.
Henshaw, P. S. (1957). *Radiology* 69: 30–6.
Hergersberg, M. (1991). *Experientia* 47: 1171–85.
Hershko, A. (1988). *J. Biol. Chem.* 263: 15237–40.
Hershko, A. (1991). *Trends Biochem. Sci.* 16: 265–8.
Hightower, L. E. (1991). *Cell* 66: 191–7.
Hill, C. S., Packman, L. C. & Thomas, J. O. (1990). *EMBO J.* 9: 805–13.
Hill, C. S., Rimmer, J. M., Green, B. N., Finch, J. T. & Thomas, J. O. (1991). *EMBO J.* 10: 1939–48.
Hinchliffe, J. R. (1981). In *Cell Death in Biology and Pathology* (E. D. Bowen & R. A. Lockshin, eds.), 35–78. Chapman & Hall, New York.
Hiromi, Y. & Gehring, W. (1987). *Cell* 50: 963–74.
Hodges, P. & Scott, J., (1992). *Trends Biochem. Sci.* 17: 77–81.
Hoehn, H., Bryant, E. M., Johnston, P., Norwood, T. K. & Martin, G. M. (1975). *Nature* 258: 608–9.
Hoeller, M., Westin, G., Jiricny, J. & Schaffner, W. (1988). *Genes Dev.* 2: 1127–35.
Hoey, T., Dynlacht, B. D., Peterson, M. G., Pugh, B. F. & Tjian, R. (1990). *Cell* 61: 1179–86.
Hofbauer, R. & Denhardt, D. T. In press.
Hoffman, O. W. (1974). *J. Molec. Biol.* 86: 349–62.
Hoffmann, M. (1991). *Science* 253: 136–7.
Hofstetter, H., Kressman, A. & Birnsteil, M. L. (1981). *Cell* 24: 573–85.
Hohmann, P. & Cole, R. D. (1971). *J. Molec. Biol.* 58: 533–40.
Hohmann, P., Tobey, R. A. & Gurley, L. R. (1976). *J. Biol. Chem.* 251: 3685–92.
Holliday, R. (1969). *Nature* 221: 1224–8.
Holliday, R. (1984). *Monographs in Developmental Biology* 17: 60–77.
Holliday, R. (1986). *Exptl. Cell Res.* 166: 543–52.
Holliday, R. (1987). *Science* 238: 163–9.

Holliday, R. (1988). *Perspectives in Biology and Medicine* 32: 109–123.
Holliday, R. (1990). *J. Gerontol.* 45: 36–41.
Holliday, R., Huschtscha, L. I., Tarrant, G. M. & Kirkwood, T. B. L. (1977). *Science* 198: 366–72.
Holliday, R. & Tarrant, G. M. (1972). *Nature* 238: 26–30.
Holmes, G. E. & Holmes, N. R. (1986). *Molec. Gen. Genet.* 204: 108–14.
Holt, J. T., Venkat-Gopal, T., Moulton, A. D. & Nienhuis, A. W. (1986). *Proc. Natl. Acad. Sci (U.S.A.)* 83: 4794–8.
Holtzman, D. M. & Mobley, W. C. (1991). *Trends Biochem. Sci.* 16: 140–4.
Honda, B. M. Dixon, G. H. & Candido, E. P. M. (1975). *J. Biol. Chem.* 250: 8681–5.
Hopfield, J. J. (1974). *Proc. Natl. Acad. Sci. (U.S.A.)* 71: 4135–9.
Horbach, G. J. M. J., Prinvew, H. M. G., van der Knoef, M., van Bezooijen, C. F. A. & Yap, S. H. (1984). *Biochim. Biophys. Acta* 783: 60–6.
Horbach, G. J. M. J. & van Bezooijen, C. F. A. (1991) In *Liver and Aging* (K. Kitani, ed.), 183–94. Elsevier, Amsterdam.
Horikoshi, M., Yamamoto, T., Ohkuma, Y., Weil, P. A. & Roeder, R. G. (1990). *Cell* 61: 1171–8.
Hosbach, H. A. & Kubli, E. (1979). *Mech. Age. Dev.* 10: 141–9.
Howlett, D., Dalrymple, S. & Mays-Hoopes, L. L. (1989). *Mut. Res.* 219: 101–6.
Hsiang, M. W. & Cole, R. D. (1977). *Proc. Natl. Acad. Sci. (U.S.A.)* 74: 4852–6.
Hynes, R. O. (1985). *Ann. Rev. Cell Biol.* 1: 67–90.
Hynes, R. O. (1990). *Fibronectins*, Springer-Verlag, New York.
Iacopino, A. M. & Christakos, S. (1990). *Proc. Natl. Acad. Sci. (U.S.A.)* 87: 4078–82.
Iatrou, L., Spira, A. W. & Dixon, D. H. (1978). *Dev. Biol.* 64: 82–98.
Icard. C., Beaupain, R., Diatloff, C. & Macieira-Coelho, A. (1979). *Mech. Age. Dev.* 11: 269–78.
Ichimura, S., Mita, K. & Zama, M. (1982). *Biochemistry* 21: 5329–34.
Igo-Kemenes, T. & Zachau, H. G. (1977). *Cold Spring Harbor Symp. Quant. Biol.* 42: 109–18.
Iguchi-Ariga, S. M. M. & Schaffner, W. (1989). *Genes Dev.* 3: 612–19.
Ikebe, S., Tanaka, M., Ohno, K., Sato, W., Hattori, K., Kondo, T., Mizuno, Y. & Ozawa, T. (1990). *Biochem. Biophys. Res. Commun.* 170: 1044–8.
Imamura, K., Taniuchi, K. & Tanaka, T. (1972). *J. Biochem.* 72: 4511–15.
Izumo, S., Nadal-Ginard, B. & Mahdavi, V. (1988). *Proc. Natl. Acad. Sci. (U.S.A.)* 85: 339–43.
Jackson, D. A. (1991). *BioEssays* 13: 1–10.
Jackson, D. A., McCready, S. J., & Cook P. R. (1984). *J. Cell Sci. (Suppl.)* 1: 59–79.
Jackson, P. D. & Felsenfeld, G. (1985). *Proc. Natl. Acad. Sci. (U.S.A.)* 82: 2296–300.
Jaiswal, Y. K. (1988). Ph.D. dissertation, Banaras Hindu University, India.

Jaiswal, Y. K. & Kanungo, M. S. (1990). *Biochem. Biophys. Res. Commun.* 168: 71–7.

James, T. C. & Kanungo, M. S. (1978). *Biochim. Biophys. Acta* 538: 205–11.

Jeffreys, A. J. & Flavell, R. A. (1977). *Cell* 12: 1097–1108.

Jiang, J., Hoey, T. & Levine, M. (1991). *Genes Dev.* 5: 265–77.

Johns, E. W. (1964). *Biochem. J.* 92: 55–9.

Johns, E. W. (ed.) (1982). *The HMG Chromosomal Proteins*. Academic Press, New York.

Johnson, P. F. & McKnight, S. L. (1989). *Annu. Rev. Biochem.* 58: 799–839.

Johnson, R., Chrisp, C. & Strehler, B. L. (1972). *Mech. Age. Dev.* 1: 183–98.

Johnson, R. & Strehler, B. L. (1972). *Nature* 240: 412–14.

Johnson, T. E. (1987). *Proc. Natl. Acad. Sci. (U.S.A.)* 84: 3777–81.

Jones, P. A. (1986). *Cell* 40: 485–6.

Jones, P. A. & Taylor, S. M. (1980). *Cell* 20: 85–93.

Jump, D. B. & Oppenheimer, J. H. (1983). *Mol. Cell. Biochem.* 55: 159–76.

Kadonaga, J. T., Carner, K. R., Masianz, F. R. & Tjian, R. (1987). *Cell* 51: 1079–90.

Kadonaga, J. T. & Tjian, R. (1986). *Proc. Natl. Acad. Sci. (U.S.A.)* 83: 5889–93.

Kaldor, G. & Min, B. K. (1975). *Fed. Proc.* 34: 191–4.

Kallmann, F. J. & Jarvik, L. F. (1959). In *Handbook of Ageing and Individual* (J. E. Birren, ed.), 216–63. University of Chicago Press, Chicago.

Kamakaka, R. T. & Thomas, J. O. (1990). *EMBO J.* 9: 3997–4006.

Kameshita, I., Matsuda, Z., Tanigushi, T. & Shizuta, Y. (1984). *J. Biol. Chem.* 259: 4770–6.

Kanai, Y., Miwa, M., Matsushima, T. & Sugimura, T. (1981). *Proc. Natl. Acad. Sci. (U.S.A.)* 78: 2801–4.

Kanungo, M. S. (1970). *Biochem. Rev.* (India) 41: 13–23.

Kanungo, M. S. (1975). *J. Theo. Biol.* 53: 253–61.

Kanungo, M. S. (1980). *Biochemistry of Ageing*. Academic Press, London.

Kanungo, M. S. & Gandhi, B. S. (1972). *Proc. Natl. Acad. Sci. (U.S.A.)* 69: 2035–8.

Kanungo, M. S. & Patnaik, S. K. (1975). In *Regulation of Growth and Differentiated Function in Eukaryote Cell* (G. P. Talwar, ed.), 479–90. Raven Press, New York.

Kanungo, M. S. & Patnaik, S. K. & Koul, O. (1975). *Nature* 253: 366–7.

Kanungo, M. S. & Saran, S. (1991). *Ind. J. Biochem. Biophys.* 28: 96–9.

Kanungo, M. S. & Saran, S. (1992). *Ind. J. Biochem. Biophys.* 29: 49–53.

Kanungo, M. S., Singh, A., Singh, S. & Jaiswal, Y. K. (1992). In *New Horizons in Aging Science*. Proceedings of the 4th Asia/Oceania Regional Congress of Gerontology, 1991. Tokyo University Press, Tokyo.

Kanungo, M. S. & Singh, S. N. (1965). *Biochem. Biophys. Res. Commun.* 21: 454–9.

Kanungo, M. S. & Thakur, M. K. (1977). *Biochem. Biophys. Res. Commun.* 79: 1031–6.

Kanungo, M. S. & Thakur, M. K. (1979a). *J. Steroid Biochem.* 11: 879–87.
Kanungo, M. S. & Thakur, M. K. (1979b). *Biochem. Biophys. Res. Commun.* 87: 266–71.
Kanungo, M. S. & Thakur, M. K. (1979c). *Biochem. Biophys. Res. Commun.* 86: 14–19.
Kao, C. C., Lieberman, P. M., Schmidt, M. C., Zhau, Q., Pei, R. & Berk, A. J. (1990). *Science* 248: 1646–9.
Kaptein, R. (1991). *Curr. Op. Struct. Biol.* 1: 63–70.
Karlsson, O., Edlund, T., Moss, J. B., Rutter, W. J. & Walker, M. D. (1987). *Proc. Nat. Acad. Sci. (U.S.A.)* 84: 5819–23.
Karpov, V. L., Preobrazhenskaya, O. V. & Mirzabekov, A. D. (1984). *Cell* 36: 423–31.
Kator, K., Cristofalo, V., Carpentier, R. & Cutler, R. G. (1985). *Gerontology* 31: 355–61.
Katsurada, A., Iritani, N., Fukuda, H. Noguchi, T. & Tanaka, T. (1982). *Biochem. Biophys. Res. Commun.* 109: 250–5.
Kedes, L. H. (1976). *Cell* 8: 321–31.
Kelsey, G. D., Povey, S., Bygrave, A. E. & Lovell-Badge, R. H. (1987). *Genes Dev.* 1: 161–71.
Keshet, I., Lieman-Hurwitz, J. & Cedar, H. (1986). *Cell* 44: 535–43.
Kimura, T., Mills, F. C., Allan, J. & Gould, H. (1983). *Nature* 306: 709–12.
Kincade, J. M. Jr. & Cole, R. D. (1966a). *J. Biol. Chem.* 241: 5790–7.
Kincade, J. M. Jr. & Cole, R. D. (1966b). *J. Biol. Chem.* 241: 5798–805.
Kirkwood, T. B. L. (1977) *Nature* 270: 301–4.
Kirkwood, T. B. L. & Holliday, R. (1975) *J. Theo. Biol.* 53: 481–96.
Kirkwood, T. B. L. & Holliday, R. (1979). *Proc. Roy. Soc. (London)* B205: 532–46.
Kirkwood, T. B. L. & Holliday, R. (1986). In *The Biology of Human Ageing* (A. H. Bittles & K. J. Collins eds.). Cambridge Univ. Press.
Kissinger, C. R., Liu, B., Martin-Blanco, E., Kornberg, T. B. & Pabo, C. O. (1990). *Cell* 63: 579–90.
Klevan, L., Datta Gupta, N., Hogan, M. & Crothers, D. M. (1978). *Biochemistry* 17: 4533–40.
Klug, A., Rhodes, D., Smith, J., Finch, J. T. & Thomas, J. O. (1980). *Nature* 287: 509–16.
Knippers, R., Otto, B. & Bohme, R. (1978). *Nucl. Acids Res.* 5: 2113–31.
Kolodziej, P. A., Woychik, N., Lo, S. M. & Young, R. A. (1990). *Molec. Cell. Biol.* 10: 1915–20.
Koopman, P., Povey, S. & Lovell-Badge, R. H. (1989). *Genes Dev.* 3: 16–25.
Kornberg, R. D. (1974). *Science* 184: 868–71.
Kornberg, R. D. (1977). *Ann. Rev. Biochem.* 46: 931–54.
Kornberg, R. D. & Lorch, Y. (1992). *Ann. Rev. Cell. Biol.* 8: 563–87.
Kornberg, R. D. & Thomas, J. O. (1974). *Science* 184: 865–8.
Kornblihtt, A. R., Vibe-Pedersen, K. & Baralle, F. E. (1983). *Proc. Natl. Acad. Sci. (U.S.A.)* 80: 3218–22.

Koslowsky, D. J., Bhat, G. J., Read, L. K. & Stuart, K. (1991). *Cell* 67: 537–46.

Koul, O. & Kanungo, M. S. (1975). *Exptl. Gerontol* 10: 273–8.

Krauss, S. & Linn, S. (1982). *Biochemistry* 21: 1002–9.

Krauss, S. & Linn, S. (1986). *J. Cell Physiol.* 126: 99–106.

Kreimeyer, A., Wielckens, K., Adamietz, P. & Hilz, H. (1984). *J. Biol. Chem.* 259: 890–6.

Kurochkin, S. N., Trakht, I. N., Severin, E. S. & Cole, R. D. (1978). *FEBS Lett.* 84: 163–6.

Kurtz, D. I., Russell, A. R. & Sinex, F. M. (1974). *Mech. Age. Dev.* 3: 37–49.

Lake, R. S. (1973). *Nature* 242: 145–6.

Lake, R. S. & Salzman, N. P. (1972). *Biochemistry* 11: 4817–26.

Lamb, M. J. (1977). *Biology of Ageing.* Blackie & Son Ltd., Glasgow.

Lamb, M. J. & McDonald, R. P. (1973). *Exptl. Gerontol.* 8: 207–17.

Langan, T. A. & Hohmann, P. (1974). *Fed. Pro⁻.* 33: 1597 (Abs.).

Latchman, D. S. (1990). *Gene Regulation: A Eukaryotic Perspective,* Unwin Hyman, London.

Laughon, A. (1991). *Biochemistry* 30: 11357–67.

Laybourn, P. J. & Kadonaga, J. T. (1991). *Science* 254: 238–45.

Lebkowski, J. S. & Laemmli, U. K. (1982). *J. Molec. Biol.* 156: 309–24.

Lee, H., Paik, W. K. & Borun, T. W. (1973). *J. Biol. Chem.* 248: 4194–9.

Lee, H. C., Paz, M. A. & Gallop, P. M. (1982). *J. Biol. Chem.* 257: 8912–18.

Lee, M. G., Norbury, C. J., Spurr, N. K. & Nurse, P. (1988). *Nature* 333: 676–9.

Leffak, M., Grainger, R. & Weintraub, H. (1977). *Cell* 12: 837–45.

Leibovitz, B. E. & Siegel, B. V. (1980). *J. Gerontol.* 35: 45–56.

Lennox, R. W. & Cohen, L. H. (1988). In *Histone Phosphorylation* (K. W. Adolph, ed.), vol 11: 33–56. CRC Press, Boca Raton, FL.

Levi, V., Jacobson, E. L. & Jacobson, M. K. (1978). *FEBS Lett.* 88: 144–6.

Levin, P., Janda, J. K., Joseph, J. A., Ingram, D. K., Roth, G. S., Linn, S., Kairis, M. & Holliday, R. (1981). *Proc. Natl. Acad. Sci. (U.S.A.)* 73: 2818–22.

Levine, A., Cantoni, G. L. & Razin, A. (1991). *Proc. Natl. Acad. Sci. (U.S.A.)* 88: 6515–8.

Levinger, L. & Varshavsky, A. (1982). *Cell* 28: 375–85.

Levy, A. & Noll, M. (1981). *Nature* 289: 198–203.

Lewis, C. D. & Laemmli, U. K. (1982). *Cell* 29: 171–81.

Lewis, C. M. & Holliday, R. (1970). *Nature* 228: 877–80.

Lewis, C. M. & Tarrant, G. M. (1972). *Nature* 239: 316–18.

Libby, P. R. (1968). *Biochem. Biophys. Res. Commun.* 31: 59–65.

Lifton, R. P., Goldberg, M. L., Karp, R. W. & Hogness, D. S. (1978). *Cold Spring Harbor Symp. Quant. Biol.* 42: 1047–52.

Limas, C. J. & Limas, C. (1978). *Nature* 271: 781–83.

Lindop, P. J. & Rotblat, J. (1961). *Proc. Roy. Soc. (London)* B154: 332–68.

Lindquist, S. & Craig, E. (1988). *Ann. Rev. Genet.* 22: 631–71.

Linn, S., Kairis, M. & Holliday, R. (1976). *Proc. Natl. Acad. Sci. (U.S.A.)* 73: 2818–22.

Linnane, A. W., Maruzuki, S., Ozawa, T. & Tanaka, M. (1989). *Lancet* 1 (no. 8639): 642–5.

Lipps, H. J. (1975). *Cell Diff.* 4: 123–9.

Little, J. B. (1976). *Gerontologia* 22: 28–55.

Liu, A. Y. C., Lin, Z., Choi, H. S., Sorhage, F. & Li, B. (1989). *J. Biol. Chem.* 264: 12037–45.

Lu, H., Zawel, L., Fisher, L., Egly, J. M. & Reinberg, D. (1992). *Nature,* 358: 641–5.

Lu, K. Levine, R. A. & Campisi, J. (1989). *Mol. Cell. Biol.* 9: 3411–17.

Luisi, B. F., Xu, W. X. Otwinoski, Z., Freedman, L. P. & Yamamoto, K. R. (1991). *Nature* 352: 497–505.

Lumpkin, C. K. Jr., McClung, J. K., Pereira-Smith, O. M. & Smith, J. R. (1986). *Science* 232: 393–5.

Lundblad, V. & Szostak, J. W. (1989). *Cell* 57: 633–43.

Lutter, L. C. (1978). *J. Molec. Biol.* 124: 391–420.

Lyon, S. B., Buonocore, L. & Miller, M. (1987). *Mol. Cell. Biol.* 7: 1759–63.

McCay, C. M. (1952). In *Cowdry's Problems of Ageing. Biological and Medical Aspects.* 3rd ed. (A. L. Lansing, ed.), 139–202. Williams and Wilkins, Baltimore.

McCay, C. M., Crowell, M. F. & Mognard, L. A. (1935). *J. Nutri.* 10: 63–79.

McCay, C. M., Sperling, G. & Barnes, L. L. (1943). *Arch. Biochem.* 2: 469–79.

McClure, W. R. (1985). *Ann. Rev. Biochem.* 54: 171–204.

McCormick, A. & Campisi, J. (1991). *Curr. Op. Cell Biol.* 3: 230–4.

McGhee, J. D. & Felsenfeld, G. (1979). *Proc. Natl. Acad. Sci. (U.S.A.)* 76: 2133–7.

McGhee, J. D. & Felsenfeld, G. (1980). *Ann. Rev. Biochem.* 49: 1115–56.

McGhee, J. D., Nickol, J. M., Felsenfeld, G. & Rau, D. C. (1983). *Cell* 33: 831–41.

McGinnis, W., Garber, R. L., Wirz, J., Kuroiwa, A. & Gehring, W. J. (1984). *Cell* 37: 403–8.

Mackay, W. J. & Bewley, G. C. (1989). *Genetics* 122: 643–52.

McKnight, S. L. (1991). *Sci. Amer.* 264: 32–9.

Mader, S. Kumar, V., deVernevil, H. & Chambon, P. (1989). *Nature* 338: 271–4.

Magnuson, V. L., Young, M., Schattenberg, D. G., Mancini, M. A., Chen, D., Steffensen, B. & Klebe, R. J. (1991). *J. Biol. Chem.* 266: 14654–62.

Maier, J. A. M., Voulalas, P., Roeder, D. & Maclag, T. (1990). *Science* 249: 1570–5.

Makrides, S. C. (1983). *Biol. Rev.* 58: 343–422.

Man, N. T. & Shall, S. (1982). *Eur. J. Biochem.* 126: 83–8.

Maniatis, T., Goodbourne, S. & Fischer, J. A. (1987). *Science* 236: 1237–45.

Manjula & Sundari, R. M. (1980). *J. Biosci.* 2: 243–51.

Manjula & Sundari, R. M. (1981). *Ind. J. Biochem. Biophys.* 18: 192–9.
Maren, T. H. (1967). *Physiol. Rev.* 47: 595–781.
Markert, C. L. & Moller, F. (1959). *Proc. Nat. Acad. Sci. (U.S.A.)* 45: 753–62.
Markert, C. L. & Ursprung, H. (1962). *Dev. Biol.* 5: 363–81.
Martin, G. M., Sprague, C. & Epstein, C. (1970). *Lab. Invest.* 23: 86–92.
Martin, K. J. (1991). *BioEssays* 13: 499–503.
Marushige, K. (1976). *Proc. Natl. Acad. Sci. (U.S.A.)* 73: 3937–41.
Marushige, Y. & Marushige, T. (1978). *Biochim. Biophys. Acta* 518: 440–9.
Matus, A. & Green, G. D. J. (1987). *Biochemistry* 26: 8083–6.
Mayne, R., Vail, M. S. & Miller, E. J. (1975). *Proc. Natl. Acad. Sci. (U.S.A.)* 73: 2818–22.
Mays-Hoopes, L., Cleland, G., Bochantin, J., Kalunian, D., Miller, J., Wilson, W., Wong, M. K., Johnson, D. & Sharma, D. K. (1983). *Mech. Age. Dev.* 22: 135–49.
Medawar, P. B. (ed.). (1957) *The Uniqueness of the Individual.* Methuen, London.
Medvedev, Z. A. (1990). *Biol. Rev.* 65: 375–98.
Medvedev, Zh. A., Medvedeva, M. N. & Huschtscha, L. I. (1977). *Gerontology* 23: 334–41.
Medvedev, Zh. A., Medvedeva, M. N. & Robson, L. (1978). *Gerontology* 24: 286–92.
Meehan R. R., Lewis, D. D., McKay, S., Kleiner, E. L. & Bird, A. P. (1989). *Cell* 58: 499–507.
Meijlink, K., Curran, T., Miller, A. D. & Verma, I. D. (1985). *Proc. Natl. Acad. Sci. (U.S.A.)* 82: 4987–91.
Mennecier, F. & Dreyfus, J. C. (1974). *Biochim. Biophys. Acta* 364: 320–6.
Menzies, R. A., Mishra, R. K. & Gold, P. H. (1972). *Mech. Age. Dev.* 1: 117–32.
Migeon, B. R., Axelman, J. & Beggs, A. H. (1988). *Nature* 335: 93–6.
Miller, J., MacLachlan, A. D. & Klug, A. (1985). *EMBO J.* 4: 1609–14.
Milman, N., Graudel, N. & Andersen, H. C. (1988). *Clin. Chim. Acta* 176: 59–62.
Miquel, J. (1991). *Arch. Gerontol. Geriatr.* 12: 99–117.
Miquel, J., Bensch, K. G., Philpott, D. E. & Atlan, H. (1972). *Mech. Age. Dev.* 1: 71–97.
Mirkovitch, J., Mirault, M. E. & Laemmli, U. K. (1984). *Cell* 39: 223–32.
Mirsky, A. E. & Pollister, A. W. (1946). *J. Gen. Physiol.* 30: 117–48.
Mirzabekov, A. D., Shick, V. V., Belyavsky, A. V. & Bavykin, S. G. (1978) *Proc. Natl. Acad. Sci. (U.S.A.)* 75: 4184–8.
Mitchell, P. J. & Tjian, R. (1989). *Science* 245: 371–8.
Mitra, S., Sen, D. & Crothers, D. M. (1984). *Nature* 308: 247–50.
Miwa, M., Tanaka, M., Matsushima, T. & Sugimura, T. (1974). *J. Biol. Chem.* 249: 3475–82.
Moerman, E. J. & Goldstein, S. (1991). *Molec. Cell. Biol.* 11: 3905–14.

Mohandas, T., Sparkes, R. S. & Shapiro, L. J. (1981). *Science* 211: 393–6.

Monk, M. (1986). *BioEssays* 4: 204–8.

Moreau, P., Hen, R., Wasylyk, B., Everett, R., Gaub, M. P. & Chambon, P. (1981). *Nucl. Acids Res.* 9: 6047–68.

Morin, G. B. (1989). *Cell* 59: 521–9.

Morin, G. B. (1991). *Nature* 353: 454–6.

Morris, G. E., Piper, M. & Cole, R. (1976). *Nature* 263: 76–7.

Moss, B., Gersowitz, A., Weber, L. & Baglioni, C. (1977). *Cell* 10: 113–20.

Moudgil, V. K. & Kanungo, M. S. (1973a). *Comp. Gen. Pharmacol.* 4: 127–30.

Moudgil, V. K. & Kanungo, M. S. (1973b). *Biochem. Biophys. Res. Commun.* 52: 725–30.

Moudgil, V. K. & Kanungo, M. S. (1973c). *Biochim. Biophys. Acta* 329: 211–20.

Moyzis, R. K. (1991). *Scient. Amer.* 265: 34–41.

Müller, M. M., Gerster, T. & Schaffner, W. (1988) *Eur. J. Biochem.* 176: 485–95.

Muller, R., Bravo, R. & Burckhardt, J. (1984). *Nature* 312: 716–20.

Muller, W. E. G., Zahn, R. K., & Arendes, J. (1979). *Mech. Age. Dev.* 14: 39–48.

Muller, W. E. G., Zahn, R. K., Schroder, C. H. & Arendes, J. (1979) *Gerontology* 25: 61–8.

Murano, S., Thweatt, R., Shmookler-Reis, R. J., Jones, R. A., Moerman, E. J. & Goldstein, S. (1991). *J. Biol. Chem.* 266: 14654–62.

Murcia, G. de, Menissier-de Murcia, J. & Schreiber, V. (1991). *BioEssays* 13: 455–62.

Murray, E. J., & Grosveld, F. (1987). *EMBO J.* 6: 2329–35.

Murray, V. & Holliday, R. (1981). *J. Molec. Biol.* 146: 55–76.

Murtha, M. T., Leckman, J. F. & Ruddle, F. H. (1991). *Proc. Natl. Acad. Sci. (U.S.A.)* 88: 10711–15.

Nacheva, G. A., Guschin, D. Y., Preobrazhenskaya, O. V., Karpor, V. L., Ebralidse, K. K. & Mirzabekov, A. D. (1989). *Cell* 58: 27–36.

Nebert, D. W. & Gonzalez, F. J. (1985). *Trends Pharmacol. Sci.* 4: 160–4.

Neelin, J. M. & Butler, G. C. (1961) *Can. J. Biochem.* 39: 485–91.

Nelson, D. A., Perry, M., Sealy, L. & Chalkley, R. (1978). *Biochem. Biophys. Res. Commun.* 82: 1346–53.

Nesse, R. M. (1988). *Exptl. Geront.* 23: 445–53.

Neuberger, M. S. (1983). *EMBO J.* 2: 1373–8.

Neumeister, J. A. & Webster, G. C. (1981). *Mech. Age. Dev.* 16: 319–26.

Nickol, J. & Felsenfeld, G. (1983). *Cell* 35: 467–77.

Niedzwiecki, A., & Fleming, J. (1990). *Mech. Age. Dev.* 52: 295–304.

Niedzwiecki, A., Kongpachith, A. M. & Fleming, J. E. (1991). *J. Biol. Chem.* 266: 9332–8.

Nikolov, D. B., Hu, S.-H., Lin, J., Gasch, A., Hoffmann, A., Hornikoshi, M., Chua, N. -H., Roeder, R. G. & Burley, S. K. (1992). *Nature,* 360: 40–6.

Nina, O. L., Seto, S. H. & Tener, G. M. (1990). *Proc. Natl. Acad. Sci. (U.S.A.)* 87: 4270–4.

Ning, Y., Weber, J. L., Killary, A. M., Ledbetter, D. H., Smith, J. R. & Pereira-Smith, O. M. (1991). *Proc. Natl. Acad. Sci. (U.S.A.)* 88: 5635–9.

Nissenson, R., Flouret, G. & Hechter, O. (1978). *Proc. Natl. Acad. Sci. (U.S.A.)* 75: 2044–8.

Noll, M. (1974a). *Nature* 251: 249–51.

Noll, M. (1974b). *Nucl. Acids Res.* 1: 1573–8.

Noll, M. & Kornberg, R. (1977). *J. Molec. Biol.* 109: 393–404.

Nordstedt, C., Gandy, S. E., Alafuzoff, I., Caporaso, G. L., Iverfeldt, K., Grebb, J. A., Winblad, B. & Greengard, P. (1991). *Proc. Natl. Acad. Sci. (U.S.A.)* 88: 8910–14.

Norton, V. G., Imai, B. S., Yan, P. & Bradbury, E. M. (1989). *Cell* 57: 449–57.

Norwood, T. H., Pendergrass, W. A., Sprague, C. A. & Martin, G. M. (1974). *Proc. Natl. Acad. Sci. (U.S.A.)* 71: 2231–5.

Nussinov, R. (1990). *Crit. Rev. Biochem. Mol. Biol.* 25: 185–224.

Ogata, N., Ueda, K., Kagamiyama, H. & Hayaishi, O. (1980). *J. Biol. Chem.* 255: 7616–20.

Ohashi, Y., Ueda, K., Kawaichi, M. & Hayaishi, O. (1983). *Proc. Natl. Acad. Sci. (U.S.A.)* 80: 3604–7.

Ohghushi, H., Yoshihara, K. & Kamiya, T. (1980). *J. Biol. Chem.* 255: 6205–6211.

Ohkuma, Y., Sumimoto, H., Hoffmann, A., Shimasaki, S., Horikoshi, M. & Roeder, R. G. (1991). *Nature* 354: 398–404.

Olins, A. L., Carlson, R. D., Wright, E. B. & Olins, D. E. (1976). *Nucl. Acids Res.* 3: 3271–91.

Olins, A. L. & Olins, D. E. (1974). *Science* 183: 330–2.

Oliva, R. & Mezquita, C. (1982). *Nucl. Acids Res.* 10: 8049–59.

Oliver, D., Balhorn, R., Granner, D. & Chalkley, R. (1972). *Biochemistry* 11: 3921–5.

Olovnikov, A. M. (1973). *J. Theo. Biol.* 41: 181–90.

Olson, E. N. (1990). *Genes Dev.* 4: 1454–61.

O'Meara, A. & Herrmann, R. (1972). *Biochim. Biophys. Acta* 269: 419–27.

Ono, T. & Cutler, R. G. (1978). *Proc. Natl. Acad. Sci. (U.S.A.)* 75: 4431–5.

Ono, T., Dean, R. G., Chattopadhyaya, S. K. & Cutler, R. G. (1985a). *Gerontology* 31: 362–72.

Ono, T., Okada, S., Kawakami, T., Honjo, T. & Getz, M. J. (1985b). *Mech. Age. Dev.* 32: 227–34.

Ono, T., Okada, S. & Sugahara, T. (1976). *Exptl. Gerontol.* 11: 127–32.

Ord, M. G. & Stocken, L. A. (1975). *Proc. 9th FEBS Symp.* 34: 113–25.

Orgel, L. (1963). *Proc. Natl. Acad. Sci. (U.S.A.)* 49: 517–21.

Orgel, L. (1970). *Proc. Natl. Acad. Sci. (U.S.A.)* 67: 1476–80.

Orgel, L. (1973). *Nature* 243: 441–5.

O'Shea, E. K., Klemm, J. D., Kim, P. S. & Alber, T. (1991). *Science* 254: 539–44.

Osterman, J., Fritz, P. J. & Wuntch, T. (1973). *J. Biol. Chem.* 248: 1011–18.

Otero, R. D. & Felsenfeld, G. (1977). *Proc. Natl. Acad. Sci. (U.S.A.)* 74: 5519–23.

Otting, G., Rian, Y. Q., Billeter, M., Muller, M. Affolter, M., Gehring, W. J. & Wuthrich, K. (1990). *EMBO J.* 9: 3085–92.

Owenby, R. K., Stulberg, M. P. & Jacobson, K. B. (1979). *Mech. Age. Dev.* 11: 91–103.

Owens, G. P., Hahn, W. E. & Cohen, J. J. (1991). *Molec. Cell. Biol.* 11: 4177–88.

Paik, W. K. & Kim, S. (1973). *Biochem. Biophys. Res. Commun.* 51: 781–8.

Paik, W. K. & Kim, S. (eds.) (1980). *Protein Methylation*. John Wiley & Sons, New York.

Painter, R. B., Clarkson, J. M. & Young, B. R. (1973). *Radiat. Res.* 56: 560–4.

Pandey, R. S. & Kanungo, M. S. (1984). *Molec. Biol. Rep.* 10: 79–82.

Panyim, S. & Chalkley, R. (1969). *Biochem. Biophys. Res. Commun.* 37: 1042–9.

Papanov, V. D., Gromov, P. S., Sokolov, N. A., Spitkovsky, D. M. & Tseitlin, P. I. (1978). *Biochem. Biophys. Res. Commun.* 82: 674–9.

Pardon, J. F., Worcester, D. L., Wooley, J. C., Tatchell, K., van Holde, K. E. & Richards, B. M. (1975). *Nucl. Acids Res.* 2: 2163–75.

Park, J. W. & Ames, B. H. (1988). *Proc. Natl. Acad. Sci. (U.S.A.)* 85: 7467–70.

Parker, C. S. & Topol, J. (1984). *Cell* 36: 273–83.

Partridge, L. (1989). In *Life Time Reproduction Success in Birds* (I. Newton, ed.). Academic Press, London.

Patel, R. S., Odermatt, E., Schwarzbauer, J. E. & Hynes, R. O. (1987). *EMBO J.* 6: 2565–72.

Patnaik, S. K. & Kanungo, M. S. (1976). *Ind. J. Biochem. Biophys.* 13: 117–24.

Paulson, J. R. & Laemmli, V. K. (1977). *Cell* 12: 817–28.

Pelham, H. R. B. (1986). *Cell* 46: 959–61.

Pentecost, B. T., Wright J. M. & Dixon, G. H. (1985). *Nucl. Acids Res.* 13: 4871–88.

Pereira-Smith, O. M., Fisher, S. F. & Smith, J. M. (1985). *Exptl. Cell Res.* 160: 297–306.

Pereira-Smith, O. M. & Smith, J. R. (1983). *Science* 221: 964–6.

Pereira-Smith, O. M. & Smith, J. R. (1988). *Proc. Natl. Acad. Sci. (U.S.A.)* 85: 6042–6.

Perry, M. & Chalkley, R. (1981). *J. Biol. Chem.* 256: 3313–18.

Peterson, M. G., Inostrozoa, J. K., Maxon, M. E., Flores, O., Admion, A., Reinherg, D. & Tjian, R. (1991). *Nature* 354: 369–73.

Peterson, M. G., Tanese, N., Pugh, B. F. & Tjian, R. (1990). *Science* 248: 1625–30.

Peterson, R. P., Cryar, J. R. & Gaubatz, J. W. (1984). *Arch. Gerontol. Geriat.* 3: 115–25.

Petes, T. D., Farber, R. A., Tarrant, G. M. & Holliday, R. (1974). *Nature* 251: 434–6.

Pfeifer, G. P., Tanguay, R. L., Steigerwald, S. D. & Riggs, A. D. (1990). *Genes Dev.* 4: 1277–87.

Phillips, D. M. P. (1963). *Biochem. J.* 87: 258–63.

Phillips, P. D., Kaji, K. & Cristofalo, V. J. (1984). *J. Gerontol.* 39: 11–17.

Pientá, K. J., Getzenberg, R. H. & Coffey, D. S. (1991). *Crit. Rev. Eur. Gene Exp.* 1: 355–85.

Pines, J. & Hunter, T. (1990). *New Biologist* 2: 399–401.

Plumb, M. A., Lobanenkov, V. V., Nicolas, R. H., Wright, C. A., Zaron, S. & Goodwin, G. H. (1986). *Nucl. Acids Res.* 14: 7675–93.

Poccia, D. L., Simpson, M. V. & Green, G. R. (1987). *Dev. Biol.* 121: 445–53.

Pogo, B. G. T., Allfrey, V. G. & Mirsky, A. E. (1966). *Proc. Natl. Acad. Sci. (U.S.A.)* 55: 805–12.

Pogo, B. G. T., Pogo, A. O., Allfrey, V. G. & Mirsky, A. E. (1968). *Proc. Natl. Acad. Sci. (U.S.A.)* 59: 1337–44.

Pollwein, P., Master, C. L. & Beyreuther, K. (1992). *Nucl. Acids Res.* 20: 63–8.

Polson, C. D. A. & Webster, G. C. (1982). *Exptl. Gerontol.* 17: 11–17.

Popp, R. A., Bailiff, E. G., Hinsch, G. P. & Conrad, R. A. (1976). *Interdisciplinary Topics in Gerontology* 18: 125.

Price, G. B., Modak, S. P. & Makinodan, T. (1971). *Science* 171: 917–20.

Printz, D. B. & Gross, S. R. (1967). *Genetics* 55: 451–67.

Prior, C. P., Cantor, C. R., Johnson, E. M., Littau, V. C. & Allfrey, V. G. (1983). *Cell* 34: 1033–42.

Prunell, A., Kornberg, R. D., Lutter, L. C., Klug, A., Levitt, M. & Crick, F. H. C. (1979). *Science* 204: 855–8.

Ptashne, M. (1986). *Nature* 322: 697–701.

Ptashne, M. (1989). *Scient. Amer.* 260: 25–31.

Ptashne, M. & Gann, A. F. A. (1990). *Nature* 346: 329–31.

Pu, W. T. & Struhl, K. (1991). *Proc. Natl. Acad. Sci. (U.S.A.)* 88: 6001–5.

Rabinovitch, P. S. & Norwood, T. H. (1980). *Exptl. Cell Res.* 130: 101–9.

Radna, R. L., Caton, Y., Jha, K. K., Kaplan, P., Li, G., Traganos, F. & Ozer, H. L. (1989). *Mol. Cell. Biol.* 9: 3093–6.

Rahman, Y. E. & Peraino, C. (1973). *Exptl. Gerontol.* 8: 93–100.

Rao, K. S., Martin, G. M. & Loeb, L. A. (1985). *J. Neurochem.* 45: 1273–8.

Rao, S. S. & Kanungo, M. S. (1974). *Ind. J. Biochem. Biophys.* 11: 208–12.

Rath, P. C., Jaiswal, Y. K. & Kanungo, M. S. (1989). Abstract in *14th Internatl. Cong. Gerontol. Mexico,* June 18–23.

Rath, P. C., & Kanungo, M. S. (1989). *Biochem. Biophys. Res. Commun.* 157: 1403–9.

Rath, P. C. & Kanungo, M. S. (1989). *FEBS Lett.* 244: 193–8.

Ratha, B. K. & Kanungo, M. S. (1977). *Biochem. Biophys. Res. Commun.* 76: 925–9.

Rattle, H. W. E., Langan, T. A., Danby, S. E. & Bradbury, E. M. (1977). *Eur. J. Biochem.* 81: 499–505.

Razin, A., Cedar, H. & Riggs, A. D. (1984). *DNA Methylation.* Springer-Verlag, New York.

Razin, A., Feldmesser, E., Katri, T. & Szyf, M. (1985). In *Biochemistry and Biology of DNA Methylation* (G. L. Cantoni & A. Razin eds.), 239–53; Alan R. Liss, New York.

Razin, A. & Riggs, A. D. (1980). *Science* 210: 604–16.

Rechsteiner, M. (1991). *Cell* 66: 615–18.

Reeves, R. & Candido, E. P. M. (1978). *FEBS Lett.* 91: 117–20.

Reeves, R. & Chang, D. (1983). *J. Biol. Chem.* 258: 679–87.

Reinberg, D., Horikoshi, M. & Roeder, R. G. (1987). *J. Biol. Chem.* 262: 3322–30.

Reiss, U. & Gershon, D. (1976). *Eur. J. Biochem.* 63: 617–23.

Reiss, U. & Rothstein, M. (1975). *J. Biol. Chem.* 250: 826–30.

Reitz, M. S. & Sanadi, D. R. (1972). *Exptl. Gerontol.* 7: 119–29.

Reveillaud, I., Niedzniecki, A., Bensch, K. G. & Fleming, J. E. (1991). *Molec. Cell. Biol.* 11: 632–40.

Reynolds, W., Smith, R. D., Bloomer, L. S. & Gottesfeld, J. M. (1982). In *The Cell Nucleus* (H. Busch & L. Rothblum, eds.) Vol. II, 63–87. Academic Press, New York.

Reznick, A. Z., Rosenfelder, L., Shpund, S. & Gershon, D. (1985). *Proc. Natl. Acad. Sci. (U.S.A.)* 82: 6114–8.

Riabowol, K., Schiff, J. & Gilman, M. Z. (1992). *Proc. Natl. Acad. Sci. (U.S.A.)* 89: 157–61.

Riabowol, K., Vosatka, R. J., Ziff, E. B., Lamb, N. J. & Framisco, J. F. (1988). *Molec. Cell. Biol.* 8: 1670–6.

Richardson, A. (1985). In *Review of Biological Research on Aging* (M. Rothstein, ed.), vol. 2, 395–419. Alan R. Liss, New York.

Richardson, A., Birchenall-Sparks, M. C., Staecker, J. L., Hardwick, J. P. & Liu, D. S. H. (1982). *J. Gerontol.* 37: 666–72.

Richardson, A., Butler, J. A., Rutherford, M. S., Semsei, I., Gu, M.-Z., Fernandes, G., & Chiang, W.-H. (1987). *J. Biol. Chem.* 262: 12821–5.

Richardson, A., Rutherford, M. S., Birchenall-Sparks, M. C., Roberts, M. S., Wu, W. T. & Cheung, H. T. (1985). *Molecular Biology of Aging* 29: 229–42.

Richardson, A. & Semsei, I. (1987). In *Review of Biological Research on Aging* (M. Rothstein, ed.), Vol. 3, 467–83. Alan R. Liss, New York.

Richmond, T. J., Finch, J. T., Rushton, B., Rhodes, D. & Klug, A. (1984). *Nature* 311: 532–7.

Richter, C. (1988). *FEBS Lett.* 241: 1–5.

Riethman, H. C., Moyzis, R. K., Meyne, J., Burke, D. T. & Olson, M. V. (1989). *Proc. Natl. Acad. Sci. (U.S.A.)* 86: 6240–4.

Riggs, A., Singer-Sam, J. & Keith, D. H. (1985). In *Biochemistry and Biology of DNA Methylation* (G. Cantoni & A. Razin, eds.), 211–22. Alan R. Liss, New York.

Riggs, M. G., Whittaker, R. G., Neumann, J. R. & Ingram, V. M. (1977). *Nature* 268: 462–4.

Rikans, L. E. & Notley, B. A. (1981). *Exptl. Gerontol.* 16: 253–9.

Ritossa, F. (1962). *Experientia* 18: 571–3.

Rittling, S. R., Brooks, K. M., Cristofalo, V. J. & Baserga, R. (1986). *Proc. Natl. Acad. Sci. (U.S.A.)* 86: 3316–20.

Robin, E. D. & Wong, R. (1988). *J. Cell Physiol.* 136: 507–13.

Robbins, P. D., Horowitz, J. M. & Mulligan, R. C. (1990). *Nature* 346: 668–71.

Robertson, O. H. (1961). *Proc. Natl. Acad. Sci. (U.S.A.)* 47: 609–21.

Robinson, S. J., Nelkin, B. D. & Vogelstein, B. (1982). *Cell* 28: 99–106.

Rockstein, M., Chesky, J. A. & Sussman, M. L. (1977). In *Handbook of the Biology of Aging* (C. E. Finch & L. Hayflick, eds.), 1–34. Van Nostrand Reinhold New York.

Rohme, D. (1981). *Proc. Natl. Acad. Sci. (U.S.A.)* 78: 5009–13.

Romero, D. P. & Blackburn, E. H. (1991). *Cell* 67: 343–53.

Rose, M. R. (1984). *Evolution* 38: 1004–10.

Rose, M. R. (1985). *Theor. Popup. Biol.* 28: 342–85.

Rose, M. R. & Graves, J. L. (1990). *Rev. Biol. Res. Ageing* 4: 3–14.

Rosen, R. (1978). *Intern. Rev. Cytol.* 54: 161–91.

Rosenfeld, M. G. (1991). *Genes Dev.*, 5: 897–907.

Ross, S. & Scott, G. (1939). *Brit. J. Radiol.* 12: 440–44.

Roth, G. S., Karoly, K., Britton, V. J. & Adelman, R. C. (1974). *Exptl. Gerontol.* 9: 1–12.

Rothstein, M. (1979). *Mech. Age. Dev.* 9: 197–202.

Rothstein, M. (1987). In *Modern Biological Theories of Ageing* (H. R. Warner et al., eds.). Raven Press, New York.

Rothstein, M., Coppens, M. & Sharma, H. K. (1980). *Biochim. Biophys. Acta* 614: 591–600.

Roy, A. K., & Chatterjee, B. (1985). In *Molecular Aspects of Medicine* (H. Baum et al., eds.), 3–88. Pergamon, Oxford.

Roy, A. K., Chatterjee, B., Demyan, W. F., Milin, B. S., Motwani, N. M., Nath, U. & Schiop, M. J. (1983a). *Rec. Prog. Horm. Res.* 39: 425–61.

Roy, A. K., Nath, T. S., Motwani, N. M. & Chatterjee, B. (1983b). *J. Biol. Chem.* 258: 10123–7.

Ruiz-Carrillo, A., Wangh, L. J. & Allfrey, V. G. (1975). *Science* 190: 117–128.

Ruiz-Carrillo, A., Wangh, L. J. & Littan, V. C. & Allfrey, V. G. (1976). *Arch. Biochem. Biophys.* 174: 273–90.

Ryan, T. M., Behringer, R. R., Martin N. C., Townes, T. M., Palmiter, R. D. & Brinster, R. L. (1989). *Genes Dev.* 3: 314–23.

Ryseck, R. P., Hirai, S. I., Yaniv, M. & Barvo, R. (1988). *Nature* 334: 535–7.

Sacher, G. A. (1956). *Radiology* 67: 250–7.

Sacher, G. A. (1959). In *Ciba Foundation Colloquia on Ageing. V. Life Span of Animals* (G. E. W. Wolstenholme & M. O'Connor, eds.), 115–33. Churchill, London.

Sacher, G. A. (1980). *Ann. Rev. Gerontol. Geriat.* Part I, pp. 3–25.

Sacher, G. A. & Staffeldt, E. F. (1974). *Amer. Nat.* 108: 593–615.

Sager, R. (1989). *Science* 246: 1406–12.

Salditt-Georgieff, M., Sheffery, M., Krauter, K., Darnell, J. E., Rifkind, R. & Marks, P. A. (1984). *J. Molec. Biol.* 172: 437–50.

Saltzman, A. G. & Weinmann, R. (1989). *FASEB J.* 3: 1723–33.

Samis, H. V., Falzone, J. A. & Wulff, V. J. (1966). *Gerontology* 12: 79–88.

Samis, H. V., Wulff, V. J. & Falzone, J. A. (1964). *Biochim. Biophys. Acta* 91: 223–32.

Samuels, M., Fire, A. & Sharp, P. A. (1982). *J. Biol. Chem.* 257: 14419–27.

Sano, H. & Sager, R. (1982). *Proc. Natl. Acad. Sci. (U.S.A.)* 79: 4584–8.

Saunders, J. W. (1966). *Science* 154: 604–12.

Sawadogo, M. & Roeder, R. G. (1984). *J. Biol. Chem.* 259: 5321–6.

Schaffner, W., Kunz, G., Daetwyler, H., Telford, J., Smith, H. O. & Birnsteil, M. L. (1978). *Cell* 14: 655–71.

Scharf, J., Dovrat, A. & Gershon, D. (1987). *Arch. Clin. Exptl. Ophthalmol.* 225: 133–6.

Schwabe, J. W. R. & Rhodes, D. (1991). *Trends Biochem. Sci.* 16: 291–6.

Schwabe, J. W. R., Nieuhaus, D. & Rhodes, D. (1990). *Nature* 348: 458–61.

Schwartz, L. M. (1991). *BioEssays* 13: 389–95.

Schwartz, L. M., Kosz, L. & Kay, B. K. (1990). *Proc. Natl. Acad. Sci. (U.S.A.)* 87: 6594–8.

Schwarzbauer, J. E. (1991). *BioEssays* 13: 527–33.

Schwarzbauer, J. E., Tamkun, J. W., Lemischka, I. R. & Hynes, R. O. (1983). *Cell* 35: 421–31.

Scott, M. P. & Wiener, A. J. (1984). *Proc. Natl. Acad. Sci. (U.S.A.)* 81: 4115–19.

Sealy, L. & Chalkley, R. (1978). *Nucl. Acids Res.* 5: 1863–76.

Selkoe, D. J. (1991a). *Scient. Amer.* 265: 40–7.

Selkoe, D. J. (1991b). *Nature* 354: 432–3.

Sellins, K. S. & Cohen, J. J. (1987). *J. Immunol.* 139: 3199–206.

Semsei, I., Ma, S. & Cutler, R. G. (1989). *Oncogene* 4: 465–70.

Semsei, I., Rao, G. & Richardson, A. (1989). *Biochem. Biophys. Res. Commun.* 164: 620–5.

Semsei, I., Szeszak, F. & Zs-Nagy, I. (1982). *Arch. Gerontol. Geriat.* 1: 29–42.

Sen, R. & Baltimore, D. (1986). *Cell* 47: 921–8.

Serfling, E., Jasin, M. & Schaffner, W. (1985). *Trends Genet.* 1: 224–30.

Serfling, E., Lubbe, A., Dorsch-Hasler, K. & Schaffner, W. (1985). *EMBO J.* 4: 3851–9.

Servillo, G., DellaFazia, M. A. & Viola-Magni, M. (1991). *Biochem. Biophys. Res. Commun.* 175: 104–9.

Seshadri, T. & Campisi, J. (1990). *Proc. Natl. Acad. Sci.* 247: 205–9.

Seyedin, S. M. & Kistler, W. S. (1980). *J. Biol. Chem.* 255: 5949–54.

Shainberg, A, Yagil, G. & Yaffe, D. (1971). *Dev. Biol.* 25: 1–29.

Sharma, H. K., Gupta, S. K. & Rothstein, M. (1976). *Arch. Biochem. Biophys.* 174: 324–32.

Sharma, H. K., Prasanna, H. R., Lane, R. S. & Rothstein, M. (1979). *Arch. Biochem. Biophys.* 194: 275–82.

Sharma, H. K., Prasanna, H. R., & Rothstein, M. (1980). *J. Biol. Chem.* 255: 5043–50.

Sharma, H. K. & Rothstein, M. (1978). *Biochemistry* 17: 2869–76.

Sharma, H. K. & Rothstein, M. (1980). *Proc. Natl. Acad. Sci. (U.S.A.)* 77: 5865–8.

Sharp, P. A. (1985). *Cell* 42: 397–400.

Shaw, C. R. & Barto, E. (1963). *Proc. Natl. Acad. Sci. (U.S.A.)* 50: 211–14.

Sheffery, M., Rifkind, R. A. & Marks, P. A. (1982). *Proc. Natl. Acad. Sci. (U.S.A.)* 79: 1180–4.

Sherwood, S. W., Rush, D., Ellsworth, J. L. & Schimke, R. T. (1988). *Proc. Natl. Acad. Sci. (U.S.A.)* 85: 9086–90.

Shick, V. V., Belyavsky, A. V., Bavykin, S. G. & Mirzabekov, A. D. (1980). *J. Molec. Biol.* 139: 491–517.

Shimada, T. & Nienhuis, A. W. (1984). *J. Biol. Chem.* 260: 2468–74.

Shmookler-Reiss, R. J. & Goldstein, S. (1980). *Cell* 21: 739–49.

Shock, N. W. (1959). In *Program and Papers of the Conference on Gerontology,* pp. 123–40. Duke University.

Sibbet, G. J. & Carpenter, B. G. (1983). *Biochim. Biophys. Acta* 740: 331–8.

Sierra, F., Fey, G. H. & Guigoz, Y. (1989). *Molec. Cell. Biol.* 9: 5610–16.

Sierra, F., Juillerat, M., Coeytaux, S., Ruffieux, C. & Guigoz, Y. (1991). In *Liver and Aging,* 123–135. Excerpta Medica, Amsterdam.

Silber, J. R., Fry, M., Martin, G. M. & Loeb, L. A. (1985). *J. Biol. Chem.* 260: 1304–10.

Silva, A. J. & White, R. (1988). *Cell* 54: 145–52.

Simpson, R. T. (1978). *Biochemistry* 17: 5524–31.

Simpson, R. T. & Kunzler, P. (1979). *Nucl. Acids Res.* 6: 1387–1415.

Simpson, R. T. & Whitlock, J. P. (1976). *Cell* 9: 347–56.

Singer, B. (1980) *Carcinogenesis* (B., Pullman P. O. P. Ts'o & H. Gelboin, eds.) 91–102. D. Reidel, Norwell, MA.

Singer, M. & Berg, P. (1991). *Genes and Genomes.* Blackwell, Mill Valley, CA.

Singh, A. (1989). Ph.D. thesis. Banaras Hindu University, Varanasi, India.

Singh, A., Singh, S. & Kanungo, M. S. (1990). *Molec. Biol. Rep.* 14: 251–4.

Singh, S. (1991). Ph.D. dissertation, Banaras Hindu University, India.

Singh, S. & Kanungo, M. S. (1991). *Biochem. Biophys. Res. Commun.* 181: 131–7.

Singh, S. N. & Kanungo, M. S. (1968). *J. Biol. Chem.* 243: 4526–9.

Singhal, R. L. (1967a). *J. Gerontol.* 22: 77–82.

Singhal, R. L. (1967b). *J. Gerontol.* 22: 343–7.

Singhal, R. L., Valadares, J. R. E. & Ling, G. M. (1969). *Amer. J. Physiol.* 217: 793–7.

Singhal, R. P., Kooper, R. A., Nishimura, S. & Shindo-Okada, N. (1981). *Biochem. Biophys. Res. Commun.* 99: 120–6.

Sisodia, S. S., Koo, E. H., Beyreuther, K., Unterbeck, A. & Price, D. L. (1990). *Science* 248: 492–5.

Slagboom, P. E., de Leeuw, W. J. F. & Vijg, J. (1990). *FEBS Lett.* 269: 128–30.

Smith, D. R., Jackson, I. J. & Brown, D. D. (1984). *Cell* 37: 645–52.

Smith, R. D. & Yu, J. (1984). *J. Biol. Chem.* 259: 4609–15.

Sokal, R. R. (1970). *Science* 167: 1733–4.

Solage, A. & Cedar, H. (1978). *Biochemistry* 17: 2934–8.

Söllner-Webb, B. & Felsenfeld, G. (1975). *Biochemistry* 14: 2915–20.

Solomon, M. J., Strauss, F. & Varshavsky, A. (1986). *Proc. Natl. Acad. Sci. (U.S.A.)* 83: 1276–80.

Sommer, B., Kohler, M., Sprengel, R. & Seeburg, P. H. (1991). *Cell* 67: 11–19.

Song, C. S., Rao, T. R., Demyan, W. F., Mancini, M. A., Chatterjee, B. & Roy, A. K. (1991). *Endocrinology* 128: 349–56.

Spector, W. E. (ed.) (1956). In *Handbook of Biological Data*, 182–4. W. B. Saunders, London.

Srivastava, S. K. & Kanungo, M. S. (1979). *Ind. J. Biochem. Biophys.* 16: 347–8.

Srivastava, S. K. & Kanungo, M. S. (1980). *Biochem. Med.* 23: 64–9.

Srivastava, S. K. & Kanungo, M. S. (1982). *Biochem. Med.* 28: 266–72.

Srivastava, V. K. & Busbee, D. L. (1992). *Biochem. Biophys. Res. Commun.* 182: 712–21.

Srivastava, V. K., Tilley, R. D., Haart, R. W. & Busbee, D. L. (1991). *Exptl. Gerontol.* 26: 453–66.

Stedman, E. & Stedman, E. (1943). *Nature* 152: 556–7.

Stein, G. H., Beeson, M. & Gordon, L. (1990). *Science* 249: 666–9.

Stein, G. H., Drullinger, L. F., Robetorye, R. S., Pereira-Smith, O. M. & Smith, J. R. (1991). *Proc. Natl. Acad. Sci. (U.S.A.)* 88: 11012–16.

Stein, G. H., & Yanishevsky, R. J. (1979). *Exptl. Cell Res.* 120: 155–65.

Stevenson, K. G. & Curtis, H. J. (1961). *Radiat. Res.* 15: 774–84.

Strähle, U., Schmid, W. & Schutz, G. (1988). *EMBO J.* 7: 3389–95.

Strehler, B. L. (1964). *J. Gerontol.* 19: 83–7.

Strehler, B. L. & Chang, M. P. (1979). *Mech. Age. Dev.* 11: 379–82.

Stuart, K. (1991). *Trends Biochem. Sci.* 16: 68–72.

Stumph, W. E., Baez, M., Lawson, G. M., Tsai, M. J. & O'Malley, B. W. (1982). *UCLA Symp. Mol. & Cell Biol.* 26: 87–104.

Su, W., Jackson, S., Tjian, R. & Echols, H. (1991). *Genes Dev.* 5: 820–6.

Subrahmanyam, G., Kannan, K. & Reddy, A. R. (1984). *J. Biochem. Biophys. Meth.* 10: 153–62.

Sugawara, O., Oshimura, M. Koi, M., Annab, L. A. & Barrett, J. C. (1990). *Science* 247: 707–10.

Sugimura. T. (1973). *Prog. Nucl. Acids. Res. Mol. Biol.* 13: 127–51.

Sugiyama, S., Hattori, K., Hayakawa, M. & Ozawa, T. (1991). *Biochem. Biophys. Res. Commun.* 180: 894–9.

Sun, J., Lau, P. P. & Strobel, H. W. (1986). *Exptl. Gerontol.* 21: 65–73.
Sung, M. T., & Freedlender, F. F. (1978). *Biochemistry* 17: 1884–90.
Sung, M. T., Harford, J., Bundman, M. & Vidalakas, G. (1977). *Biochemistry* 16: 279–85.
Supakar, P. C. & Kanungo, M. S. (1984). *Molec. Biol. Rep.* 9: 253–7.
Sussman, J. L. & Trifonov, E. N. (1978). *Proc. Natl. Acad. Sci. (U.S.A.)* 75: 103–7.
Suzuki, F., Watanabe, E. & Horikawa, M. (1980). *Exptl. Cell Res.* 127: 299–307.
Svaren, J. & Chalkley, R. (1990). *Trends Genet.* 6: 52–6.
Szilard, L. (1959a). *Nature,* 184: 958–60.
Szilard, L. (1959b) *Proc. Natl. Acad. Sci. (U.S.A.)* 45: 30–45.
Takahashi, R. & Goto, S. (1990). *Arch. Biochem. Biophys.* 277: 228–33.
Takahashi, S. & Zeydel, M. (1982). *Arch. Biochem. Biophys.* 214: 260–7.
Talbot, D. & Grosveld, F. (1991). *EMBO J.* 10: 1391–8.
Tamkun, J. W., Schwarzbauer, J. E. & Hynes, R. O. (1984). *Proc. Natl. Acad. Sci. (U.S.A.)* 81: 5140–4.
Tanaka, K., Nazawa, T., Okada, Y. & Kumahara, Y. (1980). *Exptl. Cell Res.* 123: 261–7.
Tanigawa, Y., Kitamura, A., Kawamura, M. & Shimoyama, M. (1978). *Eur. J. Biochem.* 92: 261–9.
Tanzi, R. E., George-Hyslop, P. & Gusella, J. F. (1991). *J. Biol. Chem.* 266: 20579–82.
Tatchell, K. & van Holde, K. E. (1976). *Biochemistry* 16: 5295–303.
Tazi, J. & Bird, A. (1990). *Cell* 60: 909–20.
Thakur, M. K., Das, R. & Kanungo, M. S. (1978). *Biochem. Biophys. Res. Commun.* 81: 828–31.
Thakur, M. K. & Kanungo, M. S. (1981). *Exptl. Gerontol.* 16: 331–5.
Theveny, B., Bailly, A., Rauch, M., Rauch, E., Delain, E. & Milgrom, E. (1987). *Nature* 329: 79–81.
Thoma, F. & Koller, T. (1981). *J. Molec. Biol.* 149: 709–33.
Thomas, G., Lange, H. W. & Hempel, K. (1975). *Eur. J. Biochem.* 51: 609–15.
Thomas, M., White, R. L. & Davis, R. W. (1976). *Proc. Natl. Acad. Sci. (U.S.A.)* 73: 2294–8.
Thompson, K. V. A. & Holliday, R. (1978). *Exptl. Cell. Res.* 112: 281–7.
Thorne, A. W., Kmiciek, D., Mitchelson, K., Sautiere, P. & Crane-Robinson, C. (1990). *Eur. J. Biochem.* 193: 701–13.
Tice, R. R. & Setlow, R. B. (1985). In *The Handbook of the Biology of Aging* (C. E. Finch & E. L. Schneider, eds.), 173–224. Van Nostrand Reinhold, New York.
Tidwell, T., Allfrey, V. G. & Mirsky, A. E. (1968). *J. Biol. Chem.* 243: 707–15.
Tolmasoff, J. M., Ono, T. & Cutler, R. G. (1980). *Proc. Natl. Acad. Sci. (U.S.A.)* 77: 2777–81.

Trigun, S. K. & Singh, S. N. (1987). *Cell Mol. Biol.* 33: 767–74.
Trigun, S. K. & Singh, S. N. (1988). *Cell Mol. Biol.* 34: 207–13.
Trounce, I., Byrne, E. & Marzuki, S. (1989). *Lancet* 1 (no. 8639): 637–9.
Ueda, K. & Hayaishi, O. (1985). *Ann. Rev. Biochem.* 54: 73–100.
Umesono, K., Murakami, K. K., Thompson, C. C. & Evans, R. M. (1991). *Cell* 65: 1255–66.
Van Bezooijen, C. F. A. (1984). *Mech. Age. Dev.* 25: 1–22.
van Holde, K. E. (1989). *Chromatin.* Springer-Verlag, New York.
van Holde, K. E., Sahashrabudhe, C. G., Shaw, B. R., Bruggen, E. F. J. Van & Arnberg, A. C. (1974). *Biochem. Biophys. Res. Comm.* 60: 1365–70.
Varshavsky, A. J., Bakayev, V. V. & Georgiev, G. P. (1976). *Nucl. Acids Res.* 3: 477–92.
Verzar, F. (1955). *Experientia* 11: 230.
Verzar, F. (1958). *Acta Anat.* 33: 215–29.
Verzar, F. & Thoenen, H. (1960). *Gerontologia* 4: 112–19.
Vijg, J., Vitterlinden, A. G., Mullaart, E., Lohman, P. A. M. & Knook, D. L. (1985). In *Molecular Biology of Ageing: Gene Stability and Gene Expression* (R. S. Sohal, et al., eds.). Raven Press, New York.
Vinson, C. R., Sigler, P. B. & McKnight, S. L. (1989). *Science* 246: 911–16.
Wahli, W. & Martinez, E. (1991). *FASEB J.* 5: 2243–9.
Walbot, V. (1991). *Trends Genet.* 7: 37–9.
Walburg, H. E., Cosgrove, G. E. & Upton, A. C. (1966). In *Radiation and Ageing* (P. J. Lindop & G. A. Sacher, eds.) 361–5. Taylor & Francis, London.
Walford, R. H. (1987). In *Modern Biological Theories of Ageing.* Raven Press, New York.
Wang, E. (1987). *J. Cell Physiol.* 133: 151–7.
Wang, E. & Lin, S. L. (1986). *Exptl. Cell Res.* 167: 135–43.
Wareham, K. A., Lyon, M. F., Glenister, P. H. & Williams, E. D. (1987). *Nature* 327: 725–7.
Warner, H. (1989). *Exptl. Gerontol.* 24: 351–4.
Warren, S. (1956). *J. Amer. Med. Assoc.* 162: 464–8.
Wasylyk, B. (1988). *Biochim. Biophys. Acta* 951: 17–35.
Watanabe, F. (1984). *FEBS Lett.* 170: 19–22.
Waterborg, J. H., Fried, S. R. & Mathews, H. R. (1983). *Eur. J. Biochem.* 136: 245–52.
Waterborg, J. H. & Mathews, H. R. (1984). *Eur. J. Biochem.* 142: 329–35.
Watson, J. D., Hopkins, N. H., Roberts, J. W., Steitz, J. A. & Weiner, A. M. (1987). In *Molecular Biology of the Gene,* Vol. I. Benjamin/Cummings, Menlo Park, CA.
Weber, F. & Schaffner, W. (1985). *EMBO J.* 4: 949–56.
Weintraub, H. (1972). *Nature,* 240: 449–53.
Weintraub, H., Dwarki, V. J., Verma, I., Danis, R., Hollenbnerg, S., Snider, L., Lassar, A. & Tapscott, S. J. (1991). *Genes Dev.* 5: 1377–86.
Weintraub, H. & Groudine, M. (1976). *Science* 193: 848–56.
Weintraub, H., Weisbrod, S., Larsen, A. & Groudine, M. (1981). In *Organi-*

sation and Expression of Globin Genes (G. Stamato Yannopoulos & A. W. Nienhuis, eds.), 175–90. Alan R. Liss, New York.

Weisshaar, B., Langner, K. D., Juetlermann, R., Mueller, U., Zock, C., Klimkait, T. & Doerfler, W. (1988). *J. Molec. Biol.* 202: 255–70.

Wellinger, R. & Guigoz, Y. (1986). *Mech. Age. Dev.* 34: 203–17.

Welshon, W. V., Lieberman, M. E. & Gorsky, J. (1984). *Nature* 307: 747–9.

Westin, G., Gerster, T., Muller, M. M., Schaffner, G. & Schaffer, W. (1987). *Nucl. Acids Res.* 15: 6787–98.

Whitlock, J. P. & Stein, A. (1978). *J. Biol. Chem.* 253: 3857–61.

Whittenmore, S. R., Ebendal, T., Larkfors, L., Olson, L., Seiger, A., Stromberg, F. & Persson, H. (1986). *Proc. Natl. Acad. Sci. (U.S.A.)* 83: 817–21.

Widom, J. (1989). *Ann. Rev. Biophys. Chem.* 18: 365–95.

Wieland, T. & Pfleiderer, H. (1957). *Biochem. Z.* 329: 112–16.

Wilkins, A. S. (1986). *Genetic Analysis of Animal Development.* Wiley-Interscience, New York.

Wilkinson, K. D., Urban, M. K. & Haas, A. L. (1980). *J. Biol. Chem.* 255: 7529–32.

Williams, G. C. (1957) *Evolution* 11: 398–411.

Williams, G. C. (1966) *Adaptation and Natural Selection,* Princeton University Press, New Jersey.

Williams, T. & Tjian, R. (1991). *Genes Dev.* 5: 670–82.

Wilson, D. L., Hall, M. E. & Stone, G. C. (1978). *Gerontology* 24: 426–33.

Wilson, P. D. (1973). *Gerontologia* 19: 79–125.

Wismer, C. T., Sherman, K. A., Zibart, M. & Richardson, A. (1988). *Central Nervous System Disorders of Aging* (R Strong et al., eds.), 189–98. Raven Press, New York.

Wittig, S. & Wittig, B. (1982). *Nature* 297: 31–8.

Wodinsky, J. (1977). *Science* 198: 948–51.

Wohlrab, H., Bronson, R. T., Lu, R. C. & Nemeth, V. (1988). *Biochem. Biophys. Res. Commun.* 154: 1130–6.

Wolfe, A. P. (1991). *Biochem. J.* 278: 313–24.

Wood, W. I. & Felsenfeld, G. (1982). *J. Biol. Chem.* 257: 7730–6.

Woodcock, C. L. F., Frado, L. L. Y. & Rattner, J. B. (1984). *J. Cell Biol.* 99: 42–52.

Wray, V. P., Elgin, S. C. R. & Wray, W. (1980). *Nucl. Acids Res.* 8: 4155–63.

Wright, W. E., Pereira-Smith, O. M. & Shay, J. W. (1989). *Mol. Cell Biol.* 9: 3088–92.

Wu, B., Heydari, A. R., Conrad, C. C. & Richardson, A. (1991). In *Liver and Aging* (K. Kitani, ed.), 197–212. Excerpta Medica, Amsterdam.

Wu, C. (1980). *Nature* 286: 854–60.

Wu, C. (1984). *Nature* 309: 229–34.

Wyatt, G. R. (1951). *Biochem. J.* 48: 581–4.

Wyllie, A. H. (1980). *Nature* 284: 555–6.

Yadav, R. N. S. & Singh, S. N. (1980). *Biochim. Biophys. Acta* 633: 323–30.

Yadav, R. N. S. & Singh, S. N. (1981). *Biochem. Med.* 26: 258–63.

Yaffe, D. & Fuchs, S. (1967). *Dev. Biol.* 15: 33–50.

Yagil, G. (1976). *Exptl. Gerontol.* 11: 73–8.

Yamamoto, K. (1988). *Nature* 334: 543–6.

Yanishevsky, R. M. & Stein, G. M. (1980). *Exptl. Cell Res.* 126: 469.

Yen, T. C., Chen, Y.-S., King, K. L., Yeh, S.-H. & Wei, Y.-H. (1989). *Biochem. Biophys. Res. Commun.* 165: 994–1003.

Yen, T.-C., Su, J.-H., King, K. -L. & Wei, Y.-H. (1991). *Biochem. Biophys. Res. Commun.* 178: 124–31.

Yisraeli, J., Frank, D., Razin, A. & Cedar, H. (1988). *Proc. Natl. Acad. Sci. (U.S.A.)* 85: 4638–42.

Yisraeli, J. & Szyf, M. (1984). In *DNA Methylation: Biological and Biochemical Significance* (A. Razin et al., eds.), 353–78. Springer, New York.

Yoshida, M., & Shimura, K. (1984). *J. Biochem.* 95: 117–24.

Young, D. & Carroll, D. (1983). *Mol. Cell Biol.* 3: 720–30.

Yu, G.-L. & Blackburn, E. H. (1991). *Cell* 67: 823–32.

Yu, G.-L., Bradley, J. D., Attardi, L. D. & Blackburn, E. H. (1990). *Nature* 344: 126–32.

Yu, J. & Smith, R. D. (1985). *J. Biol. Chem.* 260: 3035–40.

Yuan, J. & Horvitz, H. R. (1990). *Dev. Biol.* 138: 33–41.

Zakian, V. A. (1989). *Ann. Rev. Genet.* 23: 579–604.

Zawel, L. & Reinberg, D. (1992). *Curr. Op. Cell. Biol.* 4: 488–95.

Zeelon, P., Gershon, H. & Gershon, D. (1973). *Biochemistry,* 12: 1743–50.

Zhang, X.-Y., Fittler, F. & Honz, W. (1983). *Nucl. Acids Res.* 11: 4287–306.

Zlatanova, J. & Swetly, P. (1978). *Nature* 276: 276–7.

Zuckerkandl, E. (1965). *Scient. Amer.* 212: 110–18.

Zweidler, A. (1984). In *Histone Genes: Structure, Organisation and Regulation* (G. S. Stein, J. L. Stein, & W. F. Marzluff, eds.), 339–71. John Wiley, New York.

Index

Note: The letter f or t after a page citation indicates reference to a figure or a table, respectively.